W9-COS-387

Foreign Investment in American Telecommunications

Foreign Investment in
American
Telecommunications

J. Gregory Sidak

THE UNIVERSITY OF CHICAGO PRESS

Chicago and London

J. Gregory Sidak is the F.K. Weyerhaeuser Fellow in Law and Economics at the American Enterprise Institute and a senior lecturer at the Yale University School of Management. He has served as deputy general counsel to the Federal Communications Commission and senior counsel and economist to the Council of Economic Advisers in the Reagan administration.

HE
7775
S55
1997

The University of Chicago Press, Chicago 60637
The University of Chicago Press, Ltd., London
© 1997 by The University of Chicago
All rights reserved. Published 1997
Printed in the United States of America
06 05 04 03 02 01 00 99 98 97   1 2 3 4 5
ISBN: 0-226-75626-2

Library of Congress Cataloging-in-Publication Data

Sidak, J. Gregory.
    Foreign investment in American telecommunications / J. Gregory Sidak.
        p.   cm.
    Includes bibliographical references and indexes.
    ISBN 0-226-75626-2 (cloth : alk. paper)
    1. Telecommunication—United States.   2. Telecommunication—Law and legislation—United States.   3. Investments, Foreign—United States.   I. Title.
    HE7775.S55   1997
    384'.04— dc21

                                                                                                  96-39882
                                                                                                  CIP

♾ The paper used in this publication meets the minimum requirements of the American National Standard for Information Sciences—Permanence of Paper for Printed Library Materials, ANSI Z39.48-1984.

To the memory of my father

JOSEPH G. SIDAK

1917–1996

Contents

3 The Statute   77

Acknowledgments

THE SEED for this book was planted sometime in 1987, when, as deputy general counsel of the Federal Communications Commission, I first reviewed a matter in which the agency was seeking to prevent certain foreign direct investment in an American company licensed to use radio spectrum. The applicable law seemed illogical and perverse to me from my very first exposure to it. The experience of having to enforce such folly as a government regulator left such a bad taste in my mouth that this book is the product a decade later.

I would like to acknowledge and thank the talented persons who assisted me over a two-year period in the historical, legal, and economic research for this book: Rebecca Armendariz, Michael Bennett, Solveig Bernstein, Ramsen Betfarhad, Ian Connor, Jason Levine, Joseph Matelis, Sasha Mayergoyz, Elizabeth Mower, Mayank Raturi, and Hal Singer. Special recognition is due Marshall Smith, with whom I had the pleasure to work during his stint as a telecommunications analyst at the American Enterprise Institute. I could not have completed this book without his able and tireless assistance. In addition, Amanda Riepe and Keitha Macdonald meticulously typed the manuscript and formatted it for publication. I thank them both for their patience, care, and good cheer. Finally, Anthony Yoseloff proofread the manuscript with care and dispatch.

I have benefited from conversations with my AEI colleagues (particularly Claude Barfield, Cynthia Beltz, Jagdish Bhagwati,

Thomas Hazlett, and Douglas Irwin) and with executives in the telecommunications industry (particularly Keith Bernard, Patricia Diaz Dennis, Sam Ginn, James Graf, and Richard Nohe) and from the comments of participants in a roundtable discussion of a draft of this book at AEI in November of 1995. I have benefited as well as from the thoughtful reviews of anonymous referees selected by the University of Chicago Press. Melinda Ledden Sidak and Leigh Tripoli gave me valuable suggestions for improving the exposition of the manuscript. Representative Michael Oxley and his legislative director, Robert Foster, generously answered my numerous requests for the latest legislative materials concerning the Telecommunications Act of 1996.

Finally, I want to thank Geoffrey Huck of the University of Chicago Press for his commitment to this project, and Christopher DeMuth, president of the American Enterprise Institute, for his steadfast encouragement of my work since I arrived at AEI in 1992 and for his leadership of the extraordinary intellectual community where it has been possible for me to conduct the research that has produced this book.

J. GREGORY SIDAK

About the Author

J. GREGORY SIDAK is the F. K. Weyerhaeuser Fellow in Law and Economics at the American Enterprise Institute for Public Policy Research and a senior lecturer at the Yale School of Management. His research concerns telecommunications regulation, antitrust policy, and constitutional law issues concerning economic regulation. He directs the American Enterprise Institute's Studies in Telecommunications Deregulation.

Mr. Sidak served as deputy general counsel of the Federal Communications Commission from 1987 to 1989, and as senior counsel and economist to the Council of Economic Advisers in the Executive Office of the President from 1986 to 1987. As an attorney in private practice, he worked on numerous antitrust cases and federal administrative, legislative, and appellate matters concerning telecommunications and other regulated industries.

Mr. Sidak is coauthor, with William J. Baumol, of *Toward Competition in Local Telephony* (MIT Press and AEI Press 1994) and *Transmission Pricing and Stranded Costs in the Electric Power Industry* (AEI Press 1995). With Daniel F. Spulber, Mr. Sidak is coauthor of *Deregulatory Takings and the Regulatory Contract: The Competitive Transformation of Network Industries in the United States* (Cambridge University Press 1997) and *Protecting Competition from the Postal Monopoly* (AEI Press 1996). Mr. Sidak has published numerous articles on antitrust, telecommunications regulation, corporate governance, and constitutional law. He has testified before committees of the U.S. Senate and House of Representatives on regulatory and constitutional law matters, and his writings have been cited by the Supreme Court, the lower federal courts, state and federal regulatory commissions, and the Judicial Committee of the Privy Council of the House of Lords.

Mr. Sidak received A.B. and A.M. degrees in economics and a J.D. from Stanford University, where he was a member of the *Stanford Law Review*. He served as law clerk to Chief Judge Richard A. Posner during his first term on the U.S. Court of Appeals for the Seventh Circuit.

1

Beyond America's Borders

BEFORE 1996 most debate over U.S. telecommunications policy proceeded as though the known world ended at America's borders. Congress and the Federal Communications Commission focused on statutes and regulations that inhibited competition among various categories of American telecommunications firms in the U.S. market. That preoccupation was understandable, given the central role of the breakup of the Bell System and the obsolescence of the Communications Act of 1934, developments that together led to passage of the Telecommunications Act of 1996.[1]

Because of that inward focus, however, policymakers have failed to consider how American consumers could benefit from the entry of foreign telecommunications firms into the U.S. market—primarily through direct investment. Nor have they considered that the American telecommunications regulation offers consumers demonstrably inferior results compared with those enjoyed in certain other developed economies that have privatized their government telephone monopolies. This book is about those long neglected but consequential issues in American telecommunications policy.

U.S. policymakers no longer ignore entry and investment by foreign telecommunications carriers. In 1994 British Telecom (BT) received Justice Department and FCC approval to acquire 20 per-

---

1. Pub. L. No. 104-104, 110 Stat. 56 (1996).

cent of MCI, the second largest interexchange carrier (IXC) in the United States.[2] The same year, Deutsche Telekom and France Télécom announced an investment of $4.2 billion to acquire 20 percent of Sprint, America's third largest IXC.[3] It appeared in 1994 and 1995 that those foreign firms would be willing to invest even more in MCI and Sprint if not for section 310(b) of the Communications Act of 1934, which had the practical effect of capping at 25 percent the foreign ownership of an American telecommunications firm that is a licensee of radio spectrum.[4] Stated another way, in the absence of section 310(b), companies like MCI and Sprint would have greater freedom to accept foreign direct investment from additional overseas firms (such as Asian or Latin American carriers) to complement existing foreign participation, which by itself exhausted the 25 percent ownership then allowed by the FCC.

For years, liberalization of the foreign ownership restrictions was unthinkable. Although legal and economic analysis could readily expose the cost of the restrictions to American consumers, the FCC and Congress appeared to have been disinclined to liberalize foreign investment in telecommunications. Then, in 1994, perhaps because of the election of the Republican Congress, the political climate began to change noticeably. In the discussion concerning the telecommunications deregulation bills in Congress in 1995, the question became how, not whether, Congress should relax the restrictions on foreign investment. In the Telecommunications Act of 1996, however, Congress ultimately opted for incremental rather than revolutionary change in that regard and only repealed provisions in section 310(b) that had restricted the ability of foreigners to serve as officers or directors of U.S. radio licenses.[5]

---

2. Dept. of Justice, Antitrust Div., Proposed Final Judgment and Competitive Impact Statement: United States *v.* MCI Comm. Corp. and BT Forty-Eight Co., 59 FED. REG. 33,009 (1994); MCI Comm. Corp., 9 F.C.C. Rcd. 3960 (1995).

3. Andrew Adonis, *US Telecoms Alliance for France and Germany: Dollars 4Bn Stake in Sprint,* FIN. TIMES, June 15, 1994, at 1; Tom Redburn, *Sprint Forms European Alliance,* N.Y. TIMES, June 15, 1994, at D3. The Justice Department and the FCC approved the transaction in 1995. Department of Justice, Antitrust Div., Proposed Final Judgment and Competitive Impact Statement: United States *v.* Sprint Corp. and Joint Venture Co., 60 FED. REG. 44,049 (1995); Sprint Corp., Declaratory Ruling and Order, 11 F.C.C. Rcd. 1850 (1995).

4. 47 U.S.C. § 310(b)(4).

5. § 403(k), 110 Stat. at 131–32.

In scarcely more than a year, events on the international stage supplied the impetus for the revolutionary liberalization of telecommunications services that Congress had been unable to deliver in its amendments to section 310(b). In November 1996, a year after the FCC had promulgated regulations that implied that the agency would permit higher levels of investment by British companies,[6] BT announced that it would acquire the remaining portion of MCI for $22 billion—an amount that exceeded the sum of *all* previous foreign direct investment in American telecommunications.[7] Then, on February 15, 1997, the World Trade Organization reached an agreement on reducing barriers to trade in telecommunications services.[8] One element of that agreement was the liberalization of restrictions on foreign investment in telecommunications. Thus, the WTO agreement put the ball back in Congress's court with respect to reforming section 310(b).

To answer the question of how Congress would best reform the foreign ownership restrictions, we first examine their origins nearly a century ago. Chapter 2 shows that the national security concern underlying section 310(b) is narrower than is commonly believed. The desire to limit the ability of foreigners to communicate by radio can be traced to the Russo-Japanese War and the race for naval superiority before and during World War I. Radiotelegraphy acquired even greater military significance during World War I because of the advent of submarine warfare. The concern over the national security implications of wireless led, shortly after World War I, to the creation, and supervision by the U.S. Navy, of the Radio Corporation of America—a private corporation holding critical patents over radio hardware. Only when RCA's patent monopolies expired and competition in international wireless became feasible in the United States in the late 1920s did Congress perceive the need to limit foreign direct investment in wireless in the United States. In 1934 Congress enacted the statutory provision now designated as

---

6. Market Entry and Regulation of Foreign-Affiliated Entities, Rep. and Order, IB Dkt. No. 95-22, 11 F.C.C. Rcd. 3873 (1995).

7. MCI and BT Announce Largest International Merger in History, MCI Communications Corp. press release, Nov. 3, 1996 (www.mci.com/aboutmci/news/content/bt.shtml).

8. Office of U.S. Trade Representative, Statement of Ambassador Charlene Barshefsky, Basic Telecom Negotiations (Feb. 15, 1997).

section 310(b) of the Communications Act.

Nonetheless, refinements in encryption technologies soon vitiated the national security advantages that Congress expected section 310(b) to produce. The ineffectuality of section 310(b) became apparent only seven years after its enactment when, in 1941, Imperial Japanese diplomats readily transmitted encoded messages relevant to the imminent attack on Pearl Harbor from the United States to Tokyo over American-owned radio telegraph carriers. Thus, even in the relatively simple technological era in which Congress enacted section 310(b), the foreign ownership restrictions failed to protect America's national security.

In the six decades since enactment of the Communications Act, the FCC and the courts badly misinterpreted the statutory language of section 310(b). As chapter 3 shows, the FCC's reading of the operative provision of the statute has been contrary to law and has contributed to a complex and costly process that firms have repeatedly manipulated to impede their rivals' ability to use foreign direct investment to intensify competition.[9] Through a 1995 rulemaking the FCC sought to impose a principled structure on its application of section 310(b).[10] But the practical effect of that new regulatory edifice will simply be to deter foreign direct investment in U.S. radio licenses, for the rule will create more opportunities for incumbent American firms strategically to use the regulatory process to deter competitive entry. Contrary to the FCC's most recent policies, foreign direct investment is per se in the public interest, and the agency should bear the burden of proof when asserting otherwise.

In chapter 4 we descend from the lofty overlook of history and economics to the gritty practice of telecommunications law. The regulatory esoterica that have resulted from the myriad FCC decisions interpreting section 310(b) have skewed business decisions for the structuring of ownership and control in American telecommunications firms. The foreign ownership restrictions do not apply to wire-based telecommunications. Despite the national security origins of section 310(b), network security and survivability are not now and never have been the focus of the statute. The restrictions conse-

---

9. In 1997 congressional opponents of foreign investment in American telecommunications would again resort to that incorrect reading of the statute to attempt to upset the WTO negotiations.

10. Market Entry and Regulation of Foreign-Affiliated Entities, *supra* note 6.

quently do not impair the ability of a foreigner to own wireline assets. Section 310(b) therefore distorts investment choices from wireless to wireline technologies and needlessly raises transactions costs. Nevertheless, foreign investors and American telecommunications firms can usually achieve their desired economic outcomes under the foreign ownership restrictions; but those economic actors must do so at an artificially higher cost than if the restrictions did not exist. Foreign investors, in other words, have been constrained to resort to second-best ownership and control structures to comply with section 310(b). The business structures that the FCC has considered permissible in the past are unlikely to suffice as foreign direct investment takes place on a larger scale in transactions such as the merger of BT and MCI or the alliance among Sprint, France Télécom, and Deutsche Telekom.

For an industry as large as it is, there had been relatively little foreign direct investment in the U.S. telecommunications industry until the BT-MCI merger announced in November 1996. Chapter 5 argues, however, that the nature of such investment is changing in a manner that makes global telecommunications networks more valuable to consumers and hence more essential as an organizational form for competitive telecommunications firms. Foreign direct investment is a predictable characteristic of the ownership and control of such global networks, a characteristic intended to reduce the monitoring costs in a joint venture comprising firms from different countries, cultures, and regulatory environments. Section 310(b), however, can act as a barrier to new entry or enhanced competition in the United States.

Given the benefits in the United States and abroad from foreign direct investment in telecommunications, we ask in chapter 6 why the restrictions in section 310(b) persist. Few informed persons today believe that concerns over national security continue to motivate the FCC's current enforcement of the foreign ownership restrictions. Instead, to the extent that section 310(b) is today anything more than a chauvinist device to keep foreigners from buying American companies in a particular industry, the statute has been appropriated as a crude tool with which to secure reciprocal market access for American carriers seeking to make direct investments overseas. In an attempt to refine that policy instrument, Congress considered, in 1994 and 1995, repealing section 310(b) or amending it to provide specifically for the FCC to condition foreign invest-

ment on the existence of reciprocal market access for American
direct investment in the telecommunications industry of the nation in
question. Yet, there has been no effort to analyze the relative merits
of repeal and reciprocity in light of the current economic literature
on classical trade theory and strategic trade theory. Nor has there
been any effort to use that body of economic thought to analyze the
FCC's 1995 rule imposing a reciprocity standard as an exercise of
agency interpretation of its existing statutory authority under section
310(b). Economic analysis, however, shows that the FCC's 1995
rule is counterproductive. Moreover, other government agencies
have greater expertise to make trade policy, and it is unlikely that
the FCC could replicate their expertise without considerable expenditures of time and resources.

Contrary to the usual assessment that outright repeal of section
310(b) would be politically naïve and counterproductive to the
interests of U.S. telecommunications firms, the adoption of a reciprocity standard could easily harm American consumers and producers. Continued adherence to the FCC's traditional (but erroneous) interpretation of section 310(b) as imposing a 25 percent ceiling on foreign investment is likely to *reduce* market access for
American telecommunications firms. The collapse of the World
Trade Organization talks on telecommunication services in 1996,
when the United States walked out, surely could not be interpreted
as evidence that section 310(b) was working as a crowbar to open
overseas markets to investment and competitive entry by U.S. telecommunications firms.[11] As experience from the 1997 WTO negotiations confirmed, section 310(b) in its current form is unlikely to
secure market access for American firms because it gives foreign
governments a pretext for restricting American direct investment in
their telecommunications markets. Repealing section 310(b) would
be more likely to remove that pretext than would the FCC's strategy
of turning section 310(b) into an explicit reciprocity standard. Furthermore, given the abysmal experience of countless FCC policies
that have regulated market entry since 1934, one can only expect
that adding a reciprocity analysis to section 310(b) would encourage

---

11. Paul Lewis, *Telecommunications Talks Postponed as U.S. Balks*, N.Y.
TIMES, May 1, 1996, at D4; Bhushan Bahree, *U.S. Takes a Hard Line on Telecom*,
WALL ST. J., Apr. 30, 1996, at A3.

incumbent American firms to use the process to impede entry. That result would hardly encourage foreign governments to liberalize their restrictions on foreign direct investment in telecommunications.

Chapter 7 employs a completely different set of analytical tools to formulate the *coup de grâce* for the foreign ownership rules: Section 310(b) violates the First Amendment by restricting the ability of foreigners to speak through the electronic media or by restricting the ability of Americans to hear the views of foreigners.[12] Under current constitutional law, different First Amendment analysis applies to the mass media than to common carriers. Content-based restrictions, however, receive strict scrutiny, and the News Corp. case involving Rupert Murdoch and the Fox Network illustrates how section 310(b) can be used in a way that is *not* content neutral.[13] Under several different lines of analysis, the existing case law would support the conclusion that the foreign ownership restrictions violate the First Amendment. Furthermore, the Supreme Court's 1994 *Turner Broadcasting* decision[14] and Bell Atlantic's successful First Amendment attack on the cable–telephone company entry ban[15] may signal a sea change in the First Amendment protection of electronic speech generally. In the meantime, an embarrassing inconsistency exists between the FCC's interpretation of section 310(b) and the Supreme Court's 1990 analysis in *Metro Broadcasting*[16] of the importance of promoting diversity of ownership and diversity of expression. Consequently, in the name of protecting Americans from the threat of foreign ownership and control of broadcasting, the FCC has rebuffed attempts by Mexican and Taiwanese nationals to invest in radio and television stations planning

---

12. My experience has been that many economists dismiss such analysis as extraneous legalisms. In doing so, they seem not to recognize that, by ignoring or simplifying the actual constraints that legal rules impose on regulatory institutions, they sacrifice the superior rigor that they presume economic analysis to have over legal analysis.

13. Fox Television Stations, Inc., Mem. Op. and Order, 10 F.C.C. Rcd. 8452 (1995).

14. Turner Broadcasting Sys., Inc. *v.* FCC, 114 S. Ct. 2445 (1994).

15. Chesapeake & Potomac Tel. Co. of Va. *v.* United States, 42 F.3d 181 (4th Cir.), *cert. granted*, 115 S. Ct. 2608 (1995), *vacated and remanded*, 116 S. Ct. 1036 (1996).

16. Metro Broadcasting, Inc. *v.* FCC, 497 U.S. 547 (1990).

to offer Spanish-language and Chinese-language programming to minority audiences in the United States.

Chapter 8 examines the World Trade Organization agreement on telecommunications services that was concluded in Geneva on February 15, 1997. That agreement ensured that the controversy over foreign investment in American telecommunications would continue. As of the spring of 1997, it seemed likely that Congress would amend or repeal section 310(b) to comply with the WTO agreement. Consequently, the successful completion of the WTO talks also had the effect in the United States of renewing a debate, nearly a century in duration, over foreign investment in American telecommunications.

In sum, the message of this book is simply this: The United States must exercise leadership in reducing barriers to foreign direct investment in telecommunications. To reiterate, there are five reasons why Congress should seek to do so by simply repealing section 310(b). First, as an exhaustive assessment of the legislative history makes clear, the statute no longer serves its national security purpose—if it ever did. Second, the statute, as interpreted and enforced by the FCC, has worked a result that is contrary to law and to sound public policy. Third, the statute denies American consumers the benefits of lower prices and more robust innovation that would flow from the full participation of the world's largest telecommunications firms in the U.S. market. Fourth, section 310(b) hinders American direct investment abroad; the statute thereby reduces American producer welfare and forecloses opportunities for the increased competitiveness of American telecommunications firms on a global scale. Fifth, section 310(b), as it has been applied by the FCC, unconstitutionally restricts the freedom of electronic speech. By repealing the foreign ownership restrictions, Congress would increase the likelihood that its efforts to enhance competition among telecommunications firms would bear fruit not only domestically, but also in the growing markets that extend beyond America's borders.

2

The Legislative History

THE RADIO ACT OF 1912[1] is often considered the origin of the foreign ownership restrictions in the Communications Act of 1934.[2] In fact, the legislative history of such restrictions must begin with the outbreak of the Russo-Japanese War in 1904. The Japanese Navy's use of wireless communications to crush the Russian Fleet ushered in a new age of communications awareness, elevating the strategic control of wireless communication to a paramount issue of national security. To secure that vital control, Congress, at the U.S. Navy's urging, imposed alien ownership restrictions on wireless communica-

---

1. Act of Aug. 13, 1912, ch. 287, § 2, 37 Stat. 302 (1912).

2. *See, e.g., Trade Implication of Foreign Ownership Restrictions on Telecommunications Companies: Hearings on Section 310 of the Communications Act of 1934 Before the Subcomm. on Commerce, Trade, and Hazardous Materials of the House Commerce Comm.*, 104th Cong., 1st Sess. 15 (1995) (statement of Reed E. Hundt, Chairman, Federal Communications Commission); *Hearing on Telecommunications Policy Reform: Hearing on S. 253 Before the Subcomm. on Telecommunications of the Sen. Comm. on Commerce, Science, and Transportation*, 104th Cong., 1st Sess. 236 (1995) (testimony of Eli M. Noam); Ian M. Rose, Note, *Barring Foreigners from Our Airwaves: An Anachronistic Pothole on the Global Information Highway*, 95 COLUM. L. REV. 1188, 1194 (1995); Moving Phones Partnership L.P. *v.* FCC, 998 F.2d 1051, 1056 (D.C. Cir.) *cert. denied*, 114 S. Ct. 1369 (1994); John J. Watkins, *Alien Ownership and the Communications Act*, 33 FED. COMM. L.J. 1, 1, 4 (1980).

tions companies operating within the United States to protect American military interests during time of war or international unrest. The scope of those alien ownership restrictions manifested itself to greater and lesser degrees during the defining regulatory periods of 1904 to 1912, 1912 to 1934, and 1934 to the present day.

Ironically, as a means to achieving national security, foreign ownership restrictions in U.S. telecommunications policy have been ineffective. While intuitively appealing, such restrictions were neither properly conceived nor well tailored to achieve their intended purpose—control of the transmission to unfriendly recipients of information vital to U.S. interests, particularly in time of war. In fact, although restrictions on the foreign ownership of U.S. wireless companies successfully limited access to important radio hardware early in the twentieth century, they were largely ineffective in curbing the threats to national security from encrypted messages sent by foreigners over U.S carriers or from the dissemination of alien propaganda. Indeed, as the history of governmental control over wireless companies in both world wars documents, despite such alien ownership restrictions, Germany and Japan made use of American wireless communications systems to compromise U.S. national security.

With those considerations in mind, we shall now trace the development of foreign ownership restrictions in the U.S. wireless communications industry, the degree to which national security and military concerns animated their implementation, and the failure of those restrictions in ultimately securing their strategic objectives.

PIONEERS OF THE ETHER

The earliest days of radio communications in the United States were typified by the sporadic deployment of that new technology by the military, by the military's persistent attempts to exclude amateur and commercial use of radio by civilians, and by the unwillingness of Congress and the executive branch to let the military succeed, other than in wartime, in monopolizing radio.[3] The U.S. Navy, for reasons

3. For historical analysis of this period, see JONATHAN W. EMORD, FREEDOM, TECHNOLOGY, AND THE FIRST AMENDMENT 138–45 (Pacific Research Institute for Public Policy 1991); ANDREW F. INGLIS, BEHIND THE TUBE: A HISTORY OF BROADCASTING TECHNOLOGY AND BUSINESS 28–56 (Focal Press/Butterworth Publishers 1990); SUSAN J. DOUGLAS, INVENTING AMERICAN BROADCASTING

discussed below, disliked the use of radio by *any* civilians, domestic or foreign. In time, the fact that some of the most logical companies to spearhead the commercial development of radio for civilian purposes happened to be British corporations would provide the Navy, through its recitation of national security concerns over foreign control of radio licensees, with a convenient way to slow the growth of civilian use of radio.

That the Navy proved to be the fulcrum upon which radio communications regulation in America turned is not surprising. Superior communications confers a strategic and often decisive advantage on a combatant. As early as the American Civil War, the Union army had strung telegraph wires from aerial observers in balloons, which, wrote General Ulysses S. Grant, created a communications network linking "each division, each corps, each army and . . . my headquarters."[4] With the strategic implications of electronic communications in mind, the U.S. military (and more specifically the Navy) seemed a natural consumer of the emerging technology of wireless, which would enable ships at sea to communicate ship-to-ship and ship-to-shore. As Susan Douglas noted in her definitive history of radio, *Inventing American Broadcasting, 1899–1922*:

> To wireless inventors eagerly looking for customers, none seemed more promising than the U.S. Navy. Fresh from its victories in the Philippines and Cuba, and in the midst of a successful renovation and modernization, the U.S. Navy still lacked a critical tool: a reliable and versatile method of communications which could keep ships in touch with one another and with shore. Other

---

1899–1922 (Johns Hopkins University Press 1987); LLEWELLYN WHITE, THE AMERICAN RADIO (University of Chicago Press 1971); ERIK BARNOUW, A TOWER IN BABEL: A HISTORY OF BROADCASTING IN THE UNITED STATES TO 1933 (Oxford University Press 1966); A. WALTER SOCOLOW, THE LAW OF RADIO BROADCASTING (Baker, Voorhis & Co. 1939); TYLER BERRY, COMMUNICATIONS BY WIRE AND RADIO (Callaghan & Co. 1937); GLEASON T. ARCHER, HISTORY OF RADIO TO 1926 (American Historical Society 1938; reprinted Arno Press 1971); CLINTON B. DE SOTO, TWO HUNDRED METERS AND DOWN: THE STORY OF AMATEUR RADIO (American Radio Relay League 1936). Many of the obscure original documents concerning the early regulation of radio are reprinted in DOCUMENTS IN TELECOMMUNICATIONS POLICY (John M. Kittross ed., Arno Press 1977) [hereinafter DOCUMENTS].

4. *Quoted in* R. ERNEST DUPUY & TREVOR N. DUPUY, THE HARPER ENCYCLOPEDIA OF MILITARY HISTORY: FROM 3500 B.C. TO THE PRESENT 900 (HarperCollins Publishers, 4th ed. 1993).

navies were acquiring such a tool, and the United States Navy
could not afford to fall behind.[5]

Thus, just as the U.S. Navy needed inventors and entrepreneurs to
bring it wireless communication, so did those pioneers of the ether
need the Navy—their largest, and most natural, potential client. That
interdependence initiated the Navy's direct involvement not only in
the development of wireless communication in the United States, but
also in the regulatory schemes that sought to restrict alien ownership
of American wireless companies and culminated in the formation of
the Radio Corporation of America. The history of that interrela-
tionship is necessary to an understanding of the legislative history of
U.S. telecommunications regulation, particularly with respect to the
foreign ownership restrictions contained in section 310(b) of the
Communications Act of 1934.

RETICENCE AND MONOPOLY

Given the strategic importance of integrating wireless communication
into the U.S. Navy's command-and-control structure, it is ironic that
the Navy was at first slow to embrace the powerful technology of
radiotelegraphy. The Navy's initial reaction to that emergent technology was a curious mix of reticence and fondness for monopoly. The
Navy's reticence toward adopting the new technology was rooted in
its traditional, if not anachronistic, conceptions of a captain and his
ship. To the old Navy, wireless communication threatened the autonomy of a captain's command while at sea; and the captains treated
wireless accordingly, either shutting it down completely once under
way or simply ignoring calls from shore.[6]

The counterbalance to that antipathy toward wireless among
naval commanders was their awareness of radiotelegraphy's enormous
strategic potential and the perils of failing to assimilate the new
technology. That awareness would increasingly manifest itself in a
desire to monopolize wireless technology under direct naval control.
Riven by those conflicting attitudes toward wireless communication,
the Navy's early assimilation of radiotelegraphy was predictably

---

5. DOUGLAS, *supra* note 3, at 102.

6. *Id.* at 134.

equivocal. The Navy wanted to ignore the new technology and at the same time ensure that no one else could use it to superior strategic advantage. That untenable position ultimately collapsed upon itself and so impeded not only the Navy's incorporation of wireless technology, but also the efforts of entrepreneurs trying to realize its full commercial and military potential.

FLIRTATION AND RIVALRY WITH MARCONI

Despite its ambivalence at the turn of the century, the Navy initially appeared eager to embrace wireless technology. Three years after British inventor and entrepreneur Guglielmo Marconi received the first patent for the radiotelegraph in 1896,[7] the U.S. Navy asked him to demonstrate his device. At the completion of Marconi's demonstration, the Navy was inclined to purchase his system—that is, until Marconi revealed his excessive price and unduly restrictive terms, which dictated not only the number of units the Navy must purchase, but also required the payment of annual royalties.[8]

Particularly galling to the Navy was Marconi's requirement that the Navy restrict its communications to only those ships and stations that owned and operated Marconi systems.[9] John D. Oppe, vice president and general manager of the American Marconi Company, a wholly owned subsidiary of British Marconi, detailed that restriction (along with seven others only slightly less demanding) in a strikingly presumptuous letter to the Navy's Bureau of Equipment:

> Except in time of emergency or war or in the case of war vessels, the Bureau shall not use the Marconi wireless apparatus fitted at their stations for the interchange of signals with vessels or stations not equipped with apparatus provided by the Marconi company.[10]

Marconi's terms were not animated by simple avarice. He was keenly

---

7. *See* WHITE, *supra* note 3, at 11.

8. DOUGLAS, *supra* note 3, at 111–12.

9. *Inter-Departmental Board [on] Wireless Telegraphy: Inter-Departmental Board Appointed by the President to Consider the Entire Question of Wireless Telegraphy in the Service of the National Government, Washington, D.C., 1904,* in 1 DOCUMENTS, *supra* note 3, at 23.

10. *Id.*

aware that he could not hope to own the ether and lease its use. Nonetheless, as the pioneer of that new frontier and the first to demonstrate its commercial potential, he believed that he rightfully deserved to reap his share of wireless's economic rewards.[11] The only means to ensure his "rightful" bounty, Marconi realized, was to control the gateway to its use by creating his own self-contained and exclusionary operating system.[12] As the dominant player in the emerging market of wireless, Marconi believed that if he were initially successful in imposing his exclusionary operating system on the majority of wireless users, then he would effectively control the airwaves.

Marconi's scheme, however, faced two obstacles. First, his restrictive terms and high costs rankled many of his potential customers as imperious and greedy. Second, and more critically, if Marconi succeeded in his plan, the British, who already controlled a majority of the world's submarine cables, would control the ether as well.[13] The prospect of conceding effective control over so strategic a military asset as communications was unacceptable to the other world powers. Those considerations led the Navy to spurn the American Marconi Company and go it alone.[14] Selected to spearhead the Navy's initial efforts was Francis M. Barber, a retired naval officer with extensive international experience and an expertise in electrical engineering. His assignment was to procure and assemble an in-house radiotelegraphy system that would, in his words, "be able to drive the American Marconi Company out of business."[15]

As those sentiments reveal, considerations beyond simply price motivated the Navy's rejection of Marconi's system. Both parties were keenly aware that in the balance hung the question of who would control the new technology. Marconi's draconian terms of control were driven by his desire to earn monopoly rents on his new technology; the Navy's rejection of Marconi's system not only was a refusal to accede to his price and terms, but also reflected the Navy's own growing ambition to monopolize wireless communica-

---

11. DOUGLAS, *supra* note 3, at 106.

12. *Id.*

13. *Id.*

14. *Id.* at 115.

15. *Quoted in id.* (letter from Commander F. M. Barber to Admiral R. B. Bradford, Chief, Bureau of Equipment, Apr. 2, 1902, National Archives, box 85).

tion. Wrote Susan Douglas: "Ironically, [the Navy] wanted exactly what they condemned Marconi for pursuing: a monopoly of the airwaves."[16] Thus began the fierce rivalry between the Navy and American Marconi for control over the radio waves that would produce alien ownership restrictions and create the Radio Corporation of America.

THE NAVY'S PATCHWORK

The Navy initially relied upon European and American equipment cobbled together not according to any assimilative technological plan, but according to what was least expensive.[17] Consequently, the Navy's radio communications systems became a patchwork of different manufacturers' equipment, cannibalized and reconfigured to meet the Navy's idiosyncratic operational needs—needs often at odds with the efficient use of the new technology.[18] As a result, the Navy ended up with a fragmented wireless network that was only crudely functional.

By 1904, despite having twenty radio stations along the nation's coast and thirty-four ships equipped or being equipped with radiotelegraphs, the Navy still had no standardized operating platform.[19] Each radiotelegraphy station used its own peculiar operating system such that radio operators, on being transferred to a new station or ship, often had to learn anew the systems and their transmitting procedures.[20] To overcome that problem, many a radio operator improvised his own radio unit, which he carried with him, hooking it into whatever existing system he found at his new station and removing it when he left.[21] Thus, four and a half years after Marconi's demonstration, the Navy had succeeded not in displacing Marconi, but in creating a system that was woefully inadequate, particularly when

---

16. *Id.* at 134.

17. *Id.* at 136.

18. *Id.*

19. EMORD, *supra* note 3, at 139 (citing JOHN O. ROBINSON, SPECTRUM MANAGEMENT POLICY IN THE UNITED STATES: AN HISTORICAL ACCOUNT 6 (Federal Communications Commission, Office of Plans and Policy, Apr. 1985); *Inter-Departmental Board*, *supra* note 9, at 7).

20. DOUGLAS, *supra* note 3, at 136.

21. *Id.*

compared with those possessed by the world's other naval powers. In 1902, *Electrical World*—then one of the leaders of the technical press—published its somber assessment of the Navy's efforts to integrate wireless: "As matters stand now, we would be at a great disadvantage in this respect if attacked by any reasonable power."[22] Nor did matters improve over the next several years.

THE RUSSO-JAPANESE WAR
AND THE ROOSEVELT BOARD

The potential consequences of the Navy's bumbling efforts were put in stark relief by a confluence of events beginning in 1904 that underscored wireless's immense military and commercial potential. On February 4, 1904, the Russo-Japanese War began when a surprise attack by Japanese torpedo boats destroyed Russia's fleet anchored in the harbor of Port Arthur.[23] In the culminating Battle of Tsushima in May 1905—described by military historians as "the greatest naval battle of annihilation since Trafalgar"—the Japanese sank or captured all but six ships in Russia's entire fleet, including three battleships.[24] Efforts to explain Japan's stunning naval victories centered on the superiority of Japan's Marconi wireless system over the Russian Navy's German-made equipment.[25] Commercially, the Russo-Japanese War was equally significant in demonstrating wireless telegraphy's manifold applications, as journalists for the first time used radiotelegraphy to file their war reports from halfway around the world.[26]

The lessons of the Russo-Japanese War concerning the strategic importance of wireless communication were not lost on one particularly interested observer: President Theodore Roosevelt. Surveying the early stages of the war, Roosevelt was struck by two realizations concerning wireless technology. First, the war powerfully demonstrated the strategic significance of having an effective wireless communications system, particularly with respect to maintaining a strong

---

22. *Id.* at 119 (*quoting* 40 ELECTRICAL WORLD, no. 10, at 354 (1902)).

23. DUPUY & DUPUY, *supra* note 4, at 1008–9.

24. *Id.* at 1014.

25. W. P. JOLLY, MARCONI 148 (Stein & Day 1972); DOUGLAS, *supra* note 3, at 123.

26. INGLIS, *supra* note 3, at 48.

navy—then viewed as the principal index of international power.[27] Second, Roosevelt was concerned that the radio interference generated by the press and other private wireless entities transmitting in the theater of operations, would threaten the military's ability to communicate effectively during wartime conditions.[28] Therefore, he appointed a presidential commission with pronounced allegiances to the Navy to analyze the "entire question of wireless telegraphy"—the Interdepartmental Board of Wireless Telegraphy, or, as it was widely known, the Roosevelt Board.[29] So swift was Roosevelt's action to improve America's wireless position that, by the time his Nobel prize–winning efforts brought Russia and Japan to peace in Portsmouth, New Hampshire, in August of 1905, the board had already been at work for a year studying the problems the U.S. government encountered in using wireless.[30]

One military danger that the Roosevelt Board explicitly mentioned was the potential for torpedo boats to be equipped with radios.[31] The attack on Port Arthur had heightened the understanding of how small, versatile attack vessels could be far more effective predators if radio communications could coordinate their actions and inform them of the movements of enemy ships.

Not surprisingly, the government-oriented Roosevelt Board concluded that to prevent radio interference in ship-to-ship and ship-to-shore communications, the wireless communication industry should be brought "under full governmental supervision,"[32] with the Navy being charged with the responsibility of overseeing the national integration of wireless communication:

> The Board believes it to be in the interest not only of governmental, but public economy and efficiency, to permit the naval stations to handle the public service, for in the present state of the art but one station is desirable for the public interests in such places.[33]

---

27. DOUGLAS, *supra* note 3, at 123.
28. *Id.* at 123–24.
29. EMORD, *supra* note 3, at 139.
30. INGLIS, *supra* note 3, at 51; *Inter-Departmental Board, supra* note 9.
31. *Id.* at 19.
32. *Id.* at 11.
33. *Id.* at 10.

To that end, the board recommended that the Navy build a nationwide wireless network using a standardized system.[34] The board presented several rationales to justify that government takeover of wireless. The first was that

> [t]his method of placing private stations under full government supervision is desirable in order to regulate them for their mutual and the public welfare, as well as from considerations of national defense. Aside from the necessity of providing rules for the practical operation of such stations, it seems desirable that there should be some wholesome supervision of them to prevent the exploitation of speculative schemes based on a public misconception of the art.[35]

To that vague, paternalistic justification the board added a second equally dubious rationale for government supervision of radio—namely, that such a takeover was necessary "[t]o prevent the control of wireless telegraphy by monopolies or trusts,"[36] despite the absence of any evidence that such a threat existed at the time.[37]

ENTREPRENEURS AND GOVERNMENT INTERVENTION

Roosevelt accepted the board's recommendations and ordered his secretary of the Navy, Paul Morton, to execute them.[38] The board's efforts were undone by several factors, however. First, its proposals provoked a hostile reaction from the public, which looked with suspicion upon government intervention in private industry.[39] Second, the rapidly evolving nature of the technology itself outstripped the board's efforts to bring it under government supervision. Most notably, in 1906, Canadian-born inventor and entrepreneur Reginald A. Fessenden, sending Christmas greetings to the U.S. Navy's ship radio operators, was the first to transmit the human voice by radio

---

34. *Id.*
35. *Id.*
36. *Id.* at 11.
37. EMORD, *supra* note 3, at 139.
38. *Id.*
39. DOUGLAS, *supra* note 3, at 125–26.

waves.[40] Although those greetings heightened the Navy's desire to end private interference in wireless communications, they foreshadowed an explosion in private commercial and amateur radio use.

The civilian public was intrigued by the new medium and the accessible nature of the materials required to construct one's own radio transmitter. Between 1907 and 1912 the electromagnetic spectrum became increasingly occupied by radio users of all types roaming the meter band in search of contact. The Navy's wireless operators were favorite targets. The Navy, fearing that this uncontrolled proliferation of radio use would degrade maritime communications, again strongly lobbied Congress for comprehensive federal regulation of wireless communication, insisting that the ether be declared government property. Joining the Navy in that call for nationalizing the wireless industry was Charles Nagel, secretary of commerce and labor. Nagel warned that the unrestrained use of the radio waves would create a dissonant "babel," frustrating the government's efforts to put the new technology to productive use.[41]

The Navy's arguments for military control over the wireless industry were met with skepticism, given the Navy's own spotty history of integrating wireless into its command structure. Despite the Roosevelt Board's study, and despite the Navy's own awareness of the strategic role that wireless communications promised to play in military warfare, the Navy's record in integrating the new technology continued to be significant largely for its futility. In 1909 the Bureau of Equipment was still receiving reports of Navy wireless shore stations so neglected and obsolete as to be inoperative.[42] One report even requested that a station be moved from the deteriorating pigeon coop in which it had first been installed to more suitable premises.[43] Needless to say, such reports did not enhance Congress's confidence that the Navy would be a diligent steward of wireless's integration into America's military and commercial infrastructures.

Nor did the nation's general entrepreneurial ethos aid the Navy's cause. The public, and its representatives in Congress, viewed with suspicion any effort to impose government control over private enterprise, particularly a nascent industry as promising and dynamic as

---

40. EMORD, *supra* note 3, at 139.
41. *Id.* at 140.
42. DOUGLAS, *supra* note 3, at 137.
43. *Id.* at 137.

wireless communication. The press almost universally decried the board's recommendations. The *New York Times* opined that the proposals needlessly threatened a government takeover of "an art which is yet only in an embryo state of development."[44] *Electrical World*, singling out the Navy for derision, denounced the board's recommendations:

> The Navy Department is particularly disqualified at the present time from becoming the custodian of wireless. . . . Such a policy cannot be too strongly condemned, not only because it involves an extension of military authority over what in times of peace is a purely commercial function, but because of the deadening effect on the development of the art that would inevitability result in bureaucratic control. . . . That such development would occur under military domination none, we believe, will seriously assert.[45]

The Navy would continue for the next several years to press for more government control of the ether, but it would have to wait until World War I for Congress finally to intervene decisively in the nation's wireless industry and begin restricting foreign direct investment. In the meantime, Congress's early efforts to regulate radiotelegraphy would continue to be most remarkable for their narrowness of scope and circumspection.

THE FIRST FEDERAL TELECOMMUNICATIONS LEGISLATION

Undaunted by the earlier defeat of its efforts to declare the airwaves government property, the Navy seized upon sporadic instances of interference to continue to lobby for complete federal supervision of the wireless industry. Testifying in 1910 before the Senate with respect to the proposed Wireless Ship Act, the Navy painted a grim picture of the wireless situation unless it was brought under government control:

---

44. N.Y. TIMES, Sept. 10, 1906, at 6.
45. 43 ELECTRICAL WORLD, no. 23, at 1068 (1904); 44 ELECTRICAL WORLD, no. 9, at 319 (1904).

Calls of distress from vessels in peril on the sea go unheeded or are drowned out in the etheric bedlam produced by numerous stations all trying to communicate at once. . . . It is not putting the case too strongly to state that the situation is intolerable, and that it is continually growing worse.[46]

Secretary of Labor and Commerce Nagel echoed the Navy's call for nationalizing the radio spectrum and warned, "The ether is common property, and with the cheapest apparatus unrestrained trivial messages can create Babel."[47] Nagel also raised the threat of the economic exploitation of that common resource, ominously prophesying that, unless the ether was brought under government control, "in time one company will absorb the other and establish a monopoly."[48] The Navy and Nagel, in short, presented an apocalyptic vision of wireless if it were allowed to develop free of centralized government control. That vision, however, failed to convince Congress that radical moves toward nationalization and central planning were necessary.

Instead, Congress enacted the Wireless Ship Act in 1910, which was directed at a far narrower goal—the better protection of life at sea.[49] To that end, the act required only that radio equipment and operators be present on all ships leaving U.S. ports with fifty or more persons on board. That first tentative effort at wireless regulation illuminated not only Congress's small appetite for regulating the new industry, but also the country's still limited perceptions of radiotelegraphy's power and potential. Congress and most Americans in 1910 perceived radio exclusively as a facilitator of wireless telegraphy and telephony—a tool of point-to-point communications, not point-to-multipoint. Though acknowledged for its profound economic and military significance—enabling ship-to-shore and ship-to-ship communications, as well as intercontinental communications that bypassed the existing network of undersea cables—wireless's other innovative

---

46. S. REP. NO. 659, 61st Cong., 2d Sess. 4 (1910) (testimony of Lieutenant Commander Clelland Davis, Bureau of Equipment, U.S. Navy).

47. *Selections from Reports of the Department of Commerce and Labor, 1911, in* 1 DOCUMENTS, *supra* note 3, at 669.

48. *Id.* at 673.

49. Act of June 24, 1910, 36 Stat. 629 (1910). *See* STEPHEN BROOKS DAVIS, THE LAW OF RADIO COMMUNICATION 32 (McGraw-Hill 1927).

applications, such as radio broadcasting, would not emerge for another decade.

The Wireless Ship Act was not the only pioneering legislation concerning telecommunications that Congress enacted in 1910. The Mann-Elkins Act of 1910 gave the Interstate Commerce Commission jurisdiction over interstate telegraphy and telephony and made telephone companies common carriers, thus requiring them to provide service at just and reasonable rates on nondiscriminatory terms.[50] Communications regulation was not, however, the primary focus of the Mann-Elkins Act. Its original purpose was to strengthen the ICC's regulatory authority over railroads and to establish a specialized court to review ICC orders.[51] In an ironic twist, both Representative James R. Mann and Senator Stephen B. Elkins opposed the amendment regulating telephone-telegraph carriers, arguing that it should be dealt with by separate legislation specifically focused on the needs and problems of those industries, rather than be subsumed under railroad regulation.[52]

The American Telegraph & Telephone Company and other independent telephone companies embraced and supported that new regulation.[53] Although the ICC's record as a regulator of telecommunications would, in retrospect, look rather anemic and lead Congress to transfer the agency's jurisdiction over interstate telecommunications to a new Federal Radio Commission in 1927, that body of railroad regulators would, for the first few years at least, use its powers under the Mann-Elkins Act to leave its mark on American telecommunications. With the Department of Justice, the ICC investigated the allegedly monopolistic enterprises of AT&T.[54] Under the threat of antitrust action by the ICC, AT&T agreed to the so-called Kingsbury Commitment in 1913.[55] The agreement required AT&T to provide interconnection to, and to stop acquiring, indepen-

50. Commerce Court (Mann-Elkins) Act, ch. 309, § 7, 36 Stat. 539 (1910). *See also* MICHAEL K. KELLOGG, JOHN THORNE & PETER W. HUBER, FEDERAL TELECOMMUNICATIONS LAW 15 (Little, Brown & Co. 1992).

51. MAX PAGLIN, A LEGISLATIVE HISTORY OF THE COMMUNICATIONS ACT OF 1934, at 6 (Oxford University Press 1989).

52. *Id.*

53. *Id.*

54. *Id.* at 8.

55. *Id.* at 7.

dent telephone companies.[56] The Kingsbury Commitment and its blanket restriction on AT&T's acquisition of independent carriers was terminated in 1921.[57] In its place Congress enacted the Willis-Graham Act, conferring authority on the ICC to approve telephone transactions and thereby grant antitrust immunity to those approved.[58] In that capacity the ICC made its largest impression on the field, but otherwise influenced regulatory policy only negligibly.[59]

THE RADIO ACT OF 1912

Following the sinking of the *Titanic* in 1912, Congress revisited the issue of radio interference and wireless's regulation in general. The *Titanic* disaster propelled radiotelegraphy into the fore of public consciousness, as wireless played a dramatic role in relaying the terrible events as they unfolded to a disbelieving nation. In the disaster's aftermath there was considerable speculation that amateur radio use interfered with the other ships' and stations' ability to hear the *Titanic*'s distress signal and thereby hampered rescue efforts. In response, Congress passed the Post-*Titanic* Radio Communications Act—or as it more simply became known, the Radio Act of 1912[60]—prohibiting the use of wireless for radio communication without a license issued by the secretary of commerce and labor.[61]

The new statute imposed other regulatory restrictions as well. It required that applicants designate a specific wavelength on which they proposed to operate.[62] It limited wavelength use to 300 and 600 meters.[63] It prohibited private or commercial shore stations from using their transmitters during the first fifteen minutes of each hour to prevent interference with naval vessels that used that period to

---

56. *See id.* at 80–81; PETER TEMIN, THE FALL OF THE BELL SYSTEM: A STUDY IN PRICES AND POLITICS 9–11 (Cambridge University Press 1987).

57. PAGLIN, *supra* note 51, at 8.

58. Willis-Graham Act, ch. 20, 42 Stat. 27 (1921) (codified at 47 U.S.C. § 221(a)). For discussion of this legislation, see KELLOGG, THORNE & HUBER, *supra* note 50, at 148; PAGLIN, *supra* note 51, at 8.

59. PAGLIN, *supra* note 51, at 8.

60. Radio Communications Act of Aug. 13, 1912, Pub. L. No. 62-264, 37 Stat. 302 (1912).

61. *Id.* § 1.

62. *Id.* § 2.

63. *Id.* § 4, reg. 1.

transmit their signals.[64] It empowered the secretary of commerce and labor to change meter-band limitations and to revoke licenses for "good cause," though not to deny any applicant a license.[65] And it gave the president the power, which Woodrow Wilson would soon exercise, to seize any radio apparatus in time of war.[66]

The Radio Act of 1912 also introduced foreign ownership restrictions into U.S. communications regulation. With the eager tutelage of the U.S. Navy, Congress by 1912 clearly comprehended the economic and military implications of radio. The new statute mandated that the secretary of commerce and labor grant radio licenses to appropriate applicants, though it gave him discretion only to select a proper wavelength.[67] Section 2 provided that "such license shall be issued only to citizens of the United States or Porto Rico, or to a company incorporated under the laws of some State or Territory or of the United States or Porto Rico."[68] At the Navy's behest, Congress inserted the citizenship requirement to prevent foreign agents in the United States from transmitting messages by radio to other nations, especially in time of war or other international tension.[69] Additionally, contemporary marine law, which served as a model for the 1912 legislation because radio was at that time largely a marine operation, required ship masters to be U.S. citizens.[70]

---

64. *Id.* § 4, reg. 11.

65. *Id.* § 4, reg. 1.

66. *Id.* § 2.

67. Hoover *v.* Intercity Radio Co., 286 F. 1003 (D.C. Cir. 1923). For a discussion of that decision, see THOMAS G. KRATTENMAKER & LUCAS A. POWE, REGULATING BROADCAST PROGRAMMING 5–7 (MIT Press & AEI Press 1994).

68. § 2, 37 Stat. at 1.

69. *Radio Communication: Hearings on H.R. 15357 Before the House Comm. on the Merchant Marine and Fisheries*, 62d Cong., 2d Sess. 70 (1912) (statement of Lieutenant Commander David W. Todd, U.S. Navy) [hereinafter *H.R. 15357 Hearings*]; *Radio Communication: Hearings on S. 3620 and S. 5334 Before the Subcomm. of the Senate Comm. on Commerce*, 62d Cong., 2d Sess. 9, 36 (1912) (statement of Lieutenant Commander Todd) [hereinafter *S. 3620 and S. 5334 Hearings*].

70. *S. 3620 and S. 5334 Hearings*, *supra* note 69, at 36–37 (statement of E. T. Chamberlain, Commissioner of Navigation, Department of Commerce and Labor); S. REP. NO. 698, 62d Cong., 2d Sess. 12–13 (1912).

NATIONAL SECURITY, PROPAGANDA, AND TRADE

The inclusion of foreign ownership restrictions in the Radio Act of 1912 stemmed from Congress's genuine concerns that foreign control of radio in the United States could compromise national security and be used for propaganda purposes to influence the American citizenry during times of conflict. Those fears seem overstated nearly a century later, but they were real at the time and thus not a legislative subterfuge by which to effect a protectionist policy on foreign direct investment. That is not to say that trade concerns never entered the debate over the inclusion of alien ownership restrictions. They did and were the subject of controversy. But the manner in which key members of Congress debated trade issues clarified that the legislative intent for enacting the foreign ownership restrictions in Radio Act of 1912 was *not* to effect a policy against foreign direct investment in the U.S. wireless industry. Rather, Congress made the judgment that some sacrifice in terms of the free flow of capital into the nascent U.S. wireless industry was the necessary price to pay for the improvement in national security that the Navy believed its proposed foreign ownership restrictions would produce. Congress accepted that bargain knowing that it would lessen opportunities for investment abroad by U.S. communications firms and slow the growth of the wireless industry in the United States.

In the House, Representative Mann from Chicago—the cosponsor two years earlier of the Mann-Elkins Act and the Republicans' minority floor leader—denounced the foreign ownership restriction.[71] The alien ownership restriction in the 1912 legislation, Mann argued, foreclosed "close interchange between nations and people" and was unjustified on national security grounds.[72] He specifically objected to the legislation's application against Canadians: "We want permission over there to operate radio stations. Why should we say they should not have permission here?"[73]

Not surprisingly, Mann was joined by the American Marconi Company—then the nation's preeminent company pioneering the

---

71. BIOGRAPHICAL DIRECTORY OF THE AMERICAN CONGRESS, 1774–1989 at 1419 (Government Printing Office 1989) [hereinafter BIOGRAPHICAL DIRECTORY].

72. 48 CONG. REC. 10,503 (1912) (statement of Rep. James R. Mann).

73. *Id.*

development of radio technology and, more significantly, a foreign-owned subsidiary of British Marconi—in objecting to those restrictions as impeding commerce.[74] John Bottomley, testifying on behalf of American Marconi, specifically objected to the legislation's licensing provisions as unduly restrictive and unnecessary. The licensing scheme, he argued,

> does not provide that we can put a station down and demand a license. It would be within their option to give us a license or not, just as they saw fit. I do not understand why wireless should be singled out to be licensed and legislated for any more than land lines or telephone companies, or anything of that sort. The wireless companies are in a somewhat initiatory state. They do not want to be hampered by these restrictions . . . . We are bitterly opposed to the adoption of any bill which hinders our work and this licensing feature is one we object to.[75]

Despite those objections, Congress kept the citizenship requirement in the Radio Act of 1912, marking the beginning of Congress's regulatory restrictions on foreign ownership in the wireless communications industry.

National security concerns trumped those voiced in defense of free trade in capital. Ironically, concerns regarding the bill's negative impact on commerce and trade proved far more prescient than those predicated on the need to protect the United States from foreign radio threats. While the bill proved to be of questionable efficacy as a prophylactic against foreign influence and interference in domestic affairs, it succeeded and continues to succeed as protectionist trade policy. It is in this latter form that foreign ownership restrictions now gain their significance, a form far distant from the purposes for which they were initially enacted.

---

74. *S. 3620 and S. 5334 Hearings*, *supra* note 69, at 35–37 (statement of John Bottomley, representing Marconi Wireless Telegraph Co. of America).

75. *Id.* at 17–18.

THE COLLAPSE OF FOREIGN OWNERSHIP
RESTRICTIONS UNDER THE WICKERSHAM OPINION

Within weeks of their enactment, Attorney General George W. Wickersham had to construe the alien ownership restrictions in the new Radio Act. His task foreshadowed the difficulty that enforcing alien ownership restrictions would pose in the future.

A New York corporation, Atlantic Communication Company, applied for a license to operate transatlantic radio equipment on Long Island. Atlantic Communication, however, was the subsidiary of the German telecommunications firm Telefunken.[76] The secretary of commerce and labor thus faced a sticky question that the text of the Radio Act of 1912 failed to address: If a U.S. corporation is the subsidiary of a foreign company, does the Radio Act prohibit the secretary of commerce and labor from granting the U.S. corporation a radio license? No, concluded the attorney general. The statute did not delegate to the secretary of commerce and labor any discretion in granting a license when an applicant came within the class of persons or corporations eligible for licensing.[77] Because Atlantic Communication was duly incorporated under New York law, the secretary was *required* to issue the company a radio license.[78]

The attorney general's straightforward reading of the Radio Act exposed in an instant the new statute's inability to counter the threat to which its alien ownership restrictions were directed: foreign influence and control over wireless companies operating in the United States. Within months of the enactment of the Radio Act on August 13, 1912, Congress's first attempt at proscribing foreign influence in domestic wireless communications had been blunted. Attorney General Wickersham surely recognized that his opinion would have the effect of drawing attention to a loophole in the foreign ownership restrictions through which a truck could be driven. In what might have been a gesture to allow Congress to save face, the attorney general noted hypothetically that the Radio Act still empowered the president to close any station in time of war, public peril, or disaster.[79] Absent such compelling circumstances, however, foreign

---

76. DOUGLAS, *supra* note 3, at 269.
77. *Radio Communication—Issue of Licenses*, 29 OP. ATT'Y GEN. 579, 582 (1912).
78. *Id.* at 580.
79. *Id.* at 582.

governments and companies were able to own and control wireless companies in the United States through the simple artifice of creating holding companies incorporated in the United States.

Attorney General Wickersham surely was not the first lawyer in Washington in 1912 to interpret correctly the straightforward language of the Radio Act's foreign ownership restrictions. It is no more believable that the secretary of commerce and labor was surprised to learn that, even after passage of the Radio Act, he lacked power to bar foreigners from investing in the U.S. wireless industry. Any corporate lawyer would have immediately seen that the incorporation of an American subsidiary would enable one to circumvent the Radio Act's foreign ownership restrictions. Why then did the secretary of commerce and labor simply not issue Atlantic Communications its license rather than seek an opinion beforehand from the attorney general saying that the secretary had no choice? Why, moreover, did Congress enact such a porous provision?

We cannot be sure of the answers to those questions, but it is useful to recall the political setting in 1912. Since his election in 1908, after serving as Theodore Roosevelt's vice president, President William Howard Taft had so infuriated his former boss that Roosevelt had entered the presidential race in 1912 as an independent and founder of the Bull Moose Party. The three-way race benefited the Democratic nominee, Woodrow Wilson, who took office as president in March 1913. As we have seen, in 1912 the Republican minority leader in the House, James Mann, unsuccessfully opposed the citizenship requirement in the Radio Act on the grounds that it would induce Canada and other nations to deny U.S. firms the right to hold radio licenses. Although the Republicans had lost the debate in Congress on unrestricted foreign investment in wireless, at least they still controlled the executive branch, for the time being, when the Radio Act was enacted in August. A third-term President Roosevelt would presumably defer to the Navy on questions concerning the wireless industry, as he had done in 1904, and a President Wilson would presumably have a more decidedly Progressive agenda that favored government control over unfettered capitalism.

In that setting, an opinion from the attorney general on the corporate subsidiary question could be produced quickly—long before the results of the presidential election might require the Taft administration to clean out their desks. In fact, the request from the secretary of commerce and labor for an advisory opinion from the attorney general was dated October 22, or two weeks from the election. The attorney general's reply was dated November 22, well after the election.

Alternatively, Taft's administration could have denied Atlantic Communication a license and waited for the company to establish through litigation the same legal proposition that the attorney general's opinion could be expected to announce. But that route would have had the obvious disadvantage of frustrating the interpretation that Taft's administration, we hypothesize, wanted to promote. It would have the additional disadvantages of being slower, of spilling into a new presidential term with a new chief executive who might instruct his attorney general to change the government's legal theory in the case, and of introducing the uncertainty of relying on judges, rather than Attorney General Wickersham, to answer that question of law.

This scenario grows more complicated, for in the same letter of October 24, 1912, the secretary posed a second question to the attorney general. If the attorney general determined that the secretary had no discretion to deny Atlantic Communication a license, then, asked the secretary,

> can the application for the license described be denied until by reciprocal arrangement with Germany, American capital is guaranteed the right of investing in and controlling corporations organized under German laws to operate coast stations in Germany for trans-Atlantic radio communication?[80]

In other words, the secretary was asking whether he had the legal authority to condition foreign direct investment in the U.S. wireless industry on market access abroad for American investors. As we shall see in chapter 7, the Federal Communications Commission was still asking the same question eighty-three years later with respect to its powers under section 310(b) of the Communications Act of 1934. ·

The FCC of 1995 surely did not like the answer that Attorney General Wickersham gave to that second question in 1912:

> This is answered by what has been said as the mandatory character of the licensing provisions of the act. An arrangement

---

80. *Id.*

somewhat similar to that indicated in your question was required by the President as a condition to the landing of foreign-owned cables. But that case is not analogous. Action by the Executive was justified there because Congress had not legislated, and it was recognized that the power to impose conditions at all was subject to subsequent congressional action. Here, Congress has acted and has covered the subject, and, as above stated, you have no discretion but to carry out the provisions of the statute. Therefore, your second question must also be answered in the negative.[81]

Wickersham had delivered a stunning victory for the outgoing free traders in Taft's Republican administration. Not only would foreign capital be allowed to continue flowing into the U.S. wireless industry, but such foreign direct investment could not be made conditional on the existence of reciprocal opportunities to invest in the foreigner's home market.

Within two years of the enactment of the Radio Act of 1912 and the Wickersham opinion, the inadequacy of the statute's foreign ownership restrictions in protecting national security was dramatically and damagingly illustrated. Europe would be torn by war, and a new president would be forced to condemn Atlantic Communication's wireless station, just as Attorney General Wickersham had conjectured in 1912.[82] But the government's appropriation of wireless facilities did not occur until strong circumstantial evidence indicated that Germany had used its American-based wireless facilities to compromise the national security of the United States before its entry into World War I.

ENTER THE CORPORATE LIONS

The timidity that marked Congress's regulation of the wireless industry between 1910 and America's entrance into World War I belied the bold advances, shifting alignments of power, and restructurings that transformed the industry during the same period and set the stage for what would follow. If the first era of radiotelegraphy is properly characterized by the ascent of the brilliant entrepreneurs and inventors—Marconi, Fessenden, and Lee De For-

---

81. *Id.* at 582–83 (citations omitted).

82. DAVIS, *supra* note 49, at 49.

est—then the next era is characterized by their eclipse. Although those personalities continued to play significant roles in the development of radio technology, sovereignty over the future of wireless shifted to those who owned the key patents. The new titans were corporations such as AT&T, General Electric, and Westinghouse, which wrestled with one another, and with the U.S. Navy, to acquire patents.

In its first decade of development, wireless remained a phenomenon of uncertain potential. On one side was a collection of mercurial inventors and on the other an imperious U.S. Navy—the former trying to push their new technology to the limit, the latter acting to restrain it. World events and technological improvements pushed wireless increasingly into the mainstream, where other corporate players began to take interest. The *Titanic* disaster marked the emergence of wireless as more than a curiosity; it had become a vital part of modern telecommunications. In the aftermath of the tragedy, Congress required in the Radio Act of 1912 that *all* ships have wireless on board.[83] Radiotelegraphy further burnished its image by playing vital roles in subsequent emergencies at sea and on land—leading to the rescue of stricken ships and providing vital communication when natural disasters such as blizzards crippled wireline telephone and telegraphy.[84]

Not surprisingly, the corporate entity most interested in the evolution of wireless technology was AT&T. Though publicly dismissive of radiotelegraphy in its early stages, AT&T President Theodore Vail and head researcher J. J. Carty carefully tracked wireless's development.[85] In the wake of continued technological advancements and public relations successes, wireless no longer seemed cabined to such fringe communications needs as those that exist in the military or at sea. By 1913, Marconi was predicting that long-distance and transoceanic voice transmission would soon be possible by radio.[86] Others forecasted a future where everyone would communicate through his own wireless set.[87] Fanciful or not, Vail faced the real possibility that AT&T could find itself in competition with another long-distance network with which it would have to intercon-

---

83. § 1, 37 Stat. at 3.
84. Douglas, *supra* note 3, at 242.
85. *Id.*
86. *Id.*
87. *Id.* at 242.

nect, or, as in the case of competition between telephony and telegraphy, perhaps be undercut by the rival technology.[88] To Vail, that possibility was unacceptable. His vision for AT&T was for it to be a national monopoly—a universal long-distance voice transmission system complete unto itself. As Vail described his vision:

> One system with a common policy, common purpose and common action; comprehensive, universal, interdependent, intercommunicating like the highway system of the country, extending from every door to every other door, affording electrical communication of every kind, from every one at every place to every one at every other place.[89]

Consequently, any emergent technology that threatened the integrity of AT&T's dominion needed to be brought under its control. The challenge, of course, was how to gain that control.

Though still not a technological innovator, AT&T and, more specifically, Carty, were keenly aware of the advances being made in wireless and realized that the solution to the problems vexing voice transmission lay in a continuous wave technology that amplified transmissions. The necessary device to provide such amplification was called a "repeater." Carty predicted at the time:

> Whoever can supply and control the necessary telephone repeater will exert a dominating influence on the art of wireless telephony. . . . A successful telephone repeater, therefore, would not only react most favorably upon our service where wires are used, but might put us in a position of control with respect to the art of wireless telephony should it turn out to be a factor of importance.[90]

In 1912, after being frustrated in its own efforts to develop such a repeater, AT&T began experimenting with De Forest's audion tube, with which the inventor had achieved significant success in voice amplification.[91] Bell scientists improved on De Forest's audion tube,

---

88. ITHIEL DE SOLA POOL, TECHNOLOGIES OF FREEDOM 31 (Belknap Press/Harvard University Press 1983).

89. *Quoted in id.* at 29–30 (original source unidentified).

90. *Quoted in* DOUGLAS, *supra* note 3, at 243.

91. *Id.*

exhausting the gas from it and transforming it into a vacuum tube.[92] While AT&T now had its repeater and patents on its improvement, however, the basic rights still belonged to De Forest.[93]

The subsequent events leading to AT&T's acquisition of De Forest's patents are rife with subterfuge, drama, and acrimony.[94] When the dust settled in 1917, AT&T had gained control over perhaps the most significant invention of the era—and control over the future of wireless telegraphy. AT&T's acquisition of De Forest's patents also signified the beginning of a new phase in the evolution of wireless technology. As Douglas wrote of the acquisition: "The transfer of technological control from independent inventor to corporate research lab was complete."[95]

AT&T was not the only corporation to secure control over continuous wave technology during that period. While AT&T possessed the rights to the vacuum tube, General Electric developed and patented an alternative technological means of continuous wave transmission: the alternator.[96] GE and AT&T, however, had different objectives. Whereas AT&T wanted to control continuous wave technology to protect its monopoly over long-distance voice telecommunications, GE had no ambition to build its own communications network.[97] Principally a manufacturer, GE viewed the alternator as a product to be marketed and sold to those interested in continuous wave transmission.[98] The end use of its product was a matter of indifference to GE.

By 1915, American corporations owned the patents to continuous wave technology—the key to wireless voice transmission. Conspicuously absent from the group of companies pursuing continuous wave technology were American Marconi and its parent, British Marconi. In 1912, American Marconi dominated the U.S. wireless communication marketplace.[99] Its technological supremacy, however, was a different matter. Marconi failed to see until too late that the

---

92. *Id.*
93. *Id.* at 243–44.
94. *See id.*
95. *Id.* at 247.
96. *Id.* at 252.
97. *Id.* at 253.
98. *Id.*
99. *Id.*

future of wireless transmission lay in continuous wave transmission.[100] Instead, American Marconi clung too long to its spark technology, which was effective in telegraphic transmissions but badly outmoded when used for wireless telephony.

The outbreak of World War I derailed Marconi's belated efforts to catch up, as the war effort increasingly absorbed the company's resources.[101] The war also disrupted negotiations by American Marconi to purchase GE's alternator-based continuous wave technology and left the Marconi companies temporarily without access to the next generation of wireless technology.[102] As a result, despite American Marconi's market dominance, its technological position in 1915 was tenuous. The British and American Marconi companies for the first time found themselves at a technological disadvantage to their American competitors.

THE ASCENT OF
DANIELS AND HOOPER

During the same period, the U.S. Navy also underwent significant transformation in both its acceptance and integration of wireless. Beginning in 1912, two naval officers began to figure predominantly in the drive to integrate radio into the operations structure of the Navy: Josephus Daniels and Stanford C. Hooper. Daniels served as Woodrow Wilson's secretary of the Navy from 1913 to 1921.[103] During his tenure, he tirelessly advocated not only the Navy's integration of wireless, but also the Navy's complete takeover of wireless. Though never able to bring wireless under direct Navy control, Daniels played a critical role in the implementation of foreign ownership restrictions.

Stanford C. Hooper is widely considered the father of naval radio.[104] Tutored in the art of wireless communication since childhood, Hooper passionately advocated integrating wireless into the Navy.[105]

---

100. *Id.* at 254.
101. *Id.* at 254–55.
102. *Id.* at 254.
103. *Id.* at 258.
104. L. S. HOWETH, HISTORY OF COMMUNICATIONS—ELECTRONICS IN THE UNITED STATES NAVY xiv (Bureau of Ships and Office of Naval History 1963).
105. DOUGLAS, *supra* note 3, at 260.

With ingenious methods and unflagging insistence, Hooper was ultimately able to overcome the Navy's institutional resistance to radio.[106] After successfully integrating radio communications into the fleet's operations between 1912 and 1914, he centralized and standardized the Navy's network of shore stations.[107] By the beginning of World War I, Daniels and, primarily, Hooper had transformed the Navy's wireless network from an ineffective hodgepodge to an integrated system that significantly enhanced the Navy's fighting capabilities. The Navy's monopolistic ambitions to control the radio spectrum, however, remained unrealized, though not forgotten.

The final element of transformation within the wireless industry in the critical period from 1910 to 1915 was the realignment of the relationship between the Navy and private enterprise. While wireless's first stage was characterized by mutual suspicion and antagonism between the clashing cultures of the Navy and the entrepreneurial inventors of radio technology, increased synergy and reliance marked the second stage. The corporate cultures of AT&T and GE had replaced such temperamental entrepreneurs as De Forest and Fessenden. The Navy, for its part, had grown in its appreciation of wireless's potential. Hooper and Daniels symbolized a new generation of naval officer who was comfortable with radio and willing to work with industry to advance the technology, especially once World War I began.[108]

THE EFFECTS OF WORLD WAR I

World War I profoundly changed America's perception, both militarily and civilly, of the wireless industry and dramatically affected the course of its regulation. To the American public, wireless was, before the war, a curious abstraction that had pervaded society. As with many new technologies, there was a lag between its introduction and the subtle realization of its transformative impact. World War I made tangible the many uses of radio. From a civil perspective, wireless no longer was simply a novel form of communication, but a potential tool of propaganda and political influence.[109] Militarily,

---

106. *Id.* at 261–66.
107. *Id.* at 266.
108. *Id.* at 267–68.
109. *Id.* at 268.

wireless's threat to national security became real. It had become possible to transmit from one continent to another vital information, such as ship movements; and if such information could be transmitted instantly across the Atlantic, it also could be transmitted instantly to the newest and deadliest naval weapon prowling the Atlantic, the German *Unterseeboot*. In fact, U-boats possessed effective wireless systems used to communicate with their bases for orders and intelligence reports.[110]

This civil and military sensitivity to wireless's manifest strategic significance rose with America's desire to avoid the European war. Britain declared war on Germany on August 4, 1914. The next day, President Woodrow Wilson issued a proclamation of American neutrality, which included censorship and neutrality regulations on wireless stations operating within the continental United States.[111] Those regulations, authorized by the emergency powers granted the president two years earlier by the Radio Act of 1912,[112] prohibited private wireless licensees from transmitting or delivering any nonneutral messages or from acting with any bias toward a belligerent during the hostilities.[113] Wilson delegated responsibility for enforcing those restrictions to the Naval Radio Service, with the secretary of the Navy empowered with broad discretion to enforce the neutrality regulations as he saw necessary.[114] It was a charge that Secretary Daniels and the Navy zealously pursued, for it was their opportunity at last to control the airwaves.[115] What began as a mandate for censorship soon evolved into a concerted effort for complete naval control of wireless.

---

110. JOHN TERRAINE, THE U-BOAT WARS, 1916–1945, at 30–33 (G. P. Putnam's Sons 1989).

111. Exec. Order No. 2011 (Aug. 5, 1914), *reprinted in* 17 A COMPILATION OF THE MESSAGES AND PAPERS OF THE PRESIDENTS, 1798–1915, at 7962.

112. § 2, 37 Stat. at 2.

113. Exec. Order No. 2011, *supra* note 111.

114. *Id.*

115. HOWETH, *supra* note 104, at 227.

NEUTRALITY, U-BOATS,
AND THE BULLARD BILL

Wilson's imposition of those neutrality and censorship regulations was problematic from the start because two of the primary belligerents—Germany and Britain—operated several of the largest and most powerful radio stations on the Atlantic coast of the United States.[116] Those stations were powerful enough not only to communicate with military and merchant ships at sea, but also to reach their native lands to consult with diplomats and military commanders.[117] To ensure that Wilson's neutrality regulations were followed, the Navy barred the transmission of all coded messages and dispatched censors to foreign-controlled, long-distance stations to monitor all incoming and outgoing messages.[118]

For Germany, the neutrality restrictions were particularly onerous. Britain, at the outbreak of hostilities with Germany, dredged up the latter's transatlantic cables and severed them, leaving Germany with only wireless as a sovereign means of communication with the United States.[119] Unlike Britain, Germany consequently had no way to communicate confidentially with its diplomats in the United States unless it was allowed to transmit in code. The United States, acknowledging that disadvantage, allowed Germany to transmit in code so long as the American censors were given copies of the code books and the encrypted messages did not concern military matters.[120]

Despite those measures, the foreign-controlled stations along the Atlantic generated suspicion within the U.S. military and civilian communities. Given the ability of those long-range wireless stations to monitor ships leaving U.S. ports and to communicate that information to ships at sea, repeated allegations surfaced that those stations, particularly the German ones, were eluding the censors through clever schemes that enabled the stations to transmit vital military information.[121] National paranoia concerning the mysterious Sayville, New

---

116. DOUGLAS, *supra* note 3, at 269.
117. *Id.* at 269.
118. *Id.* at 227–28.
119. BARBARA W. TUCHMAN, THE ZIMMERMANN TELEGRAM 10–11 (Macmillan Co. 1966).
120. DOUGLAS, *supra* note 3, at 270.
121. *Id.* at 270–72.

York, station grew as Germany relied more on submarine warfare. On a single day in September 1914, Britain lost three cruisers and 1,400 sailors to attacks by one U-boat.[122] The following month, Britain lost another cruiser and a battleship.[123] Then Germany directed its submarine campaign on neutral and civilian vessels. On February 19, 1915, a Norwegian ship was sunk.[124] During that time, Telefunken's station at Sayville was particularly clouded in controversy and a prime suspect of having transmitted information concerning ship movements—if not directly to the U-boats, then indirectly to Germany. Adding to that suspicion, Telefunken upgraded the Sayville station in April 1915, nearly trebling its power from a 35-kilowatt alternator to a 100-kilowatt one and constructing an aerial consisting of three 500-foot towers.[125]

Between April and May of 1915, the monthly gross tonnage of British merchant shipping lost to enemy action nearly quadrupled.[126] On May 1, an American ship was torpedoed.[127] Six days later, when the British liner *Lusitania* was sunk off Ireland, 1,198 on board, including 124 Americans, perished.[128] On August 19, another British liner, the *Arabic*, was sunk with four Americans on board; then, on September 1, Germany announced a cessation of its unrestricted submarine war.[129] By February 21, 1916, however, Germany announced that it would extend its submarine campaign to armed merchantmen.[130] On March 24, a U-boat sank the *Sussex*, a French ferry on which a number of Americans were traveling across the English Channel.[131] On April 19, President Wilson gave Germany an ultima-

---

122. DUPUY & DUPUY, *supra* note 4, at 1035.

123. *Id.*

124. *Id.* at 1049–50.

125. DOUGLAS, *supra* note 3, at 272–73.

126. TERRAINE, *supra* note 110, at 766.

127. DUPUY & DUPUY, *supra* note 4, at 1049–50.

128. *Id.*

129. *Id.* at 1050.

130. *Id.* at 1056.

131. *Id.* The depredations of the radio-equipped U-boats came to symbolize how, in the course of accomplishing their immediate military objectives, the new technologies of warfare, so powerfully indiscriminate in their carnage, threatened to destroy much of what European civilization had achieved. For example, one passenger on the *Sussex* was Spain's greatest living composer, Enrique Granados. He had journeyed to New York for the Metropolitan Opera's first performance of his masterpiece, the opera *Goyescas*. The composer was so acclaimed by American critics that President Wilson

tum: The United States would sever diplomatic relations unless Germany were "immediately [to] declare and carry into effect its abandonment of the present method of warfare against passenger and freight carrying vessels."[132] Germany's response to the ultimatum, delivered on May 5 by wireless from Berlin to the Sayville station, was to order its submarines not to sink merchant ships "without warning and without saving human lives unless the ship attempt to escape or offer resistance."[133] In less than a year, Germany rescinded that order.

The public's suspicion had turned to certainty that the U-boats were receiving reports of ship movements from German-controlled wireless stations in the United States. As an alternative response to the submarine war, an interdepartmental committee chaired by Navy Captain W. H. G. Bullard drafted a radio regulation bill, introduced in Congress on December 22, 1916, that proposed for the first time that the government restrict the ownership of the principals of domestic licensee corporations.[134] The proposed restrictions would prohibit alien officers and impose a maximum one-third limit on alien directors and stockholders.[135] In addition, the bill would prohibit the granting of radio licenses to foreign governments or their representatives.[136]

---

requested him to give a recital at the White House before returning to Spain. To meet the president's request, Granados had to cancel a voyage that would have taken him directly to Spain. Instead, he sailed for England. Although a lifeboat plucked Granados from the frigid water of the English Channel, he dove back into the water to rescue his wife. The two were last seen clinging to a small raft. *Granados May Be Safe, Hope That Composer and His Wife Are on a Hospital Ship*, N.Y. TIMES, Apr. 1, 1916; *Sussex Survivor Returns*, N.Y. TIMES, May 23, 1916, at 2. Six weeks later, the Metropolitan Opera held an all-star performance to benefit the Granados's six orphans, the youngest only three years old. *$11,000 for the Granados, Six Great Artists at Benefit for Late Composer's Children*, N.Y. TIMES, May 8, 1916, at 9.

132. Special Message (Apr. 19, 1916), *reprinted in* 17 A COMPILATION OF THE MESSAGES AND PAPERS OF THE PRESIDENTS 8121; Note to Germany on the Sinking of the French Steamship *Sussex* in the English Channel (Apr. 19, 1916), *reprinted in id.* at 8125.

133. Germany's Reply to President Wilson's Note on the Sinking of the *Sussex* (May 5, 1916), *reprinted in id.* at 8127, 8129.

134. H.R. 19350, §§ 7, 9, 64th Cong., 2d Sess. (1916).

135. *Id.* § 7.

136. *Id. See also Hearings on H.R. 19530 Hearings Before the House Comm. on Merchant Marine and Fisheries*, 65th Cong., 1st Sess. 77–78 (1917) (statement of

Congress held hearings on Bullard's bill in January 1917. Congress did not reserve its hostility for German-controlled companies. American Marconi, which had an alien officer and was one-third British-owned, was questioned about the extent to which it was controlled from abroad.[137] American Marconi's representatives, with good reason, believed the company to be the target of the proposed legislation.[138] American Marconi was the largest owner of American radio stations and, therefore, the Navy's chief rival for dominance of American wireless. John W. Griggs, president of the Marconi Wireless Company of America, defended private enterprise in the wireless industry and denounced the Bullard bill as a coercive maneuver by the Navy to force American Marconi to sell out:

> Let us see just what it is and what the effect of it is to be on the Marconi Co., and whether it is wise, whether it is necessary, and whether it is just. It has been admitted here by Commander Todd and Capt. Bullard—admitted, as I have read the statements here—that the object of this bill is to coerce the Marconi Co. into letting go of its business, particularly its coastal stations. The proposition is to give the Navy Department unlimited authority to do commercial business in competition with these gentlemen who have put their money into a mercantile venture, and to so conduct the Government end of it that eventually in five years, we would be glad to sell out. Now, I am not making that charge against the Navy; that is what they say their purpose is.[139]

Congress never voted on Bullard's foreign ownership bill. World events in early 1917 soon overtook it. Nonetheless, the Bullard bill became the blueprint for eventual legislation restricting foreign ownership.

Captain W. H. G. Bullard).

137. *Id.* at 369–70 (questioning of David Sarnoff, Commercial Manager, American Marconi Co., by Rep. Edmonds); *see also id.* at 346–48 (statement of Commander David W. Todd, U.S. Navy, regarding monopolistic practices of British Marconi and alleging that American Marconi was its subsidiary).

138. *Id.* at 177 (statement of John W. Griggs, President, Marconi Wireless Co. of America).

139. *Id.* at 172.

NAVY SEIZURES OF WIRELESS
BEFORE APRIL 1917

In the environment of suspicion and paranoia before America's entry into World War I, Secretary Daniels seized every opportunity to bring wireless further under the Navy's control. At the war's outset, Daniels seized the German station at Tuckerton, New Jersey, pursuant to President Wilson's executive order directing the Navy to take appropriate or more high-powered stations on the Atlantic coast to provide a terminal for a U.S. circuit with Europe.[140] Daniels also succeeded in September 1914 in temporarily shutting down American Marconi's most powerful and important station at Siasconset, on Nantucket Island, after the company refused to comply with the censorship restrictions and had its case thrown out of court.[141] Later, playing on the U-boat hysteria and growing reports of nonneutral German wireless transmissions from Sayville, the Navy on July 9, 1915—two months after the sinking of the *Lusitania*—took control of that German station and began operating it under Navy direction.[142] The Germans were no longer allowed to transmit or receive radio messages except through stations that the Navy controlled.

Daniels seized upon the alleged treachery of the Germans at Sayville as evidence of the perils of allowing control of wireless to remain outside direct government control. In the *Annual Report of the Secretary of the Navy* in 1916, Daniels argued:

> It is becoming increasingly evident that no censorship of radio stations can be absolutely effective outside of complete government operation and control. . . . The government must in the end follow the lead of almost all other governments and obtain control of all coast radio stations and operate them, in conjunction with naval stations, for commercial work in times of peace.[143]

Daniels would have to wait until the outbreak of World War I to get his wish.

---

140. HOWETH, *supra* note 104, at 229.
141. DOUGLAS, *supra* note 3, at 271.
142. *Id.* at 273.
143. 1916 ANNUAL REPORT OF THE SECRETARY OF THE NAVY 27 (1917).

The events at Sayville, however, were already exposing the flawed predicate of Daniel's argument—if the government controlled the hardware, it necessarily could control the content of what was transmitted. At Sayville, the Navy effectively controlled the station and even the content of its transmissions but was apparently still unable to prevent the Germans from transmitting vital military information. Considerable suspicion and circumstantial evidence suggested that the Germans were able, through ingenious transmissions techniques, to elude the American censors and transmit strategically sensitive information.[144] Indeed, the monthly loss of merchant tonnage to U-boat attacks immediately *rose* after the Navy's seizure of the Sayville station. In July 1915 Britain lost merchant vessels totaling 52,847 tons; but in August, after the U.S. Navy had taken control of the Sayville station, Britain lost 148,464 tons.[145]

In addition, nationalized control over radio stations was no guarantee against the wireless dissemination into a country of information that would be harmful to national security. Immediately upon the outbreak of World War I, Germany began transmitting its version of the war to anyone with a receiver who cared to listen, be they amateurs, press, or government.[146] As the Germans demonstrated, wireless provided an effective means of propaganda over which other nations had virtually no control. Ultimately, control over the airwaves in the name of national security proved more difficult to achieve than merely seizing control of the hardware. The true danger to national security lay in the *content* of the transmissions, a far more difficult element of wireless to regulate and one that legislation to that point had not addressed.

NATIONALIZATION OF WIRELESS
DURING WORLD WAR I

In early 1917, despite Woodrow Wilson's commitment to neutrality, American relations with Germany deteriorated rapidly. On January 31, 1917, Germany announced its resumption of unrestricted

---

144. DOUGLAS, *supra* note 3, at 269–75.
145. TERRAINE, *supra* note 110, at 766. The world total figures for lost merchant tonnage also rose sharply, from 109,640 in July to 185,866 in August. *Id.*
146. DOUGLAS, *supra* note 3, at 275.

submarine warfare.[147] In reaction to that renewed threat to neutral ships and U.S. passengers, President Wilson addressed a joint session of Congress on February 3 and, reminding its members of a U-boat's sinking of the *Sussex* without warning the year before, announced that he would break diplomatic relations with Germany.[148] Additionally, all persons of German extraction employed at the Tuckerton and Sayville radio stations were dismissed.[149] Several months later they were arrested along with Dr. Karl Frank, head of the Telefunken's U.S. subsidiary, Atlantic Communication, in a wartime sweep of alleged German spy rings.[150] Still, Wilson adhered to his increasingly untenable policy of neutrality. Even Wilson's February 26 request for Congress to enact the Armed Ship Bill, enabling American ships to carry arms for protection in their neutral activities on the high seas, was an effort more to deter Germany from engaging in an "overt act" that would force America into the war than to protect U.S. maritime interests.[151]

Wilson's policy of neutrality suffered a devastating blow by the uncovering of the "Zimmermann telegram." Intercepted by the British, the Zimmermann telegram detailed a proposed German defensive alliance with Mexico and Japan in the event of America's entry into the war.[152] Though Wilson received the telegram on February 24, he did not release it to American newspapers until March 1, when it sparked national outrage.[153] Compounding the Germans' diplomatic perfidy in Wilson's eyes was the knowledge that German Foreign Minister Zimmermann used the U.S. State Department cable to transmit the encoded telegram to Ambassador Bernstorff in the United States, a cable that Wilson had naïvely allowed Germany to use in the hopes of negotiating a diplomatic end to the war.[154] That incident illustrated again the necessity of controlling not so much the transmis-

---

147. DUPUY & DUPUY, *supra* note 4, at 1060.

148. Diplomatic Relations with Germany Severed (Feb. 3, 1917), *reprinted in* 17 A COMPILATION OF THE MESSAGES AND PAPERS OF THE PRESIDENTS 8206.

149. DOUGLAS, *supra* note 3, at 274.

150. *Id.*

151. Letter Asking Congress for Authority to Supply Merchant Ships with Defensive Arms (Feb. 26, 1917), *reprinted in* 17 A COMPILATION OF THE MESSAGES AND PAPERS OF THE PRESIDENTS 8209; *see also* TUCHMAN, *supra* note 119, at 170.

152. *Id.* at 175.

153. *Id.* at 175–76.

154. *Id.* at 172.

sion medium itself, but its content, which would be only imperfectly addressed when Congress eventually imposed foreign ownership restrictions on wireless companies.

On April 6, 1917, the United States entered the Great War. Empowered by section 2 of the Radio Act of 1912 to shut down any radio station "in time of war or public peril or disaster" or to "authorize the use or control of any such station or apparatus by any department of the government,"[155] Wilson the same day authorized the Navy to commandeer *all* domestic radio stations.[156] In other words, even though the United States had just declared war against Germany and would shortly send soldiers to reinforce the British Army, the U.S. Navy was to seize British-controlled radio stations as well as German-controlled stations. On April 7, the Navy acted, taking over fifty-three commercial stations, most of them owned by American Marconi, and shutting down an additional twenty-eight.[157]

The Navy proved an adept steward of wireless during wartime. The Navy dedicated its formidable resources to the integration, technological development, and standardization of wireless communications in the United States.[158] By the end of the war, the Navy's wireless strategy had dramatically improved the U.S. wireless communication network and established the Navy's complete control over radio—not only over its hardware, but also over access and content.[159] Under naval control, radio also proved itself to be an invaluable tool in advancing America's diplomatic, political, and ideological interests, both nationally and internationally.[160] Daniels and others believed that they had built a compelling case to keep wireless

---

155. 37 Stat. 302, § 2 (1912).

156. Exec. Order (Apr. 6, 1917), *reprinted in* 17 A COMPILATION OF THE MESSAGES AND PAPERS OF THE PRESIDENTS 8241; Exec. Order (Apr. 30, 1917), *reprinted in id.* at 8254. *See also Emergency Control of Systems of Communications: Hearings on H.J. Res. 309 Before the House Comm. on Interstate and Foreign Commerce,* 65th Cong., 2d Sess. 7–9 (1918) (additional statement of Josephus Daniels, Secretary of the Navy) (more than fifty stations taken over by the Navy).

157. *Government Control of Radio Communication: Hearings on H.R. 13159 Before the House Comm. on the Merchant Marine and Fisheries,* 65th Cong., 3d Sess. 7–9 (1918) (statement of Secretary Daniels) [hereinafter *H.R. 13159 Hearings*]. *See also* 1917 ANNUAL REPORT OF THE SECRETARY OF THE NAVY 44 (1918).

158. HOWETH, *supra* note 104, at 237–59.

159. *Id.; see also* DOUGLAS, *supra* note 3, at 279–80.

160. *Id.* at 288.

under naval control after the war. But again, the Navy badly mis-judged the attitudes of the American people and Congress.

THE NAVY'S INFLUENCE
AFTER WORLD WAR I

As World War I drew to a close, the Navy lobbied Congress to na-tionalize ownership of all radio in peacetime, arguing that the govern-ment could handle commercial business as it had during the war.[161] Leading the argument to maintain the government monopoly of wireless after the war was Commander Stanford Hooper. Hooper as-serted that radio "is a natural monopoly; either the government must exercise that monopoly by owning the stations or it must place the ownership of these stations in the hands of some one commercial concern and let the government keep out of it."[162] The statement presaged the Navy's role in forming, after its failed attempts at keeping radio under its direct control, the Radio Corporation of America. RCA's creation would provide the Navy with the surrogate means by which to keep radio not only under monopoly control, but also under its own indirect influence.

Congress was somewhat receptive to the Navy's arguments. In 1918 Representative Joshua Alexander introduced a bill, vigorously supported by the Navy and Secretary Daniels, that would nationalize all radio transmitters and give the Navy permanent control over their use and licensing.[163] The public opposed the bill, as it had President Theodore Roosevelt's efforts to bring wireless under government control more than a decade earlier. Although during the war the Navy had achieved admirable success in managing wireless, the same could not be said for the government's control over industries such as tele-phone and other utilities. Long-distance telephone rates increased

---

161. *H.R. 13159 Hearings, supra* note 157, at 5, 9–10 (testimony of Secretary Daniels); *see also Authorizing Use of Radio Stations Under Control of Navy Department for Commercial Purposes: Hearings [on H.R. 8783] Before the Comm. on the Merchant Marine and Fisheries,* 66th Cong., 1st Sess. 38–41 (1919) (statement of Secretary Daniels).

162. BARNOUW, *supra* note 3, at 53 (*quoting Government Control of Radio Communication: Hearings Before the House Comm. on Merchant Marine and Fisheries,* 65th Cong., 3d Sess. 10–11 (1918)).

163. *See* POOL, *supra* note 88, at 111.

dramatically under government wartime control, yet AT&T still lost money.[164] Government control over America's railroads fared no better.[165] The *New York Times* editorialized:

> Not for any temporary and not for any permanent cause, or merely assumed cause, should the government be allowed to put its bungling and paralyzing hand upon private business. . . . [T]he country does not pine for nationalization.[166]

When Congress balked at Alexander's bill and rebuffed Daniels's other efforts to establish naval control over wireless, the Navy was forced to explore alternative means of dominating wireless.

EVENTS LEADING TO THE FORMATION
OF THE RADIO CORPORATION OF AMERICA

The Navy's failure to establish postwar control over the wireless industry left the door open to its rivals—most significantly, American Marconi. Seizing upon the opportunity to reestablish itself in the marketplace, American Marconi resumed its negotiations with GE—interrupted by the war—to purchase Alexanderson alternators and thereby obtain the continuous wave technology that it needed for long-distance voice transmission.[167] Although GE refused to assign to American Marconi exclusive rights to the alternator, the tentative agreement provided for the purchase of twenty-four alternators, a purchase order that would occupy GE production capacity for several years and thereby produce in effect an exclusive supply agreement.[168]

The execution of such an agreement raised two intolerable possibilities for the Navy. First, it would end the Navy's control over the wireless industry in the United States. Second, and more dangerous from the Navy's perspective, a foreign-owned U.S. subsidiary would exercise significant control over American airwaves. The Alexanderson alternator was widely regarded as the most powerful and best radio system available.[169] It was unthinkable in terms of

164. DOUGLAS, *supra* note 3, at 281.
165. *Id.*
166. *Id.* at 284 (quoting N.Y. TIMES, July 25, 1919, at 10).
167. *Id.* at 285.
168. *Id.*
169. TOM LEWIS, EMPIRE OF THE AIR: THE MEN WHO MADE RADIO 141–42

national security that such valuable technology developed in the United States should fall under the control of a company largely owned and subsidized by the British government.[170] Consequently, the Navy moved decisively in early 1919 to preempt that threat to its mission of ensuring American dominance over domestic wireless communications.

Interceding on the Navy's behalf were W. H. G. Bullard, who had been promoted to admiral since drafting the prewar bill attempting to restrict foreign ownership, and Commander Stanford Hooper. On April 8, 1919, Bullard and Hooper privately urged GE to end its negotiations with American Marconi. They stressed the critical importance of keeping the postwar wireless communications network under American control. GE's proposed contract with American Marconi, they warned, would enable Britain to create a worldwide radio monopoly comparable to what Britain had already achieved in submarine cables.[171] If GE went through with its sale to the Marconi companies, Britain would dominate radio communications to and from the United States.[172]

Appeals to nationalism, however, were not the only tool that Bullard and Hooper used to persuade GE to forgo its agreement with American Marconi. Faced with the futility of trying to retain control over wireless, the Navy struck upon an alternative means to keep wireless not only under U.S. sovereignty, but under indirect naval control: the formation of a new, all-American company—what would be called the Radio Corporation of America. Bolstering the possibilities of that new company was the fact that during the war the Navy had acquired licenses to valuable radio patents.[173] The Navy's patents in conjunction with those controlled by GE alone would be sufficient to form an American company that could exercise significant, if not yet exclusive, control over wireless in the United States.[174] Through

---

(Edward Burlingame Books 1991).

   170. *Id.*

   171. 67 CONG. REC. 5489, 5493–94 (1926) (statement of Rep. Free).

   172. *Cable-Landing Licenses: Hearings on S. 4301 Before a Subcomm. of the Sen. Comm. on Interstate Commerce*, 66th Cong., 3d Sess. 333–34 (1920) (statement of Owen Young, Chairman of the Board, Radio Corporation of America) [hereinafter *S. 4301 Hearings*].

   173. LEWIS, *supra* note 169, at 142–43.

   174. *Id.*

a buyout of American Marconi, the new radio company could control America's long-distance and point-to-point wireless networks.[175] The implication, of course, was that this new company would, in place of American Marconi, buy GE's alternators and generally provide an attractive new venture in which GE would have a significant stake.[176] GE, however, still wavered.

Finally, Bullard played his trump card. Taking Owen Young, GE's president, aside, Bullard told him that President Wilson, struggling to preserve his Fourteen Points, had himself asked for Young's help in blunting the British drive for domination of global wireless communication.[177] The Navy's appeal was not to be understood simply as a business transaction, but as a patriotic act to preserve a resource of vital national interest. Young understood the strategic importance of wireless and agreed to the Navy's plan.[178] On April 9, GE informed American Marconi that negotiations between the two companies were formally terminated.[179]

Young then handled the more delicate task of convincing American Marconi that it was in its best interests to sell out to the all-American corporation that GE and the Navy were forming. The unspoken threat behind those overtures to American Marconi was that, if it remained independent and chose instead to compete with the new GE-Navy corporation, the federal government would make it difficult for such a foreign-influenced company to compete with an all-American one.[180] As Young delphically put it to E. J. Nally, American Marconi's vice president, in response to the latter's query as to Washington's attitude if American Marconi continued to operate as a foreign-owned subsidiary: "I cannot say, but I will say this, Mr. Nally: the American Marconi interests are greatly menaced because of the English holdings in the Company and the attitude of the Government toward such holdings."[181] It became apparent that American Marconi's best course was to allow GE to acquire its American

---

175. DOUGLAS, *supra* note 3, at 285.

176. JOSEPHINE YOUNG CASE & EVERETT NEEDHAM CASE, OWEN D. YOUNG AND AMERICAN ENTERPRISE 176 (David R. Godine 1982).

177. LEWIS, *supra* note 169, at 143.

178. *Id.*

179. DOUGLAS, *supra* note 3, at 286.

180. LEWIS, *supra* note 169, at 144.

181. CASE & CASE, *supra* note 176, at 183.

operations—including its patent licenses—and fuse them into one corporation with those patent licenses already owned or controlled by GE and the Navy.[182] British Marconi found itself checkmated: If it did not agree to the buyout of American Marconi, it would not be able to purchase the Alexanderson alternators that it needed for its own British and continental operations.[183] On September 5, 1919, British Marconi reluctantly consented to sell its American interests—364,826 shares of American Marconi stock.[184]

In retrospect, the Navy's asserted threat of a potential British monopoly of wireless was naïve, if not disingenuous. The United States could have responded to that hypothetical British monopoly simply by licensing more radio spectrum for international wireless telephony and telegraphy. Furthermore, as a legal matter, the Sherman Act not only had by then been on the statute books for twenty-nine years,[185] but also had been used by the Supreme Court on a single day in 1911 to break up both the Standard Oil and American Tobacco trusts.[186] Then, three years later, Congress supplemented the Sherman Act by enacting the Clayton Act of 1914,[187] which addressed incipient diminutions in competition caused by mergers or acquisitions. Finally, if there was any doubt about the extraterritorial scope of antitrust subject matter jurisdiction under either the Sherman or Clayton Act,[188] that doubt was removed by a new statute. Concern in 1915 about the possibility of international predatory pricing by European firms following the end of the war[189] prompted Congress to

182. *Id.*

183. *Id.* at 146.

184. ARCHER, *supra* note 3, at 173.

185. 26 Stat. 209 (1890).

186. Standard Oil Co. of N.J. *v.* United States, 221 U.S. 1 (1911); United States *v.* American Tobacco Co., 221 U.S. 106 (1911). *See* ROBERT H. BORK, THE ANTITRUST PARADOX: A POLICY AT WAR WITH ITSELF 33–41 (Free Press 1978; rev. ed. 1993).

187. 38 Stat. 730 (1914).

188. American Banana Co. *v.* United Fruit Co., 213 U.S. 347 (1909). The Supreme Court broadened the extraterritoriality of American antitrust laws eighteen years later. United States *v.* Sisal Sales Corp., 274 U.S. 268 (1927).

189. 1915 SECRETARY OF COMMERCE ANNUAL REPORT 42; Henry C. Emery, *The Problem of Anti-Dumping Legislation, in* OFFICIAL REPORT OF THE THIRD NATIONAL FOREIGN TRADE CONVENTION 73, 81 (1916); *see also* JACOB VINER, DUMPING: A PROBLEM IN INTERNATIONAL TRADE 242–46 (1923); WILLIAM SMITH CULBERTSON, COMMERCIAL POLICY IN WAR TIME AND AFTER (D. Appleton & Co. 1923).

enact the Antidumping Act of 1916.[190]

THE CONSOLIDATION OF RCA

On October 17, 1919, the Radio Corporation of America, with the patents of GE and Marconi, was incorporated in the state of Delaware.[191] GE purchased the assets of the American Marconi company on behalf of RCA, a merger that put American Marconi out of business.[192] The Navy, for its part, managed to insert into RCA's articles of incorporation three provisions that not only restricted alien ownership, but also guaranteed continued naval involvement and influence over radio in the United States. One prohibited the appointment or election of a corporate officer or director who was not a U.S. citizen.[193] A second provision limited foreign equity ownership and voting rights to 20 percent of the outstanding shares.[194] The third permitted the U.S. government to participate in the administration of RCA's affairs, as the directors might vote advisable.[195] In essence, through RCA the Navy succeeded in achieving what it could not achieve for itself: an American-controlled institution with a monopoly over domestic wireless operations and under tangible, albeit indirect, Navy influence.

Control over the patents of American Marconi and GE, however, was not sufficient to provide RCA with universal dominion over wireless. Although foreign hegemony had been eliminated from the U.S. radio industry, no American company had enough patents to provide a complete, integrated radio network. The myriad of inventors and the frenetic purchase of patents by various corporations produced a technological interdependence whereby one company's patented technology required the use of another's to operate.[196] Thus,

---

190. Formally enacted as part of the Revenue Act of 1916, ch. 463, § 803, 39 Stat. 798 (1916) (codified at 15 U.S.C. § 72). The 1916 legislation is analyzed in J. Gregory Sidak, *A Framework for Administering the 1916 Antidumping Act: Lessons from Antitrust Economics*, 18 STAN. J. INT'L L. 377 (1982).

191. LEWIS, *supra* note 169, at 146.

192. *S. 4301 Hearings, supra* note 172, at 335–36 (statement of Owen Young).

193. FEDERAL TRADE COMMISSION, REPORT ON THE RADIO INDUSTRY 19 (Government Printing Office 1924) [hereinafter FTC RADIO REPORT].

194. *Id.*

195. *Id.*

196. DOUGLAS, *supra* note 3, at 289.

RCA's ultimate success grew from its ability to construct an interlocking technology network with other corporations through extensive cross-licensing and market recognition agreements. To that end, the patents "were used to clarify the boundaries of industries and licenses were granted for particular uses rather than particular patents."[197] The first such agreement was between AT&T and RCA.

AT&T controlled the patent rights to De Forest's audion—the vacuum tube that enabled long-distance voice transmission—and thus the technological linchpin of any wireless network.[198] But, use of the audion depended upon another invention, the Fleming valve, to which RCA held the rights.[199] Aware of the futility of noncooperation, both companies agreed to cross-license their patents and divide the market into their respective spheres of influence.[200] RCA established exclusive rights to use the pooled patent licenses for international wireless telegraphy and ship-to-shore communication.[201] AT&T retained control over wireless telephony, including exclusive rights to "all land radio telephony for toll purposes."[202] Those exclusionary rights addressed manufacturing as well. GE established control over the manufacture of amateur apparatus, vacuum tubes, and radio receivers.[203] AT&T exercised exclusive rights to manufacture wireless telephone transmitters.[204]

Westinghouse presented a second obstacle to RCA's efforts to consolidate control over wireless. After the war, Westinghouse devised an aggressive strategy to develop the radio market, one that placed it squarely in competition with RCA.[205] Though outflanked by RCA in gaining control over many crucial technology licenses, Westinghouse acquired a company that possessed the exclusive rights to crucial broadcasting technology—most notably, Howard Armstrong's regeneration and superheterodyne inventions that amplified and

---

197. GERALD W. BROCK, THE TELECOMMUNICATIONS INDUSTRY: THE DYNAMICS OF MARKET STRUCTURE 166 (Harvard University Press 1981).

198. DOUGLAS, *supra* note 3, at 289.

199. *Id.*

200. BROCK, *supra* note 197, at 166.

201. *Id.*

202. *Id.*

203. DOUGLAS, *supra* note 3, at 289.

204. *Id.*

205. LEWIS, *supra* note 169, at 151–52.

filtered weak incoming signals.[206] But Westinghouse's control over such vital patents was not enough. It still had only a tenuous hold on many inventions that it was using, including the vacuum tube technology licensed to RCA.[207] RCA's position was little better. Its plans to develop and market the radiola—a home radio receiver—hinged on getting access to Westinghouse's regeneration and superheterodyne patents.[208] Again, as with the AT&T licensing and manufacturing agreements, RCA and Westinghouse came together out of mutual necessity. In the end, through cross-licensing agreements RCA gained use of Westinghouse's patents in return for one million shares of RCA stock[209] and 40 percent of RCA's orders for radio components.[210]

Subsequent agreements with other companies solidified RCA's position. Two years after its formation, RCA had successfully shifted control over wireless technology, and the ether itself, away from military control to corporate. In pooling patent rights through cross-licensing agreements and then dividing them by market, those agreements provided all the parties with an effective barrier to entry into their respective markets and a unified monopoly of patents that would make it nearly impossible for an outside company to compete.[211] Through such interlocking corporate licenses and agreements, the RCA corporate trust possessed the means to control access not only to the technology of wireless communication, but also to the ether itself. RCA's mandate was to create world leadership for the United States in the manufacture and sale of radio apparatus. By 1921, it had succeeded in doing so.[212]

For the Navy, RCA was a triumph as well. In essence, RCA was a corporate reincarnation of the military monopoly that existed during the war. It preserved the monopoly that the Navy believed essential

---

206. *Id.*
207. *Id.* at 154.
208. *Id.*
209. *Id.*
210. DOUGLAS, *supra* note 3, at 290.
211. BROCK, *supra* note 197, at 167.
212. *S. 4301 Hearings*, *supra* note 172, at 336; *To Regulate Radio Communication: Hearings on H.R. 7357 Before the House Comm. on the Merchant Marine and Fisheries*, 68th Cong., 1st Sess. 162–63, 165, 170 (1924) (statement of David Sarnoff, Vice President and General Manager, Radio Corporation of America) [hereinafter *H.R. 7357 Hearings*].

to managing radio in the postwar environment, and it ensured that radio technology remained owned and controlled by Americans. Perhaps even more important, the Navy managed to preserve some influence over radio by writing into the RCA's corporate charter the provision that at least one Navy officer would sit "by invitation" on the board of directors.[213] Indeed, one of the first actions of the board of directors was to invite President Wilson to nominate a naval officer of a rank superior to captain to sit on the board and represent the government's views concerning the management of radio in the United States.[214] The Navy responded by nominating Admiral Bullard, who served on RCA's board from 1920 to 1931.[215]

After the end of World War I, the Navy restored all high-power radio stations to their private owners. In early 1920 a Senate bill proposed stringent alien-control provisions.[216] The bill authorized a Navy officer to attend licensees' board and stockholder meetings, where he could challenge votes exceeding a proposed 20 percent alien-ownership limitation. A radio licensee could have no alien officers or directors, nor could aliens, their representatives, or companies "dominated or controlled by alien interests" hold radio licenses.[217] The bill, plainly patterned after RCA's corporate charter, was not enacted, though its influence on the present section 310(b) is evident. In particular, that 1920 bill was the first to use the term "representative of an alien."

THE BOOM IN RADIO LICENSEES

Although the cross-licensing agreements that created RCA succeeded in delineating spheres of interests among the competing powers in the radio industry in the postwar era, they were structured according to limited prewar conceptions of radio's use. The agreements presumed that radio's primary purpose was to establish long-distance, point-to-point communication between specific senders and receivers. Consequently, the agreements dealt only with pooling patents and dividing markets according to narrowly defined corporate interests. Even as

---

213. FTC Radio Report, *supra* note 193, at 19.
214. Howeth, *supra* note 104, at 359.
215. *Id.*
216. S. 4038, § 6, 66th Cong., 2d Sess. (1920).
217. *Id.*

RCA was being formed, however, amateurs across America were already expanding radio into a new frontier of vast potential: broadcasting.

The potential of broadcasting and the failure of the original agreements to delineate respective corporate rights in relation to that new frontier prompted controversy over which corporation or corporations would control broadcasting. AT&T quickly exploited that ambiguity by establishing its proprietary interest.[218] RCA reacted predictably to AT&T's efforts to occupy the field by challenging AT&T legally through the arbitration clause of their original agreement.[219] After considerable litigation, the parties to the original agreement negotiated a new agreement in 1926 to replace their outmoded 1920 agreement.[220] The new agreement set out the fields of interest more clearly, specifically in relation to the phenomenon of broadcast radio. In essence, AT&T conceded broadcasting to RCA in return for greater dominance in point-to-point communications.[221]

The explosion of radio broadcasting in the 1920s also forced the government to restructure its regulatory scheme concerning wireless communications. The Radio Act of 1912 was primarily directed at maritime use, although its language was broad enough to encompass all radio. In substance, the act required that radio users be licensed by the secretary of commerce.[222] And although the statute authorized the secretary to specify the wavelength, it undercut the practical significance of such discretion in licensing by also allowing a station, at its own election, to use wavelengths other than those designated by the secretary.[223] In 1912 such free-roaming license created few problems. Chatter on the ether was confined largely to marine and amateur use, and enough channels were available to prevent undue interference among those radio users.[224]

The emergence of broadcasting in 1921 radically changed the occupancy of the ether. Applications for licenses multiplied. Without any statutory authority to provide specified wavelengths for broadcast-

218. POOL, *supra* note 88, at 34–35.
219. BROCK, *supra* note 197, at 169.
220. *Id.*
221. *Id.*
222. DAVIS, *supra* note 49, at 40.
223. *Id.*
224. *Id.*

ing, and despite the need to provide such channels to avoid inter-ference among the various services, the secretary of commerce selected two bands—360 meters, and later 400 meters—as sufficient for broadcasting's purposes.[225] No effort was made to assign separate channels for each station.[226] That simplistic method of assigning spectrum resources for radio broadcasting soon became severely out-moded.

In response to the exploding phenomenon of radio broadcasting, Secretary of Commerce Herbert Hoover called a conference on radio telephony to generate a legislative response to the problems presented by broadcasting.[227] The first conference met in February 1922 and produced a report containing a variety of recommendations for radio licensing.[228] The report did not address foreign ownership, but instead focused on whether the electromagnetic spectrum was a public re-source that the government should regulate. Hoover held successive conferences over the next three years that addressed the same issue and each time advocated the same policy—the ether, as a "public medium," should be strictly regulated by the federal government.[229] Each year, Congress rebuffed the recommendations.[230]

During that period, broadcasting was growing so rapidly that by 1923 several hundred stations were already vying to be heard on the two wavelengths assigned for nationwide use.[231] Interference among broadcasters became so pervasive that in the resulting confusion the stations effectively cancelled each other out. Also in 1923, the U.S. Court of Appeals for the D.C. Circuit ruled in *Hoover* v. *Intercity Radio Company* that, under the Radio Act of 1912, the secretary's licensing powers did not include the discretion to withhold licenses,

---

225. *Id.*

226. *Id.*

227. EMORD, *supra* note 3, at 147 (citing MINUTES OF OPEN MEETINGS OF THE DEPARTMENT OF COMMERCE CONFERENCE ON RADIO TELEPHONY (Feb. 27–28, 1922)).

228. *Id.*

229. PROCEEDINGS OF THE FOURTH NATIONAL RADIO CONFERENCE AND RECOMMENDATIONS FOR REGULATION OF RADIO, CONFERENCE CALLED BY HERBERT HOOVER, SECRETARY OF COMMERCE, WASHINGTON, D.C. 7 (Nov. 9–11, 1925) [hereinafter FOURTH NATIONAL RADIO CONFERENCE].

230. EMORD, *supra* note 3, at 152.

231. DAVIS, *supra* note 49, at 40.

even on the grounds of preventing interference.[232] The secretary's only discretionary power lay "in selecting a wavelength, within the limitations prescribed in the statute, which, in his judgment, [would] result in the least possible interference."[233] To exercise that power, the court continued, the secretary would first have to devise an effective frequency allocation scheme that avoided the pitfall of interference.[234]

That task must have seemed beyond the ability of the secretary and his beleaguered Commerce Department. By 1924 the department, unable to gauge either broadcasting's market or its potential, had all but conceded that it was helpless in the face of the new phenomenon:

> The broadcast listener is an unknown quantity. Dependable figures indicating the number of persons deriving pleasure and benefit from this new and fascinating service can not be furnished. Its effect can not be forecast, nor its value estimated. An accurate expression of its views is unobtainable.[235]

Charged by the D.C. Circuit to construct a regulatory framework by which to assign radio frequencies in a noninterfering manner, Secretary Hoover resorted again to his radio conferences to devise a solution.[236]

In 1925 Hoover's fourth radio conference again concluded that radio required new comprehensive legislation to ensure adequate regulatory control.[237] The changing nature of radio presented an increasing number of issues (ranging from the financial qualifications for licensees to the protection of broadcast rights) that exceeded the regulatory reach of the Radio Act of 1912.[238] In 1926 an Illinois federal court reinforced that conclusion when it ruled in *United States* v. *Zenith Radio Corp.* that, while the 1912 statute gave the secretary of commerce discretion to assign licenses, it did not authorize him to

---

232. 286 F. 1003, 1007 (D.C. Cir. 1923).

233. *Id.*

234. *Id.* at 1005–6.

235. COMMISSIONER OF NAVIGATION, 1924 ANNUAL REPORT TO THE SECRETARY OF COMMERCE 22 (1924).

236. PAGLIN, *supra* note 51, at 9.

237. *Id.* (citing FOURTH NATIONAL RADIO CONFERENCE, *supra* note 229, at 8–9).

238. *Id.* at 10.

devise a new regulatory scheme by which to exercise that power.[239] Without any prescribed legislative standard, the court reasoned, the secretary's design of such a scheme would not be discretionary but arbitrary.[240] Consequently, Secretary Hoover abandoned all efforts to instill order in the airwaves and confined the Commerce Department's role to that of a registration bureau.[241]

The conventional wisdom is that the *Zenith* decision plunged the ether into chaos. Radio became a Leviathan struggle of all against all. New stations erupted into the ether on frequencies selected capriciously. Existing stations surfed the ether in search of clean air, each modulating broadcast frequency, power, and times to make themselves heard above the din.

Though Congress had flirted for several years with bills to strengthen regulatory control over the airwaves, the tumult in 1926 and the resultant public dissatisfaction demanded the legislature's greater attention. After years of resisting government regulation of radio, Congress, it is commonly believed, had to act to reverse the intolerable levels of interference produced by an absence of regulation. In his message to Congress on December 7, 1926, President Calvin Coolidge called for new legislation to regulate radio:

> Due to the decision of the courts, the authority of the department under the law of 1912 has broken down; many more stations have been operating than can be accommodated within the limited number of wavelengths available; further stations are in course of construction; many stations have departed from the scheme of allocation set down by the department, and the whole service of this most important public function has drifted into such chaos as seems likely, if not remedied, to destroy its great value. I must urgently recommend that this legislation should be speedily enacted.[242]

Congress responded by passing the Radio Act of 1927, which implemented a new regulatory framework for radio.

---

239. 12 F.2d 614 (N.D. Ill. 1926).

240. *Id.* at 618.

241. PAGLIN, *supra* note 51, at 9.

242. Message to Congress (Dec. 7, 1926), *quoted in* DAVIS, *supra* note 49, at 54.

FABRICATED CHAOS?

An alternative theory of the enactment of the Radio Act of 1927, propounded by economist Thomas Hazlett, is that the interference problems of the mid-1920s did not reflect market failure, but rather the conscious decision of government officials to prevent the emergence of an efficient market for rights in radio propagation.[243] Hazlett asserts that Secretary Hoover's Commerce Department precipitated the chaos because it wished to retain control over radio as a "public medium," and the emerging broadcast industry saw that the new legislation would shield existing licensees from further competition. During the debate of the Radio Act of 1927, Senator Key Pittman of Nevada claimed that private parties lobbying for the bill exacerbated the interference problem to secure monopoly protection to be afforded by the legislation:

> Why was it that just recently broadcasting concerns of the West all changed their wave lengths, sometimes a hundred degrees, to have them conflict, and on the next day said, "If you do not pass this bill, you will have the same condition for another year"? Mr. President, I do not believe that I am naturally suspicious, but . . . [this] bill is fair to only one institution. It is fair to the monopoly that will be created under it. The monopoly that may be created under it is practically free of control.[244]

Hazlett argues that the spectrum was functioning properly under Hoover's licensing policies until 1926 and that property rights in the spectrum were well defined, freely alienable, and largely secure.[245] Congress therefore fully understood in 1927 that a system of property rights in the broadcast spectrum was feasible.[246] Congress chose, however, to allocate spectrum through a political process rather than through markets, and it restricted competition by limiting the supply of frequencies available for radio broadcasting below the level then technically feasible.[247] Moreover, that federal regulation, which

---

243. Thomas W. Hazlett, *The Rationality of U.S. Regulation of the Broadcast Spectrum,* 33 J.L. & ECON. 133, 145 (1990).

244. 63 CONG. REC. 4111 (1927).

245. Hazlett, *supra* note 243, at 143–47.

246. *Id.* at 158–63.

247. *Id.* at 152–58.

expressly preempted state law, was enacted three months after an Illinois court in November 1926 recognized a broadcaster's common law property right to eject trespassers, by force of injunction, from the frequency on which it operated.[248] Secretary Hoover feared that such rights would become vested and escape government content controls, an objective that dominated his radio conferences and directed their initiatives for government regulation of the airwaves. Hazlett argues that, after Hoover was continually rebuffed by Congress in his efforts to impose greater regulatory controls on the spectrum, he undermined the existing regulatory structure to produce chaos in the spectrum and thus force Congress's hand.

At first, Hoover responded to Congress's indifference by trying to establish content controls through his own office's powers. According to Hazlett, Hoover obtained an understanding from the broadcast industry's leaders to accept content controls in return for restrictions on new market entrants.[249] Consequently, from November 1925 to April 1926 Hoover refused to issue any more licenses.[250] In April 1926 the *Zenith* decision denied the secretary of commerce any discretion to regulate the airwaves in a manner not specified in the Radio Act of 1912,[251] and, at Hoover's request, Acting Attorney General William Donovan confirmed the correctness of the court's ruling.[252]

Then, abandoning all attempts at regulation, Hoover provoked a crisis aimed at precipitating congressional action. The secretary, Hazlett wrote, "saw his *Zenith* 'defeat' and the ensuing confusion, which he had predicted, as a predicate to achieving his policy agenda."[253] The gambit proved successful. Chaos ensured, as expected, and forced Congress to enact, in the Radio Act of 1927, the comprehensive federal regulatory controls that Hoover had long sought.

---

248. This case, Tribune Co. *v.* Oak Leaves Broadcasting Station (Cir. Ct., Cook County, Ill., Nov. 17, 1926), appears to be publicly available today only in the *Congressional Record*, where it was inserted in its entirety several weeks after being handed down. 68 CONG. REC. 216, 219 (1926). For a discussion of *Oak Leaves*, see Hazlett, *supra* note 243, at 149–52.

249. *Id.* at 152–54.

250. *Id.*

251. United States *v.* Zenith Radio Corp., 12 F.2d 614 (N.D. Ill. 1926).

252. 35 OP. ATT'Y GEN. 126 (1926).

253. Hazlett, *supra* note 244, at 159.

If Hazlett's theory is correct, then Congress in 1927 enacted the most intrusive regulatory controls to that time imposed on the use of spectrum—not in response to genuine market failure, but in response to conscious efforts by the federal government to prevent a market from functioning. Those controls continue to exist today in only slightly altered form through the Communications Act of 1934 and its amendments. "The entrusting to federal regulators of power over life and death of American broadcasters slipped through Congress and remains public policy today," Hazlett argues, "due to a fundamental misunderstanding."[254]

The period of broadcast cacophony had several implications for the regulation of foreign direct investment. First, it was clear that the Navy had lost its battle to deny the private sector control over wireless. The enormous growth in sales of home radios during the mid-1920s had created a permanent constituency of household listeners that would oppose any government takeover of the radio industry in peacetime. Second, the Navy's interest in restricting foreign ownership of wireless was completely compatible with Secretary Hoover's goal of creating a government body to limit access to radio and influence its content. Indeed, the same pervasiveness and popularity of radio broadcasting that motivated Hoover to regulate also bolstered the Navy's claim, previously rather flimsy, that foreigners could disseminate propaganda by radio. Before the early 1920s, propagandists would have to settle for an audience of amateur radio operators tuning in almost randomly with rather crude receivers. (And, in any event, the radio propaganda in 1917 or 1927 could just as easily emanate from Berlin or Moscow by short wave and thus fail to implicate the foreign ownership of U.S. wireless stations in *any* respect.) Likewise, Hoover would be content to restrict foreign investment as the Navy wished, because the reciprocal flows of benefits—from the regulator to the regulated, and vice versa—upon which his scheme of regulation was predicated would function more smoothly if the economic benefits of regulating a radio licensee could be prevented from spilling over to owners in another country. Indeed, by 1995 broadcasters would be so comfortable with their regulatory bargain that they would show virtually no interest in being covered by proposals to liberalize the foreign ownership restrictions.

---

254. *Id.* at 141–42.

THE FOREIGN OWNERSHIP RESTRICTIONS
IN THE RADIO ACT OF 1927

The foreign ownership restrictions contained in the Radio Act of 1927 originated in the early efforts to reform the Radio Act of 1912. In 1922, Representative Wallace H. White, Jr., a Republican from Maine and a participant in Hoover's conferences, introduced a bill to amend the Radio Act of 1912.[255] The bill provided that no license was to be granted or transferred to any alien or his representative, a foreign government or its representative, a company organized under the laws of a foreign government, a company of which any officer or director was an alien, or a company one-fifth or more of whose voting stock was owned or controlled by such persons or entities.[256]

Though unsuccessful, Representative White's 1922 bill contained language restricting foreign ownership of wireless entities in the United States that would become the blueprint for the restrictions ultimately incorporated into section 12 of the Radio Act of 1927, which provided, among other things, that:

> The station license required hereby shall not be granted to, or after the granting thereof of such license shall not be transferred in any manner, either voluntarily or involuntarily, to (a) any alien or the representative of any alien; (b) to any foreign government, or the representative thereof; (c) to any company, corporation, or association organized under the laws of any foreign government; (d) to any company, corporation, or association of which any officer or director is an alien, or of which more than one-fifth of the capital stock may be voted by aliens or their representatives or by a foreign government or representative thereof, or by any company, corporation, or association organized under the laws of a foreign company.[257]

The legislative history gives little indication why Congress broadened the alien ownership restrictions. The paucity of evidence probably

---

255. H.R. 11964, 67th Cong., 2d Sess. (1922); *Hearings on H.R. 11964 Before the House Comm. on Merchant Marine and Fisheries*, 67th Cong., 2d Sess. 2 (1922) [hereinafter *H.R. 11964 Hearings*].

256. H.R. 11964, § 2(B), 67th Cong., 2d Sess. (1922).

257. Radio Act of 1927, ch. 169, § 12, 44 Stat. 1162 (1927).

reflects the fact that Congress's major concern when enacting the Radio Act of 1927 was to end the chaos thought to have been created by the inadequacy of the Radio Act of 1912 and to address the competing demands for spectrum created by the growth of broadcasting.[258] The legislative history that does exist emphasizes two principal purposes for the foreign ownership restrictions.

First, the foreign ownership restrictions were explained as an attempt to eliminate loopholes in the then-existing law, particularly as to domestic corporations controlled from abroad, and to render the new legislation consistent with the policies of other nations and with U.S. navigation law.[259] Second, the foreign ownership restrictions were considered a method to prevent alien activities against the United States in time of war. In a letter dated March 22, 1932, to Senator Couzens of Michigan, chairman of the Senate Interstate Commerce Committee, Secretary of the Navy Charles Adams wrote: "The lessons that the United States had learned from the foreign dominance of the cables and the dangers from espionage and propaganda disseminated through foreign-owned radio stations in the United States prior to and during the [First World] War brought about passage of the Radio Act of 1927, which was intended to preclude any foreign dominance in American radio."[260]

Some in Congress, however, doubted the need for foreign ownership restrictions. During debate on the 1927 bill, Senator Burton Wheeler—a Democrat from Montana who had been Robert La Follette's running mate on the Progressive ticket in 1924[261]—noted that section 12 was "based, presumably, upon the idea of preventing alien activities during time of war."[262] He argued that such restrictions were "unnecessary, as the war clauses gave the solution by granting power [to the President to seize all radio stations in time or threat of

---

258. DAVIS, *supra* note 49, at 79.

259. *See, e.g.*, 64 CONG. REC. 2332 (1923) (statement of Rep. White); 62 CONG. REC. 8400 (1922) (memorandum accompanying S. 3694 upon its introduction); *To Regulate Radio Communication: Hearings on H.R. 5589 Before the House Comm. on the Merchant Marine and Fisheries*, 69th Cong., 1st Sess. 23–24 (1926) (statement of Rep. White).

260. *Hearings on H.R. 8301 Before the House Comm. on Interstate and Foreign Commerce*, 73d Cong., 2d Sess. 26 (1934) [hereinafter *H.R. 8301 Hearings*].

261. BIOGRAPHICAL DICTIONARY, *supra* note 71, at 2033.

262. 68 CONG. REC. 3037 (1927).

war]."[263] Although Senator Wheeler's view failed to prevail, he drew attention to a fundamental weakness in the logic undergirding foreign ownership restrictions: The existence of the greater power, under section 2 of the Radio Act of 1912, to seize radio stations "in time of war or public peril or disaster"[264] cast serious doubt on the need, on national security grounds, for the creation in 1927 of the lesser power to restrict foreign ownership of wireless.

Congress, however, believed the national security interests involved to be sufficient to require heightened safeguards to protect the airwaves from foreign influence. Because nationalization was an unpalatable alternative, foreign ownership restrictions provided a convenient means to ensure domestic control of radio stations in the United States; domestic control, in turn, conveyed the reasonable expectation that licensees would cooperate with the government during international conflicts.[265]

It is unclear whether the foreign ownership restrictions in the Radio Act of 1927 were originally conceived to apply to broadcasting. The Navy officer most responsible for passage of the restrictions, Captain Stanford Hooper, expressed little interest in the broadcast use of radio when testifying before Congress following enactment of the new legislation.[266] The Navy's concern was control of radio for international communications. Hooper believed that the newly created Federal Radio Commission should administer radio broadcasting because each zone of the United States had individual interests, but he contended that international communications were a matter of national concern that the Navy Department could most efficiently administer.[267]

---

263. *Id.*

264. § 2, 37 Stat. 302, (1912).

265. James G. Ennis & David N. Roberts, *Foreign Ownership in U.S. Communications Industry: The Impact of Section 310*, 19 INT'L BUS. LAW 243, 244 (1991).

266. *Commission on Communications: Hearings on S. 6 Before the Sen. Comm. on Interstate Commerce*, 71st Cong., 1st Sess. 328–31 (1929) (statement of Captain Stanford C. Hooper).

267. *Id.*

The Loophole in
the Radio Act of 1927

After Congress enacted the Radio Act of 1927, a controversy arose
over licensees that complied with the statutory requirements for aliens
but were controlled by large corporations upon which no limitations
had been imposed. That loophole allowed holding companies to
circumvent the foreign ownership restrictions by using American-
owned and directed subsidiaries. Faced with that loophole, Congress
made repeated efforts to tighten the restrictions to effect section 12's
purpose.[268]

The catalyst for revisiting the foreign ownership restriction of the
1927 statute was the international conglomerate International
Telephone and Telegraph. ITT had several alien directors and various
alien officers.[269] Nonetheless, through subsidiaries ITT controlled four
U.S. companies outright—the Postal Telegraph & Cable Corporation,
the Commercial Cable Co., All-America Cables, Inc., and Mackay
Radio and Telegraph Company—and held a managing interest in
another, Commercial Pacific Cable Co.[270] ITT was technically in
compliance with section 12 of the Radio Act of 1927 because its
subsidiaries all satisfied the foreign ownership requirements.

Nonetheless, it was clear to Congress and the Navy that ITT was
dodging the intent of the foreign ownership restrictions in the Radio
Act of 1927. The Navy considered ITT's ownership structure a threat
to national security. To counter that threat, Captain Hooper, as director of naval communications, spearheaded an aggressive lobbying
effort to close the loophole in section 12 of the 1927 statute by
extending foreign ownership restrictions to include holding companies. In congressional testimony in 1932 Hooper expressed the Navy's
disapproval of ITT's actions:

> Now we find that International Telephone & Telegraph has
> circumvented the intent of the law by operating as a holding
> company, with subsidiaries, among which their radio subsidiary
> actively complies with the law. I fail to see how this can be

---

268. *See, e.g.,* S. REP. No. 1004, 72d Cong., 2d Sess. (1932); 72 CONG. REC.
8052 (1930).
269. *H.R. 8301 Hearings, supra* note 260, at 214.
270. *Id.* at 213.

proper because if a holding company owns the subsidiary it dominates every act of the subsidiary.[271]

Hooper claimed that World War I had taught that, to promote readiness for a future war, *no* alien influence in American commercial communications should be tolerated.[272] In 1934 he told Congress:

> That the communication facilities of a nation are vital to the nation's welfare is universally recognized. A natural corollary of that truth is that the communication facilities of a nation must be controlled and operated exclusively by citizens of that nation, and entirely free from foreign influence.[273]

The divulgence of military secrets to domestic companies in peacetime was necessary, Hooper argued, so that American commercial radio could be fully and efficiently converted to a war effort on short notice.[274] The only way to ensure such readiness for war on the part of America's communication network—short of government ownership, which the Navy still advocated[275]—was through establishing a synergistic relationship between the Navy and private industry. Hooper wanted a relationship predicated on the Navy's ability to entrust the U.S. wireless industry with vital military secrets. He described his vision to Congress:

> While the radio communication operated by the Navy in peace time is sufficient for peace-time need, it would be inadequate in time of war and would have to be augmented by the facilities of commercial radio companies. These additional facilities, like those normally operated by the Navy, must be able to pass from peace to war status at a moment's notice.
>
> For efficient operation in war there must be training and indoctrination in peace. Such training and indoctrination must

---

271. *Id.* at 166.

272. *To Amend the Radio Act of 1927: Hearings on H.R. 7716 Before the Sen. Comm. on Interstate Commerce*, 72d Cong., 1st Sess. 16, 31–33 (1932) [hereinafter *H.R. 7716 Hearings*]; *Federal Communications Commission: Hearings on S. 2910 Before the Sen. Comm. on Interstate Commerce*, 73d Cong., 2d Sess. 166, 170–71 (1934) [hereinafter *S. 2910 Hearings*].

273. *Id.* at 170.

274. *Id.*

275. *Id.* at 166.

involve the disclosure of military secrets. . . . Such secrets may
not be divulged to any company, or to individuals of any company
regarding which the least doubt can be entertained as to the
citizenship, patriotism, and loyalty of any of its officers or
personnel.[276]

Despite his relentless efforts, Hooper at first made little progress in
persuading Congress to embrace his vision of the military-industrial
complex.

Ironically, ITT served not only as the catalyst and the target for
those reform initiatives, but also as the greatest obstacle to their
enactment. In 1932 Americans controlled approximately 90 percent
of its outstanding shares,[277] and only four of ITT's twenty-three
directors and two of its twenty-two officers were foreigners.[278] The
remaining directors and officers were American. In short, ITT,
though an international conglomerate, was still a predominantly
American company. As such, Congress, though sensitive to the
Navy's national security arguments for nationalizing wireless
communications, was wary of enacting any measure that would harm
ITT's ability to operate internationally. Although ITT was a private
corporate entity, Congress was well aware that it was better that
international communications be controlled by an American company,
with a minority foreign ownership stake, than by another nation.
Senator Wallace H. White, Jr., from Maine, ITT's champion in the
Senate, sounded that warning during consideration of the 1932 bill to
require that all directors and officers of holding companies with
controlling interests in U.S. wireless companies be American citizens:

> I think it would be a grievous hardship for them, and I think
> of even more importance it would be a grievous harm to the
> communications interests of the United States as a whole, and the
> people of the United States if this communication company should
> be deprived of these facilities. . . . It might cost this American
> company its entire foreign setup in some countries that might be
> affected by it.
> I think we should all agree that we would much prefer that
> there were none of these foreign directors but I think that weighs

---

276. *Id.* at 170–71.
277. *S. 2910 Hearings, supra* note 272, at 126–27.
278. *Hearings on H.R. 7716, supra* note 272, at 39–40.

but a feather against the tremendous advantage of having this company maintain its radio services throughout the world and maintain for us here in this country the competitive services which would result from their system.[279]

Frank C. Page, ITT's vice president also warned the Senate in testimony on the 1932 bill that American influence in international telecommunications would diminish if Congress enacted Hooper's prohibition on foreign directors and officers and that other nations would likely retaliate:

> If we get rid of our directors, there is just as much national feeling in South America and in the rest of the world as there may be anywhere else, and it is absolutely certain . . . that those countries will retaliate against our companies in the foreign field where we are carrying on American communications. It is a problem which we would have to face if this bill . . . is passed as it is now written.[280]

Added to those concerns were the antitrust implications of forcing ITT to divest its holding in U.S. communications companies. ITT was RCA's only significant competitor. If forced to leave the market, RCA would have a virtual monopoly over international wireless. And increasing RCA's market power in that manner would increase its exposure under U.S. antitrust law,[281] which would have the counter-productive effect of undermining America's efforts to influence international radio communications. Senator White advanced that unpleasant possibility in opposing the 1932 bill:

> I think to deprive [ITT] of the licenses of its subsidiaries would be taking the most far-reaching step toward a monopoly of radiocommunications in the international field that we could take. . . . [T]he great competitor in the international field is the Radio Corporation of America. One of the underlying purposes of the 1927 law was to preserve competition in the communications field, and there were various efforts made to insure that there should be competition.

279. *H.R. 7716 Hearings, supra* note 272, at 16–17.
280. *Id.* at 42.
281. *Id.* at 33–34.

I personally feel that to write this language which is here proposed into law would be a tremendous backward step and that we would be in large measure abandoning the original conception of the United States law with respect to this matter of monopoly. We would be doing what I think is a great harm to an American communication company, and we would be very closely verging on monopolistic control over international communication.[282]

To those warnings, ITT added that enactment of the 1932 bill would erase the investment of "over 90,000 Americans" in ITT's radio enterprises.[283] During that time, ITT also dragged its feet skillfully in the face of congressional pressure to propose its own solution to the controversy.[284] The foreign ownership restrictions would remain unchanged for another two years.

SECTION 310 OF THE
COMMUNICATIONS ACT OF 1934

Although Congress in 1934 focused primarily on President Franklin Roosevelt's proposal to create a Federal Communications Commission,[285] reform of the foreign ownership provisions was ripe. Challenging each other were the Navy's national security interests and America's economic interests in the global communications market. ITT was thought to provide the United States a unique platform from

---

282. *Id.* at 16.

283. *Id.* at 42.

284. ITT's president, Sosthenes Behn, exasperated Clarence Dill, the chairman of the Senate Committee on Interstate Commerce, during testimony in 1932 to extend the 1927 act's foreign ownership restrictions to holding companies:

You [Sosthenes Behn] have known that this provision has been up here for the last 2 or 3 years. You have said before this committee previously that you were gradually working something out; now you come before us and tell us that it is absolutely impractical, that it cannot be done; that you must go out of business if anything of this kind is put in.

S. 2910 Hearings, supra note 272, at 126-27.

285. Message from the President of the United States Recommending That Congress Create a New Agency to Be Known as the Federal Communications Commission (Feb. 26, 1934), *reprinted in* 3 PUBLIC PAPERS AND ADDRESSES OF FRANKLIN D. ROOSEVELT, 1934, at 107 (Samuel Rosenman ed., 1950).

which to influence .nternational wireless communications; thus, any constraint on ITT's ability to operate in the international arena would not only harm the company, but also compromise America's strategic interests.

In section 310 of the Communications Act of 1934, Congress expanded the foreign ownership restrictions in the Radio Act of 1927 to apply to holding companies.[286] Congress hoped that its solution would satisfy the Navy while not impairing ITT's international communications business.[287] Yet the reasoning for why the amendment would mutually satisfy the Navy and ITT was obscure from the official statements of the key figures behind the 1934 legislation.

Arguing that "[t]he holding company system has made such legislation necessary,"[288] Clarence Dill, chairman of the Senate Committee on Interstate Commerce, offered as the solution "a provision that none of the officers of the company shall be foreigners, that not more than one-fifth of the capital stock shall be owned and voted by foreigners, and that not more than one-fourth of the directors shall be foreigners."[289] ITT had previously fought H.R. 7716 in 1932, which would have been harsher in the sense of forbidding *any* foreigner to be a director, on the grounds that it would impair the company's ability to operate in foreign countries, to the ultimate detriment of broader American interests. So it was not clear why by 1934, as Dill asserted, the new provision directed at holding companies—more permissive than H.R. 7716 only to the extent that it allowed 25 percent of the holding company's board to be foreign—would not impede the ability of "our international communication companies"—namely, ITT—"to compete with companies in foreign countries."[290]

Nor was it any clearer why the Navy should like that amendment. Dill's argument for why the new provision "amply safeguarded . . . the American communications service" was really an argument for why the holding company provision was unnecessary in the first

---

286. *S. 2910 Hearings, supra* note 272, at 130 (statement of Sen. White); Joint Board, J.B. no. 319 (serial no. 522) (Jan. 19, 1934), *reprinted in S. 2910 Hearings, id.* at 167–68 (Army–Navy Joint Board memorandum supporting Hooper proposals).

287. S. REP. NO. 781, 73d Cong., 2d Sess. 7 (1934).

288. 78 CONG. REC. 8825 (1934).

289. *Id.*

290. *Id.*

place: "[A]fter all, if an emergency shall arise and the country shall go to war, the president will have power under the law to seize all communication companies, and have absolute control of all communication companies with facilities in the United States."[291] Congress had already granted the president that power twenty-two years earlier when enacting section 2 of the Radio Act of 1912,[292] and President Wilson had immediately invoked the power when the United States declared war on Germany on April 6, 1917.[293] In short, Senator Dill's explanation of the incremental benefits to ITT and the Navy of the holding company provision was no explanation at all.

The foreign ownership restrictions in section 310 of Communications Act of 1934 added two notable limitations to the restrictions already present in section 12 of the Radio Act of 1927. First, section 12(d) of the 1927 statute had prohibited the grant of licenses to any company "of which more than one-fifth of the capital stock may be voted by aliens."[294] Section 310(a)(4) of the 1934 statute qualified that restriction to apply only to "one-fifth of the capital stock . . . *owned of record* or voted by aliens."[295] The added phrase "owned of record" sought to "guard against *actual* alien control, and not the mere *possibility* of alien control."[296] In Senator Dill's words, "the only thing a company can be held to is what is on the books."[297] Corporations would be permitted to rely upon record ownership without having to confirm the extent of foreign ownership through independent investigation.

Second, and more important, Congress directly addressed, in section 310(a)(5), foreign participation in holding companies. H.R. 7716 was the blueprint.[298] It originally proposed in 1932 to relax the foreign ownership restrictions in section 12 of the Radio Act of 1927 by permitting the grant of a license to a corporation with alien directors and officers if aliens held no more than one-fifth of those

---

291. *Id.*

292. 37 Stat. 302, § 2 (1912).

293. Exec. Order (Apr. 6, 1917), *reprinted in* 17 A COMPILATION OF THE MESSAGES AND PAPERS OF THE PRESIDENTS 8241.

294. § 12(d), 44 Stat. 1162 (1927).

295. § 310(a)(4), 48 Stat. 1086 (1934) (emphasis added).

296. S. REP. NO. 781, 73d Cong., 2d Sess. 7 (1934) (emphasis added).

297. *S. 2910 Hearings, supra* note 272, at 122–25.

298. *H.R. 7716 Hearings, supra* note 272.

positions. The logic of changing the provision concerning officers and directors, from total exclusion of aliens under section 12 of the 1927 Act to four-fifths exclusion of them under H.R. 7716, was to create symmetry with the existing requirement that no more than one-fifth of the holding company's capital stock be voted by aliens.[299]

Senator White criticized that aspect of the House's bill as "a distinctly backward step from the standard of what I will call Americanism, written into the 1927 law," and he urged that "the whole section should be stricken out and that the entire matter should go to conference."[300] White saw the holding company provision as being "aimed at a particular situation and a particular American corporation."[301] Its intent, he argued, was to present ITT "a most acute and embarrassing situation" that would "force either the relinquishment of the licenses by the subsidiaries of this company or the ousting of all foreign officers and directors of the parent company."[302]

As a compromise, a House amendment retained the 25 percent benchmark with respect to directors of a corporate licensee, barred aliens from serving as officers, and added language placing similar limitations on holding companies.[303] That compromise substitution

---

299. S. REP. No. 1004, 72d Cong., 2d Sess. 10–11 (1932).

300. 76 CONG. REC. 3769 (1933).

301. *Id.*

302. *Id.*

303. In relevant part, the substitute provision read:

The station license required hereby shall not be granted to or held by—

(d) Any controlling or holding company, corporation, or association, of which any officer or more than one-fifth of the directors are aliens, or of which more than one-fifth of the capital stock may be voted by aliens, their representatives, or by a foreign government or representative thereof, or by any company, corporation, or association organized under the laws of a foreign country;

(e) Any corporation or association controlled by, or subsidiary to a corporation or association, of which any officer or more than one-fifth of the directors are aliens, or of which more than one-fifth of the capital stock may be voted by aliens, their representatives, or by a foreign government or representative thereof, or by any company, corporation, or association organized under the laws of a foreign country.

H.R. REP. No. 2106, 72d Cong., 2d Sess. 2 (1933). The House report explained that

made in H.R. 7716, however, only became section 310(a) after the conference committee inserted a clause at the end of the holding company provision giving the new FCC discretion to permit alien participation in holding companies beyond the statutory benchmarks.[304] As enacted, section 310(a)(5), later renumbered section 310(b)(4), provided that no license was to be granted to or held by

> any corporation directly or indirectly controlled by any other corporation of which any officer or more than one-fourth of the directors are aliens, or of which more than one-fourth of the capital stock is owned of record or voted, after January 1, 1935, by aliens, their representatives, or by a foreign government or representative thereof, or by any corporation organized under the laws of a foreign country, if the Commission finds that the public interest will be served by the refusal or revocation of such license.[305]

The proviso would empower the FCC to grant licenses to subsidiaries of holding companies with alien officers and more than one-fourth alien directors, or with more than one-fourth alien ownership. The conference report merely noted that the legislation gave the FCC that discretion.[306] Nonetheless, the decision by Congress to grant that discretion was consistent with the larger conclusion that the legislators ultimately agreed with Senator White that foreign ownership and control of holding companies whose subsidiaries hold radio licenses could actually advance U.S. interests rather than threaten them.[307]

As we shall see in chapter 3, the officer and director provisions of section 310(b)(4) eventually engendered so little support in

---

the amendment "permit[s] a station license to be granted to or held by a company of which not more than one-fifth of the directors are aliens. It also broadens the present law so as to make the inhibition against licenses being granted to or held by aliens, or a company, corporation, or association of which any officer or more than one-fifth of the directors are aliens, or of which more than one-fifth of the capital stock may be voted by aliens, also apply to any controlling, holding, or subsidiary company, corporation, or association." *Id.* at 5–6.

304. 78 Cong. Rec. 10978 (1934).

305. 48 Stat. 1086, § 310(a)(5) (1934).

306. H.R. Rep. No. 1918, 73d Cong., 2d Sess. 48–49 (1934). *See also* 78 Cong. Rec. 10988 (1934).

307. *See* S. Rep. No. 781, 73d Cong., 2d Sess. 7 (1934).

Congress that they were repealed, without elaboration in the legislative history, upon enactment of the Telecommunications Act of 1996.

PEARL HARBOR

Within seven years after their enactment, the fortified foreign ownership restrictions in the Communications Act of 1934 revealed their irrelevance to the protection of national security. Using U.S. radio common carriers, Japanese diplomats transmitted encrypted military intelligence to and from the United States, including information about the location and activities of ships in Pearl Harbor, troop exercises, air reconnaissance, and other sensitive defense issues.[308] Japan's information network provided precise news on U.S. naval defense strategies that would assist in executing the surprise attack of December 7, 1941.[309]

On March 27, 1941, Ensign Takeo Yoshikawa arrived in Honolulu to serve as the Imperial Japanese Navy's espionage agent.[310] Rotating among the U.S. radio common carriers, he continuously reported to the Japanese Navy all ship movements in Pearl Harbor.[311] Yoshikawa's report that the U.S. fleet was still in port was the last message that the Japanese consulate in Honolulu sent before the attack. Time-stamped "1941 Dec 6 pm 6 01" by RCA's office in Honolulu,[312] his encrypted message read:

> (1) On the evening of the 5th, the battleship *Wyoming* and one sweeper entered port. Ships at anchor on the 6th were: 9 battleships, 3 minesweepers, 3 light cruisers, 17 destroyers. Ships in Dock were: 4 light cruisers, 2 destroyers. Heavy cruisers and carriers have all left. (2) It appears that no air reconnaissance is being conducted by the fleet arm.[313]

---

308. DAVID KAHN, THE CODEBREAKERS 42–53 (Weidenfeld & Nicolson 1974); HOMER N. WALLIN, PEARL HARBOR: WHY, HOW, FLEET SALVAGE AND FINAL APPRAISAL 60–74 (Government Printing Office 1968).

309. *Id.* at 63.

310. KAHN, *supra* note 308, at 13.

311. *Id.*

312. *Id.* at 52.

313. *Pearl Harbor Attack: Hearings Before the Joint Comm. on the Investigation of the Pearl Harbor Attack,* 79th Cong., 1st Sess., pt. 12, at 280 (1946) [hereinafter *Pearl Harbor Attack Hearings*]. Japan's consulate in Honolulu routinely used RCA

In the early morning of December 7, the War Department in Washington intercepted and decoded a message from Tokyo to the Japanese Embassy. Secretary of War George Marshall composed a message to be transmitted immediately to Honolulu:

> Japanese are presenting at one p.m. Eastern Standard Time today what amounts to an ultimatum also they are under orders to destroy their code machine immediately Stop. Just what significance the hour set may have we do not know but be on alert accordingly Stop.[314]

Early morning interference prevented the War Department's 10-kilowatt transmitter from establishing a circuit to Honolulu. So Marshall sent the message by Western Union to San Francisco, where it was relayed by RCA over its 40-kilowatt transmitter to Hawaii. The message had taken fifty-two minutes to transmit from Washington to Honolulu and arrived at 7:33 A.M., when the first Japanese bombers were thirty-seven miles away.[315] RCA's operator put the message in an envelope, scheduled for hand delivery, marked "Commanding General." Within minutes, the attack began. Three hours after the last Japanese bomber had left Hawaii, the decoded message was finally presented to General Walter Short, who threw it in the waste basket.[316]

After World War I, the Navy had been instrumental in creating RCA to protect against the foreign influence in wireless that the admirals feared would tip the balance of naval power against the United States. By midday on December 7, 1941, the Imperial Japanese Navy, relying on intelligence routinely sent by RCA Radiogram from Japan's spy in Honolulu, had crippled America's Pacific fleet. The War Department's own network for radio commu-

---

Radiograms to relay intelligence through "hidden words" in weather report messages. WALLIN, *supra* note 308, at 63. Vital information was also obtained from public radio broadcasts in Hawaii, from American newspapers, and from the crew and passengers of ships that had been in Honolulu in mid-November of 1941. *Investigation of the Pearl Harbor Attack: Report of the Joint Comm. on the Investigation of the Pearl Harbor Attack,* 79th Cong., 2d Sess. 54–55 (1946).

314. *Pearl Harbor Attack Hearings, supra* note 313, pt. 15, at 1640.
315. KAHN, *supra* note 308, at 61.
316. *Id.* at 64–65.

nications had failed to protect the Navy at the critical moment, and the U.S. government was reduced to sending the secretary of war's warning by Radiogram. Foreign ownership of wireless had played no role in the gravest breach of national security in American history.

CONCLUSION

From 1934 to 1995, national security was the only enunciated rationale that could be found in the legislative history for the foreign ownership restrictions contained in the Communications Act. The actual experience with the restrictions, however, raises substantial doubt that they have ever been an efficacious means of achieving the security objectives for which they were conceived. Moreover, it is far from clear that the U.S. government had in 1934, or has today, the ability to control the natural implications of telecommunications and information technologies—technologies of which early radio was merely a crude forerunner. Today, in an era of satellite networks and Internet cyberspace, the technologies of communications have surpassed the government's ability to control them. More and more, when we take regulatory aim at those technologies, we find ourselves shooting at bogeymen that are only abstractly understood and described by such terms as the "ether," the "spectrum," and the "net."

In trying to regulate technologies that exist in such netherworlds, we endeavor to play a game that changes with the regulator's every move. In the case of radio, the need to take licenses away from foreigners during World War I arose from the fact that an extraordinary confluence of world conflict and technological advances in naval warfare had made wireless unusually powerful in shifting the balance in the European war. The proximate cause of the security problem, however, was not that aliens were licensed to use radio. A number of other decisions by President Wilson and his administration had a far more immediate effect on national security.

If the aliens operating the Sayville station had violated President Wilson's neutrality order, they should have been swiftly prosecuted for their offenses and prevented from repeating them. In fact, they were not arrested until 1917. Further, section 2 of the Radio Act of 1912 authorized Wilson not only to shut down any radio station "in time of war," but also to do so "in time of . . . public peril or disaster."[317] If aliens on Long Island were indeed transmitting information about ship movements to U-boats, then Wilson himself could have prevented at least some of the destruction of life and property in 1915 and 1916 by invoking his existing statutory power nearly two years sooner than he did, rather than permitting the German-controlled Sayville station to operate under Navy censorship.

By 1917, Wilson's embarrassment with the Zimmermann telegram demonstrated that wireline communications could jeopardize U.S. national security just as much as wireless communications. Increasingly, the military potential of telecommunications depended upon the sophistication of its encryption techniques. Yet, the Communications Act of 1934 did not begin to restrict a foreigner's ability to send encoded messages over the radio frequencies licensed to an American common carrier. The threat to national security had shifted from the identity of the electronic messenger to the imperviousness of the messages he carried for others.

In the years after the surprise attack on Port Arthur in 1904, the U.S. Navy recognized the military significance of wireless. But the ultimate irony awaited those in the Navy who believed that the foreign ownership restrictions enacted in 1927 and 1934 made their battleships more secure against enemy powers. In December 1941 Japan's fleet steamed undetected toward Hawaii under a radio blackout with orders to replay Port Arthur, this time with a new technology as devastating as the U-boats of 1914—bombers and torpedo-carrying fighters launched from aircraft carriers. On the eve of the attack, the Japanese consulate in Honolulu burned its code books and transmitted its last communiqué to Tokyo, not by way of some clandestine wireless station, but by an encoded RCA Radiogram. Meanwhile, the newly amended foreign ownership restrictions stood vigilant against a risk that technology had rendered obsolete.

317. § 2, 37 Stat. 302 (1912).

3

The Statute

FOR OVER SIX DECADES the FCC has misread the most important foreign ownership provision in the Communications Act of 1934. To understand how that misinterpretation could arise, we now examine the language and scope of section 310(b). That examination will show that, over the same period, Congress has repeatedly narrowed the scope of section 310(b), and indeed did so again in the Telecommunications Act of 1996, and that it has repeatedly rejected proposals to expand section 310(b) to encompass other services and licenses. We also examine that section's relationship to other provisions in the Communications Act and other statutes that affect foreign direct investment in U.S. telecommunications.

STATUTORY TEXT

Section 310(b) of the Communications Act, as amended through 1996, restricts foreign ownership or management of four kinds of FCC radio licensees:

> No broadcast or common carrier or aeronautical en route or aeronautical fixed radio station license shall be granted to or held by—
>> (1) any alien or the representative of any alien;

(2) any corporation organized under the laws of any foreign government;

(3) any corporation of which more than one-fifth of the capital stock is owned of record or voted by aliens or their representatives or by a foreign government or representatives thereof or by any corporation organized under the laws of a foreign country;

(4) any corporation directly or indirectly controlled by any other corporation of which more than one-fourth of the capital stock is owned of record or voted by aliens, their representatives, or by a federal government or representative thereof, or by any corporation organized under the laws of a foreign country, if the Commission finds that the public interest will be served by the refusal or revocation of such license.[1]

It might appear that the restrictions on foreign investment are increasingly permissive the more removed the foreign ownership is from the actual licensed facility. Foreign individuals and foreign corporations are flatly prohibited from directly holding licenses. A domestic corporate licensee may have 20 percent foreign ownership of its capital stock. A parent corporation or holding company for a domestic corporate licensee may have 25 percent foreign ownership of its capital stock. But that analysis is too simplistic because it ignores that sections 310(b)(3) and 310(b)(4) regulate not only ownership, but also *control*: Both sections limit the amount of the licensee's capital stock that foreign investors may *vote*.

The foreign interests are additive and cumulative with respect to the FCC licensee. Each statutory limit in section 310(b) "refers to total, as opposed to individual, alien ownership interests in any one

---

1. 47 U.S.C. § 310(b). Although broadcast and common carrier licenses are familiar, the two aeronautical licenses mentioned in section 310(b) are relatively unfamiliar. "Aeronautical en route stations provide air-ground communications for the operational control (flight management) of aircraft by their owners or operators. Communications relate to the safe and efficient operation of aircraft . . . . En route stations are the means by which companies satisfy Federal Aviation Administration requirements to maintain reliable communications between each aircraft and its dispatch office, in the case of large airlines, or maintain flight following systems, in the case of small airlines and commercial aircraft operators. Aeronautical fixed stations provide point to point communications pertaining to safety, regularity and economy of flight." Market Entry and Regulation of Foreign-affiliated Entities, Notice of Proposed Rulemaking, IB Dkt. No. 95-22, 10 F.C.C. Rcd. 5256, 5296 ¶ 97 & n.82 (1995) [hereinafter *Market Entry and Regulation NPRM*].

facility."[2] Thus, even if by himself a foreign investor in an American communications business were not violating the statutory restrictions, he might violate the statute in combination with the ownership interests of other foreign investors in the same American firm. That aspect of section 310(b) could conceivably create a race among foreign investors when a licensee wholly owned by American citizens is offered for sale. Particularly if the licensee is or will be a publicly traded corporation, a foreign bidder has an incentive to acquire as quickly as possible the "maximum" equity holding in the licensee (a slippery notion, as we shall see presently) that it ultimately might desire (for example, a full 20 percent holding directly in the licensee rather than, say, 15 percent); otherwise, a different foreign investor's subsequent equity acquisitions could foreclose the first foreign investor from increasing his stake in the licensee. On the other hand, if the first foreign investor acquired the maximum equity holding under section 310(b), and if an investment made by a different foreigner subsequently violated that maximum, one would expect the FCC to follow a last-in, first-out approach to mandatory divestiture, if the agency deemed divestiture of any sort to be more in the public interest than the FCC's acquiescence to the combined bloc of foreign investment in the licensee.

SUBSEQUENT AMENDMENTS
BEFORE 1996

Between 1934 and 1996 Congress made a series of changes in section 310(b). Some of those amendments have weakened the intellectual case for retaining the remainder of section 310(b).

The first amendment, enacted in 1958, permitted the FCC to license a radio station on an aircraft to an alien or the representative of a alien, provided that the person held a U.S. pilot certificate or a foreign aircraft pilot certificate that was valid in the United States on the basis of reciprocal agreements.[3] A 1974 amendment exempted safety, special, and experimental radio services from section 310(b),[4]

---

2. Corporate Ownership Reporting and Disclosure by Broadcast Licensees, Report and Order, Dkt. No. 20521 *et al.*, 97 F.C.C.2d 997, 1009, ¶ 22 (1984) [hereinafter *Attribution of Ownership Interests*].

3. Pub. L. No. 85-817, 72 Stat. 981 (1958).

4. Pub. L. No. 93-505, 88 Stat. 1576 (1974).

so that "persons who use radio services as an incident to their business" would not be forced to go without radio communications if they lacked "the resources and skills necessary to establish subsidiary corporations" with which to comply with the foreign ownership restrictions.[5]

Several other amendments have concerned amateur radio. A 1964 amendment authorized the FCC to permit an alien licensed by his government as an amateur radio operator to operate the station licensed by his government in the United States, provided that a bilateral agreement existed between the United States and the alien's government giving similar rights to U.S. amateur radio operators, and provided further that the FCC notified appropriate government agencies of any applications for authorizations, and that such agencies furnished to the FCC information bearing on the request's compatibility with U.S. national security.[6] A 1971 amendment permitted the FCC to issue a license for amateur radio service to an alien who had been admitted to the United States for permanent residence, subject to similar requirements concerning notice and compatibility with national security.[7] The 1974 amendments also enacted the provision now designated as section 310(c), which authorized the FCC "to permit an alien licensed by his government as an amateur radio operator to operate his amateur radio station licensed by his government in the United States, . . . provided there is in effect a multilateral or bilateral agreement, to which the United States and the alien's government are parties, for such operation on a reciprocal basis by United States amateur radio operators."[8] Originally, that provision allowed only for bilateral reciprocity. But a 1990 amendment extended that right of aliens on a multilateral basis.[9]

The extensive liberalization of the foreign ownership restrictions for amateur radio licensees seriously undermined the credibility of one of the two traditional rationales for section 310(b)—the putative need to prevent wireless communications with the enemy by subversive elements within the United States. If it is possible to scrutinize the national security implications of licensing individual aliens, then

5. H.R. REP. NO. 1423, 93d Cong., 2d Sess. 2 (1974).
6. Pub. L. No. 88-313, § 2, 78 Stat. 202 (1964).
7. Pub. L. No. 92-81, § 2, 85 Stat. 302 (1971).
8. Pub. L. No. 93-505, § 2, 88 Stat. 1576 (1974) (codified at 47 U.S.C. § 310(c)).
9. Act of Sept. 28, 1990, Pub. L. No. 101-396, 104 Stat. 850.

it is surely possible to scrutinize the national security implications of licensing an alien corporation that owns more than 25 percent of the stock or voting rights of a radio common carrier or broadcaster. On the other hand, if the national security scrutiny given alien individuals applying for amateur licenses is merely perfunctory, then the credibility of the national security rationale diminishes for the rest of section 310(b). Admittedly, the power and propagation characteristics of radio signals transmitted by an alien amateur radio operator today may be inferior to those of a television station or a provider of cellular telephony service. The fact remains nonetheless that the amateur radio stations so licensed would still be able to communicate with persons outside U.S. territory, and indeed those foreign licensees may have a considerably greater ability to do so today than did the German nationals who used state-of-the-art technology in 1915 to communicate with the German fleet from their station on Long Island. It is remarkable that the national security rationale for section 310(b) continues to be cited when Congress compromised, if not repudiated, that objective when it amended the Communications Act in 1971 and 1974 to permit the FCC to license foreign amateur radio operators.

THE 1996 REPEAL OF THE RESTRICTIONS ON FOREIGN OFFICERS AND DIRECTORS

The Telecommunications Act of 1996 repealed the restrictions in sections 310(b)(3) and 310(b)(4) on foreign officers and directors.[10] Immediately before the 1996 amendments, section 310(b)(3) forbade the grant of the applicable radio licenses to "any corporation of which any officer or director is an alien." Similarly, former section 310(b)(4) forbade any such grant to "any corporation directly or indirectly controlled by any other corporation of which any officer or more than one-fourth of the directors are aliens . . . if the Commission finds that the public interest will be served by the refusal or revocation of such license."

The 1996 amendments were most noteworthy for their utter lack of explicit language or legislative history explaining Congress's rationale for liberalizing foreign investment in that manner. Adding to the mystery was the fact that, as chapter 6 documents, throughout

---

10. Pub. L. No. 104-104, § 403(k), 110 Stat. 56, 131, 132.

the congressional consideration of the 1996 statute, the proposed rewrite of section 310(b) would have been far more expansive. The House and Senate bills that eventually produced the 1996 statute would have explicitly made section 310(b) a trade statute predicated on bilateral reciprocity. The repeal of the restrictions on officers and directors would have been a minor footnote to that more ambitious amendment. Without explanation, however, the conference committee stripped the reciprocity provisions in the House and Senate bills from the Telecommunications Act of 1996. The amendments to section 310(b) appeared in section 403 of the act, a catch-all provision captioned, "Elimination of Unnecessary Commission Regulations and Functions." Congress gave no clue as to why the restriction on foreign officers and directors had become unnecessary by 1996 while the remainder of sections 310(b)(3) and 310(b)(4) had not.

WIRELINE TELEPHONY

Various licenses and services are exempt from section 310(b). Aliens may hold licenses not enumerated in section 310(b). Section 310(b), in other words, is an exhaustive list and does not preclude a foreign investor from being the transferee of other, unenumerated FCC licenses held by an American entity.

Most significantly, the foreign ownership restrictions do not extend to communications by wire. If a telephone company held no radio licenses, section 310(b) would not limit the extent to which a foreigner could invest in the company. (As discussed below, to commence common carrier service, the foreigner would still need to receive FCC authorization under section 214 of the Communications Act.[11]) As a practical matter, however, wireline telephone companies almost always employ some radio links that implicate section 310(b). Such was the case in 1992 in the acquisition of Telefónica Larga Distancia de Puerto Rico (TLD), the U.S. domestic and international long-distance carrier, by Telefónica de España.[12] As chapter 5 will explain in greater detail, Telefónica de España acquired more than 25 percent of the stock in TLD, but it did so by segregating TLD's wireline and radio common carrier activities and purchasing different

---

11. 47 U.S.C. § 214.
12. Telefónica Larga Distancia de Puerto Rico, 8 F.C.C. Rcd. 106 (1992).

percentages of each. We shall examine in chapter 4 the generic business strategies that will enable a foreign investor to maximize his investment in an American telecommunications firm while complying with section 310(b).

CABLE TELEVISION

One consequence of the 1974 amendments to section 310(b) was to permit aliens to hold licenses for microwave radio stations in the cable television relay service. Although section 310 had not directly covered cable system operators, most became indirectly subject to the ownership restrictions before 1974 by virtue of their using microwave relay stations. The 1974 deregulation of foreign investment in cable may have been unintended. In any event, the subsequent unsuccessful attempts to reverse that deregulation help to illuminate the scope of section 310(b) generally.

1976 Proposed Rulemaking

In 1976 the FCC considered and ultimately rejected a proposed rule that would have restricted foreign investment in the cable television industry.[13] The experience with cable television relay services during the mid-1970s continues to be instructive because many of the same arguments raised then have been heard again in subsequent debates over whether and how to liberalize section 310(b).

In its 1976 report and order, the FCC gave several reasons for its determination that ownership restrictions were not warranted at that time. First, the agency observed that foreign investment in cable was quite limited.[14] Next, it distinguished cable operators from broadcasters on the ground that, although cable operators could produce original programming, such operators lacked the "totality" of

13. Amendment of Parts 76 and 78 of the Commission's Rules to Adopt General Citizenship Requirements for Operation of Cable Television Systems and for Grant of Station Licenses in the Cable Television Relay Service, Report and Order, Dkt. No. 20621, 59 F.C.C.2d 723 (1976) [hereinafter *Cable Television Citizenship Requirements*]; *see also* DANIEL L. BRENNER, MONROE E. PRICE & MICHAEL I. MEYERSON, CABLE TELEVISION AND OTHER NONBROADCAST VIDEO: LAW AND POLICY § 4.04[3] (Clark Boardman Callaghan, rev. ed. 1996).
14. *Cable Television Citizenship Requirements*, 59 F.C.C.2d at 726 ¶ 9.

program content control typical of broadcasting in 1976.[15] By the late 1990s, however, such a distinction had become nonexistent: The largest cable television systems operators, Tele-Communications, Inc. and Time Warner, were in 1996 far more vertically integrated into program production and distribution than was the typical television broadcast licensee subject to section 310(b).[16]

The FCC further reasoned in 1976 that "[a]lien ownership restrictions do not apply to communicators generally, to newspapers, wire news services, non-license radio and television networks, film and television producers, cable system networks and channel lessees, and it is not clear that they should apply to a system operator solely because of his potential ability to influence, through his program origination efforts, the ideas and attitudes of cable subscribers."[17] The agency also deemed foreign ownership of cable television to present little threat to national security because "[t]housands of different systems are operated across the United States, most diverse in their operators and each individually responsible to the communities that have selected and franchised their operations."[18] "It is these local jurisdictions," the FCC reasoned, "that are initially in the best position to determine whether or not an individual operator's nationality will prevent him from satisfying his 'public interest' obligations."[19]

The FCC did not base its decision solely on the conclusion that no *harm* would arise from not extending section 310(b) to cable television. It also recognized explicitly the *benefit* of foreign investment in developing new telecommunications technologies:

> The Commission further believes that citizenship prohibitions may in some measure deter the development of cable television in the United States. Although it is hoped that the industry will continue to find sources of domestic capital to fund its growth, it remains at this time relatively undeveloped and in need of new sources of capital and technology to continue its development. Foreign

---

15. *Id.* at 727 ¶ 9.

16. *See* DAVID WATERMAN & ANDREW WEISS, VERTICAL INTEGRATION IN CABLE TELEVISION (MIT Press & AEI Press 1997); David Waterman, *Vertical Integration and Program Access in the Cable Television Industry*, 47 FED. COMM. L.J. 511 (1995).

17. *Cable Television Citizenship Requirements*, 59 F.C.C.2d at 727 ¶ 9.

18. *Id.*

19. *Id.*

interests, if permitted, could contribute to this development. *The Commission, we believe, ought not deny these resources to cable without overriding reasons of national importance. In the absence of a demonstrable harm and where benefits may result, the Commission is inclined to allow free market forces to determine the direction of capital flow within the industry.* The proposed restrictions have been termed, in the comments, a "cure for which there is no disease." The slight evidence accumulated regarding cable systems owned by aliens seems to bear this out, for now.[20]

As we shall see in chapter 7, the FCC's own reasoning in 1976 for not extending foreign ownership restrictions to cable provides the most succinct and compelling argument two decades later for liberalizing foreign direct investment in telecommunications generally rather than using section 310(b) as tool of U.S. trade policy.

Out of an abundance of caution, the FCC in its 1976 cable decision noted that it would continue to monitor the extent of foreign investment in the cable television industry and that changed circumstances could justify ownership restrictions in the future.[21] The agency, arguing that it had no jurisdiction to adopt foreign ownership restrictions for cable television operators, declined to address public comments.[22] The FCC suggested without elaboration that its general statutory responsibilities for radio authorizations and regulation of cable television provided that authority.[23]

1980 Petition for Rulemaking

In 1980, as a result of increasing Canadian investment in the U.S. cable television industry, the FCC again considered the question of foreign ownership.[24] An Illinois cable system proposed that the FCC promulgate a rule that would have provided that "no cable television system . . . shall carry the signal of any television broadcast station

---

20. *Id.* at 727 ¶ 10 (emphasis added).

21. *Id.* at 727 ¶ 11.

22. *Id.* at 728 ¶ 12.

23. *Id.* at 728 n.10.

24. Amendment of Parts 76 and 78 of the Commission's Rules to Adopt General Citizenship Requirements for Operation of Cable Television Systems and for Grant of Station Licenses in the Cable Television Relay Service, Mem. Op. and Order, 77 F.C.C.2d 73 (1980) [hereinafter *Foreign Ownership of CATV Systems*].

or any programming delivered by a radio facility licensed by the Commission if it is owned, operated or leased, by" a person subject to section 310(b).[25] The rule, of course, would have severely restricted foreign ownership of U.S. cable systems.

The cable system's rationale for such regulation was a candid blend of protectionism for American firms and greater content control for regulators. As summarized by the FCC, the U.S. cable industry's reasons for seeking regulation were

> that the amount of foreign ownership has greatly increased since the Commission last considered this issue [in 1976], that it would be "unfortunate" if U.S. cable operators are unable to operate in a considerable number of major markets because alien interests who did not invest in cable systems earlier are now able to call on foreign sources of funds to acquire many of the most desirable franchises in this country, and that cable systems now have a considerable amount of discretion in the programming carried so that the considerations here are very like those applicable to broadcasting where alien interests are severely limited and alien control totally banned.[26]

Another company endorsing the proposed rule argued that "foreign capital is not needed to support domestic cable development,"[27] and the National Cable Television Association explicitly advocated that the FCC adopt a policy of reciprocity:

> When one foreign government pursues a policy of discrimination and restriction regarding U.S. commercial interests, the agencies and establishments of the United States are empowered to impose reciprocal limitations. Such limitations are not aimed at punishing foreign nationals or restricting foreign commerce. They are intended to expand international commerce by inducing the elimination of barriers to it.[28]

The parties opposed to those proposed restrictions, which largely consisted of U.S. cable systems with Canadian investors, argued that

---

25. *Id.* at 74 ¶ 2.
26. *Id.* at 75 ¶ 5.
27. *Id.* at 75 ¶ 6.
28. *Id.* at 76 ¶ 6.

"the amount of foreign ownership that now exists or is likely to exist in the future is insignificant and that the amount of power that cable system operators have to control the content of programming which is provided to their subscribers is minimal."[29] They noted that the Canadian owners at issue were "from a country with close and friendly ties to the United States."[30] They further argued that those companies would benefit American consumers by increasing competition in the United States.[31] And they argued (probably incorrectly as a matter of constitutional law, as we shall see presently) that municipalities were competent to address the question of foreign ownership in their franchising process.[32]

The Canadian investors attacked the principle of reciprocity on three grounds. First, they argued that "national policy favors the free flow of investment between countries" and that "the FCC is not an appropriate body to either alter that policy or to consider matters relating to international trade and international relations."[33] Second, and less persuasively, they argued that Canada's restrictions on cable television ownership could not be compared to those being proposed for the United States because foreign companies dominated the industry in Canada.[34] Third, the FCC would exceed its powers by concerning itself with questions of foreign commerce rather than the question of whether the cable television service provided in the United States served the public interest.[35]

The FCC rejected the petition and reaffirmed its 1976 position that section 310 should not be read "as reflecting a general policy against foreign investment in communications enterprises in the United States."[36] The agency began its discussion by firmly rejecting the idea that it should act on the principle of reciprocity to encourage foreign governments, particularly Canada, to open their markets to investment by U.S. companies:

---

29. *Id.* at 77 ¶ 8.
30. *Id.*
31. *Id.*
32. *Id.*
33. *Id.* at 77 ¶ 9.
34. *Id.*
35. *Id.*
36. *Id.* at 81 ¶ 20.

The Commission's responsibilities relate to "interstate and foreign communications" (47 U.S.C. 1), that is to telecommunications within the United States and between the United States and foreign countries. This does not imply, however, any responsibility for investment policy with respect to communications systems in foreign countries. *We do not believe a desire for reciprocity in international investment policies by itself provides an adequate basis for action on our part.* Nor are we, in any case, in a position to know if such a policy on our part would in fact have the result intended or if, to the contrary, it would lead to increasing trade barriers in other areas.[37]

In any event, the agency concluded, without specific direction from Congress, it "is obviously a matter that does not come within the sphere of the ordinary concerns of this Commission."[38] The regulation of foreign direct investment "is a matter which we believe is appropriately considered by other branches of the government."[39]

The FCC further found that, although there clearly had been changes since the FCC considered the issue in 1976, "[t]hese changes do not persuade us . . . that additional restrictions are needed"[40] in light of the fact that the increase in the total number of foreign-owned cable systems was relatively insignificant. Moreover, the agency noted, a restrictive ownership rule would likely harm consumers:

At this time it is difficult for us to perceive how the television viewing public would benefit in any way from the regulation requested. Rather it would appear that such a restriction would merely promote the self-interests of the domestic cable television industry at the expense of additional competitive alternatives for the public in the franchising process.[41]

The FCC emphasized that "the primary responsibility for selecting cable television system operators is not the Commission's."[42] Rather, the agency explained, no foreign-owned cable system may operate in

---

37. *Id.* at 79 ¶ 13 (emphasis added).
38. *Id.*
39. *Id.*
40. *Id.* at 80 ¶ 18.
41. *Id.*
42. *Id.* at 80 ¶ 19.

any community without a franchise granted by state or local government.[43] The FCC saw no reason why state and local governments could not decide on a case-by-case basis whether to offer a franchise to a foreign company. While noting that cable operators in 1980 did exercise more editorial control over programming than in 1976, the FCC rejected as "highly speculative" the assertion that local controls could not avert threats to domestic interests.[44]

The FCC also reaffirmed its 1976 conclusion that section 310 did not require the agency to adopt a rule restricting alien ownership of cable systems, for cable television was neither broadcasting nor a common carrier radio service.[45] Nonetheless, the FCC, again with belt and suspenders fastened, reiterated its 1976 promise to keep monitoring the situation and, if circumstances took a turn for the worse, to restrict alien ownership of cable systems in the future.[46]

Implications of the 1984
and 1992 Cable Legislation

At the time of its 1980 ruling the FCC was perceived to enjoy extraordinarily broad jurisdiction over cable.[47] Both reality and perception subsequently changed. The Cable Communications Policy Act of 1984[48] divided authority over cable television between the FCC and state and local authorities.[49] The statute amended the Communications Act to provide: "Any Federal agency, State or franchising authority may not impose requirements regarding the provision or content of cable services, except as expressly provided."[50] Thus, powers that Congress did not expressly grant should be considered denied. The 1984 act authorized the FCC to enact rules restricting ownership of cable systems by "persons who own other media of mass communications which serve the same community."[51] But the statute did not

43. *Id.*
44. *Id.* at 81 ¶ 19.
45. *Id.*
46. *Id.* at 81 ¶ 21 & n.9.
47. BRENNER, PRICE & MEYERSON, *supra* note 13, §§ 2.03, 2.04.
48. Pub. L. No. 98-549, 98 Stat. 2779.
49. 47 U.S.C. § 521(3).
50. *Id.* § 544(f).
51. *Id.* § 533(c).

restrict foreign ownership of cable systems or authorize the FCC to promulgate rules that would do so.

In 1992 Congress rewrote the law regulating cable television.[52] Yet, as one of the leading sponsors of that legislation noted, Congress declined to override the FCC's decision not to subject cable to section 310(b).[53] Representative Edward Markey, then the powerful chairman of the Telecommunications Subcommittee of the House Energy and Commerce Committee, strongly urged that Congress make restrictions on foreign ownership of cable systems a part of the 1992 Cable Act.[54] The amendment would have restricted foreign ownership of cable, direct broadcast satellite (DBS), "wireless cable," and other video distribution services.[55] The bill would have amended section 310(b)

---

52. Cable Television Consumer Protection and Competition Act of 1992, Pub. L. No. 102-385, 106 Stat. 1460 (1992).

53. Edward J. Markey, *Telecommunications and Financial Services Trade Hangs on NAFTA Thread*, 1 SAN DIEGO JUSTICE J. 281, 283 (1993).

54. "I rise in strong opposition to the amendment . . . to H.R. 4850 by . . . [Mr. Lent]. [T]he Lent substitute leaves cable systems vulnerable to takeover by foreign entities. It preserves a giant loophole in our existing telecommunications law that permits foreign ownership of cable television systems, direct broadcast satellite systems, and other new video distribution technologies while prohibiting foreign ownership of telephone and broadcasting companies. There is surely no reason for us to invite a breakdown of nearly 60 years of sound and consistent telecommunications policy, or to permit foreign ownership or domination of the next generation of telecommunications technologies." 138 CONG. REC. H6558 (daily ed. July 23, 1992) (statement of Rep. Markey).

55. The bill's preamble contained findings that:

> (1) restrictions on alien or foreign ownership of broadcasting and common carriers first were enacted by Congress in the Radio Act of 1912;
>
> (2) cable television service currently is available to more than 90 percent of American households, more than 62 percent of American households subscribe to such services, and the majority of viewers rely on cable as the conduit through which they receive terrestrial broadcast signals;
>
> (3) many Americans receive a significant portion of their daily news, information, and entertainment programming from cable television systems, and such systems should not be controlled by foreign entities; and
>
> (4) the policy justifications underlying restrictions on alien ownership of broadcast or common carrier licenses have equal application to alien ownership of cable television systems, direct broadcast satellite systems, and multipoint distribution services.

by adding at the end of it the following new paragraphs:

(2)(A) No cable system (as such term is defined in section 602) in the United States shall be owned or otherwise controlled by any alien, representative, or corporation described in subparagraph (A), (B), (C), or (D) of paragraph (1) of this subsection.

(B) Subparagraph (A) of this paragraph shall not be applied—

(i) to require any such alien, representative, or corporation to sell or dispose of any ownership interest held or contracted for on or before June 1, 1990, or acquired in accordance with clause (ii); or

(ii) to prohibit any such alien, representative, or corporation that owns, has contracted on or before June 1, 1990, to acquire ownership, or otherwise controls, any cable system from acquiring ownership or control of additional cable systems if the total number of households passed by all the cable systems that such alien, representative, or corporation would, as a result of such acquisition, own or control does not exceed 2,000,000.

(3)(A) For purposes of paragraph (1) of this subsection, a license or authorization for any of the following services shall be deemed to be a broadcast station license:

(i) cable auxiliary relay services;

(ii) multipoint distribution services;

(iii) direct broadcast satellite services; and

(iv) other services the licensed facilities of which may be substantially devoted toward providing programming or other information services within the editorial control of the licensee.

(B) Subparagraph (A) of this paragraph shall not be applied to any cable operator to the extent that such operator is eligible for the exemptions contained in subparagraph (B) of paragraph (2).[56]

The Bush administration strongly opposed Representative Markey's bill on the grounds that it would invite retaliation by other nations, violate existing trade agreements, and endanger negotiations to open foreign markets to U.S. telecommunications firms.[57] The House

---

H.R. 4850, 102d Cong., 2d Sess. (1992).

56. *Id.*

57. "Such a restriction invites retaliation by other countries and violates existing international obligations. It could stifle the growing investment of U.S. firms in foreign

adopted the proposed measure, but the Senate rejected it; in the face of President Bush's veto threat, the section was not adopted in conference.[58]

Representative Markey's unsuccessful attempts to extend section 310(b) to cable television and new wireless broadband services made clear that the FCC, in the absence of specific legislation, thereafter had no lawful authority to promulgate rules restricting the ownership of cable systems by foreigners. The Communications Act, as amended by the 1992 cable legislation, also did not expressly empower local governments to consider foreign ownership as a factor in franchise proceedings.[59] In the absence of having received an explicit grant from Congress, municipalities had to be deemed to have been denied such power.

Although many aspects of the franchise process were committed to the discretion of the municipality, the Communications Act limited the franchise authority's power to consider program content in franchise proceedings. The franchise authority could not include requirements for video programming or information services, although it could enforce requirements for broad categories of video programming.[60] And, in renewal proceedings, the franchise authority could consider only the "mix, quality or level of cable services."[61] It would arguably constitute censorship in violation of those provisions (as well as the First Amendment) for a franchise authority to refuse to franchise a cable system because it was foreign-owned—that is, because of the speaker's identity. Concerns over national security or the transmission of propaganda would not provide a municipality a credible basis for refusing to franchise a foreign-owned cable system, as Congress rather than the states or their subdivisions has responsibility under the Constitution for addressing such concerns.[62] The same

---

cable systems. It also threatens negotiations to: (1) eliminate the use of trade restrictions by other countries, and (2) open foreign government procurement to U.S. telecommunications products and services, an area in which the U.S. is in an increasingly strong position." 138 CONG. REC. H6487 (daily ed. July 23, 1992) (Statement of Administration Policy, Executive Office of the President, Office of Management and Budget).

58. *See* 138 CONG. REC. H8335 (daily ed. Sept. 14, 1992).

59. 47 U.S.C. § 544(f).

60. *Id.* § 544(b).

61. *Id.* § 546(c)(1).

62. *Cf.* Perpich *v.* Dep't of Defense, 496 U.S. 334 (1990) (construing Congress's

would be true if a local government refused a franchise to a foreign-owned cable system because of concerns with trade policy, as such action would intrude on the plenary powers of Congress to regulate foreign commerce.[63] Moreover, if challenged on equal protection grounds, any discriminatory classification of aliens in the franchising of cable systems by local governments would be reviewed under strict scrutiny—and consequently would be highly susceptible to being struck down as unconstitutional.[64]

PRIVATE SATELLITE CARRIAGE

Section 310(b) covers "common carrier . . . radio licenses."[65] By implication, section 310(b) does not cover private carriers employing radio licenses. Thus, the FCC has repeatedly held that section 310(b) does not restrict licensing aliens from operating satellite earth stations on a private, noncommon carrier basis.[66] The FCC stated in *Reuters Information Services, Inc.* in 1989: "Although the applicant is 100% foreign owned and controlled, alien ownership and control of Reuters does not prohibit the FCC from licensing it to operate the subject earth station on a private, non-common carrier basis."[67] Likewise, in *Burroughs Wellcome*, the FCC licensed a wholly owned subsidiary of

---

command over National Guard).

63. "It is an essential attribute of [the power of Congress to regulate foreign commerce] that it is exclusive and plenary. As an exclusive power, its exercise may not be limited, qualified or impeded to any extent by state action. . . . The principle of duality in our system of government does not touch the authority of the Congress in the regulation of foreign commerce." Board of Trustees of Univ. of Ill. *v*. United States, 289 U.S. 48, 56 (1933).

64. State statutes that classify individuals on the basis of alienage are generally reviewed under the Equal Protection Clause of the Fourteenth Amendment using strict scrutiny. Graham *v*. Richardson, 403 U.S. 365 (1971). Such statutes are upheld only if they are necessary to achieve a compelling state interest. *Id*. at 371–72; *see also* Sugarman *v*. Dougall, 413 U.S. 634, 642 (1973); *In re* Griffiths, 413 U.S. 717 (1973); Examining Bd. of Engineers *v*. Flores de Otero, 426 U.S. 572 (1976); Nyquist *v*. Mauclet, 432 U.S. 1 (1977). The logic underlying that choice of standard of review would apply a fortiori in the case of a municipal franchising statute that discriminated against aliens lawfully residing in the United States.

65. 47 U.S.C. § 310(b).

66. *See, e.g.*, Panamsat Carrier Servs., Inc., 10 F.C.C. Rcd. 928 (1995).

67. 4 F.C.C. Rcd. 5982, 5982–83 ¶ 6 (1989).

a British corporation to operate an INTELSAT earth station.[68] Further, in *COMSAT Earth Stations*, the agency licensed a subsidiary of a Palauan corporation with mostly Palauan officers, directors, and shareholders to operate an earth station on its own territory.[69]

Clearly, the FCC has correctly interpreted section 310(b) with respect to private satellite carriage. But the mere fact that the agency drew a distinction between private carriage and common carriage is curious, for that distinction, like the exemption for alien amateur radio licensees, erodes the national security rationale for applying section 310(b) to *any* radio service. If a foreigner is intent on harming the United States, how does it possibly advance U.S. national security to forbid him to provide wireless common carriage while allowing him to provide wireless private carriage of messages by satellite to and from U.S. territory? A well-financed enemy of the United States would be perfectly happy to be licensed to transmit sensitive information by satellite on a private carriage basis and forgo the opportunity to hold out his transmission capacity for hire on a common carrier basis.

To place matters in perspective, the private carriage in which Reuters lawfully engages today through its licensed satellite earth stations in the United States is the instantaneous, worldwide transmission of vast amounts of financial data relating to commodity and currency trades.[70] In any given second, Reuters probably transmits more bits of information than the German wireless station on Long Island transmitted during the entire period that it operated before the U.S. government confiscated its facilities in 1916. Again, the point is not that Congress and the FCC are wrong not to restrict foreign ownership in radio licenses necessary for private satellite carriage. Rather, the point is that the lack of anxiety over granting such licenses to foreigners makes its impossible to explain why, for example, any anxiety should arise from the prospect of a foreigner's being given the unrestricted right to invest in a cellular telephony licensee.

68. Burroughs Wellcome Co., 4 F.C.C. Rcd. 7190 (1989).

69. COMSAT Earth Stations, Inc., 8 F.C.C. Rcd. 7607 (1993).

70. *See, e.g.*, REUTERS HOLDINGS PLC, 1995 ANNUAL REPORT 4, 13, 14, 18 (1996).

PRIVATE LAND MOBILE SERVICES

Traditionally, some land-based mobile radio services were classified as "private land mobile services" under section 332 of the Communications Act.[71] Those services included the dispatch services used by taxicabs or services that offered carriage for hire to small groups of unaffiliated users.[72] As private carriers, they did not come under section 310(b) because of the exemption for private radio licensees enacted in 1974.

The Omnibus Budget Reconciliation Act of 1993[73] amended section 332 by bumping certain land-based mobile services that had formerly been considered private carriers into the category of common carriers.[74] All providers of a new category of "commercial mobile services" would be treated as common carriers.[75] Commercial mobile services include any interconnected mobile service provided to the public or a substantial portion of the public for profit.[76] Not every mobile service provider, however, fits in that category; some will remain private carriers, exempt from section 310(b).[77] But some mobile services that do fall within the new category will come under the common carrier regime, including section 310(b), for the first time.[78] The statute did not specify which services would be reclassified. That decision was left to the FCC.

Nevertheless, new section 332(c)(6) permits the FCC to waive the application of section 310(b) to foreign ownership interests in mobile services recategorized as common carriers by the 1993 Budget Act, so long as the foreign interest lawfully existed before May 24,

---

71. 47 U.S.C. § 332.

72. *See* Inquiry Relative to the Future Use of the Frequency Band 806–960 MHz, Second Rep. and Order, Dkt. No. 18262, 46 F.C.C.2d 752 (1974).

73. Pub. L. No. 103-66, 107 Stat. 312.

74. 47 U.S.C. § 332.

75. *Id.* § 332(c)(1)(A).

76. *Id.* § 332(d)(1).

77. *Id.* § 332(c)(2); Implementation of Sections 3(n) and 332 of the Communications Act: Regulatory Treatment of Mobile Services, Second Rep. and Order, GEN Dkt. No. 93-252, 9 F.C.C. Rcd. 1411 (1994) [hereinafter *Second Mobile Services Report*].

78. Implementation of Sections 3(n) and 332 of the Communications Act: Regulatory Treatment of Mobile Services, First Rep. and Order, GEN Dkt. No. 93-252, 9 F.C.C. Rcd. 1056 (1994) [hereinafter *First Mobile Services Report*].

1993.[79] Congress's intent was to "grandfather" only the individual who held the interest on that date; the waver does not cover future foreign owners.[80] Thus, the FCC will grant waivers only on condition that the extent of foreign ownership in the service not increase above the pre–May 24, 1993 level.[81] No subsequent transfer of ownership may be made in violation of section 310(b).[82] Requests for waivers had to be filed within six months of the date of enactment of the 1993 Budget Act—by February 10, 1994.[83]

When the FCC first established guidelines for filing the waivers, it estimated that the final rules determining exactly which mobile services would be reclassified as common carriers might not be available before the filing deadline.[84] The FCC urged private land mobile services to file for a waiver if there was "any chance at all" that the service might be reclassified as common carriage.[85] Issuing its first reclassification ruling in March 1994, the FCC moved services such as some "private" paging into the commercial mobile service category.[86]

SUBSCRIPTION VIDEO SERVICES

As mentioned above, section 310(b) is, by its terms, not applicable to private use licensees. The FCC extended that rationale in 1989 to exempt subscription video services, including subscription DBS, from section 310(b).[87]

The United States Satellite Broadcasting Company (USSB) objected to the exemption of subscription DBS from section 310(b) because it had proposed to offer DBS as an advertiser-supported (free)

---

79. 47 U.S.C. § 332(c)(6).

80. *First Mobile Services Report*, 9 F.C.C. Rcd. at 1056 ¶ 2 & n.4.

81. 47 U.S.C. § 332(c)(6)(A).

82. *Id.* § 332(c)(6)(B).

83. *Id.* § 332(c)(6).

84. *First Mobile Services Report*, 9 F.C.C. Rcd. at 1057 ¶ 7.

85. *Id.* at 1057 ¶ 8.

86. *Second Mobile Services Report*, 9 F.C.C. Rcd. at 1452–53 ¶ 96–99; *see* Amendment of Parts 2 and 90 of the Commission's Rules to Provide for the Use of 200 Channels, 10 F.C.C. Rcd. 6884 (1995).

87. *See* Subscription Video Services, Mem. Op. and Order, GEN Dkt. No. 85-305, 4 F.C.C. Rcd. 4948 (1989).

broadcast service.[88] USSB would thus be subject to section 310(b); its closest competitors, other satellite video distribution systems that operated subscription services as the customer-programmers of common carriers, would not be subject to 310(b). USSB contended that "its ability to raise financing in foreign countries will be unfairly and adversely affected by the application of section 310(b) to licensees but not customer-programmers that will be competing with licensees."[89] The FCC declined to "impose regulatory burdens on nonbroadcast services, where such burdens are not required nor in the public interest, simply because broadcasters are subject to those restraints. . . . [N]either the letter nor the intent of [section 310(b)] supports application of alien ownership restrictions to the customers of common carriers, who do not own or control communications facilities."[90] The FCC also pointed out that the common carrier licensee from whom the programmer obtains access to transmission facilities is already subject to the benchmarks.[91]

At the time of the FCC's decision, subscription video services were a commercial flop. Since 1989, however, the wireless delivery of broadband has grown in commercial viability and would seem likely to continue growing if the FCC were to allocate sufficient spectrum for such services. Today, there are multiple technologies both for DBS service and for terrestrial wireless systems.[92] The two leading terrestrial services are "wireless cable," or multichannel multipoint distribution service (MMDS), which uses line-of-sight microwave radio channels in the 2.1 GHz and 2.5 GHz bands, and local multipoint distribution service (LMDS), which uses an omnidirectional antenna in each of many cells to transmit in the 27.5 to 29.5 GHz range.[93] Although intellectual consistency is not the FCC's hallmark, the logic of its 1989 decision on subscription video, if extended to those two new wireless broadband technologies, would exempt them from section 310(b).

---

88. *Id.* at 4948 ¶¶ 2–3.
89. *Id.* at 4948 ¶ 3.
90. *Id.* at 4948 ¶¶ 4–5.
91. *Id.* at 4948 ¶ 5.
92. For a survey of those services and an assessment of their commercial viability, *see* LELAND L. JOHNSON, TOWARD COMPETITION IN CABLE TELEVISION 111–48 (MIT Press & AEI Press 1994).
93. *See id.* at 128, 134–35.

Another potential terrestrial wireless service for subscription-based video is the local television broadcaster. Using digital compression, existing television broadcasters could offer multiple channels of video programming in their existing 6 MHz assignments for over-the-air broadcasting.[94] Some of those compressed channels could be offered on a subscription basis. If that scenario were to develop, television broadcasters would have an odd but powerful argument concerning foreign ownership: The frequencies carrying their "free" compressed channels are subject to section 310(b), but the frequencies carrying their subscriber-based compressed channels are not. In turn, that differential regulation of foreign investment could induce a reconfiguration of the ownership of television broadcast companies to take advantage of the availability of foreign capital to a greater extent.

STATUTORY MISINTERPRETATION
OF SECTION 310(b)(4)

For years the FCC has misread the plain language of the key subsection of the foreign ownership restrictions in the Communications Act. Section 310(b)(4) allows foreign ownership of the holding company of a communications licensee to exceed 25 percent but gives the FCC the discretion to deny or withdraw the license—"*if* the Commission finds that the public interest will be served by the refusal or revocation of such license."[95] A treatise on international telecommunications regulation compiled by the Federal Communications Bar Association in 1993 supports that reading of the statute:

> Under the terms of the statute, the Commission must find that a *refusal* of the license to a company in which alien ownership in its holding company exceeds the twenty-five percent benchmark serves the public interest. Therefore, the onus is on the Commission to prove that the relaxed public interest standard mandates a refusal of the license request.[96]

---

94. *See id.* at 137–47.

95. 47 U.S.C. § 310(b)(4) (emphasis added).

96. Tara Kalagher Giunta, *Foreign Participation in Telecommunications Projects, in* FEDERAL COMMUNICATIONS BAR ASSOCIATION, INTERNATIONAL PRACTICE COMMITTEE, INTERNATIONAL COMMUNICATIONS PRACTICE HANDBOOK, 1993, at 43, 44 (Paul J. Berman & Ellen K. Snyder eds. 1993) (emphasis in original).

Although that interpretation is the only one consistent with the statute's straightforward use of the English language, it presumes innocently that the Communications Act means what it says rather than what the FCC's lawyers say that it means.

The FCC regards its discretion under section 310(b)(4) to be broad—so broad as to authorize the agency to reverse the burden of proof that Congress specified. The FCC presumes foreign investment in an American holding company exceeding 25 percent to be unlawful, such that the applicant must affirmatively prove to the FCC's satisfaction that the agency's grant of a *waiver* of that putative ceiling on foreign ownership would *serve* the public interest in the applicant's particular facts and circumstances. In *PrimeMedia Broadcasting, Inc.*, the FCC provided a succinct misstatement of the law when it asserted in 1988 that "alien equity interests in a parent corporation . . . may only amount to 25%, *unless* the Commission finds that the public interest would be served."[97] As of 1996, the FCC still had not acknowledged the error of that statement. To the contrary, as recently as February 1995, when the FCC issued a notice of proposed rulemaking addressing in part the regulation of foreign investment, the agency failed to acknowledge that it had been erroneously applying the statute and declined to solicit any public comment on whether its extravagant claims to discretion in the enforcement of section 310(b)(4) had any basis in law.[98] *"Under the plain language of the Communications Act* and its legislative history," the FCC asserted in the notice of proposed rulemaking, "the Commission has broad discretion in applying Section 310(b)(4)."[99]

The FCC, of course, has no discretion with which to rewrite an act of Congress. In its committee report accompanying H.R. 1555, the "Communications Act of 1995," the House Commerce Committee noted that "the Commission has consistently misinterpreted section 310(b)(4) by creating a presumption that foreign investment is not in the public interest if it exceeds 25 percent of the equity of an American radio licensee."[100] The committee further stated that its proposed "amendments to section 310(b) . . . do not constitute

---

97. 3 F.C.C. Rcd. 4293, 4295 ¶ 12 (1988) (emphasis added).

98. *Market Entry and Regulation NPRM*, 10 F.C.C. Rcd. at 5263–65 ¶¶ 15–19, 5293–98 ¶¶ 92–103.

99. *Id*. at 5294 ¶ 93 (emphasis added).

100. H.R. REP. NO. 204, pt. 1, 104th Cong., 1st Sess. 120–21 (1995).

congressional acquiescence to the Commission's past misinterpretation of section 310(b)(4)."[101]

Furthermore, even if the FCC's reading of section 310(b)(4) did not exceed the agency's statutory authority, it would still harm the public interest and thus work results that could only be described as an abuse of agency discretion in violation of the Administrative Procedure Act.[102] American telecommunications firms must modernize their infrastructure and forge the global alliances necessary to compete in the provision of full-service networks on an international scale. Such investments and alliances would plainly benefit American consumers. Viewed ex ante, opening American telecommunications markets to greater foreign investment should be at least as likely to incline other nations to do likewise as to cause them to maintain or raise barriers to incoming U.S. investment. For those reasons, there should be a strong presumption that it would *disserve* the public interest for the FCC to refuse or revoke a license of a firm whose holding company is owned more than 25 percent by a citizen of any nation friendly to the United States. Put differently, it is difficult to hypothesize *any* legitimate, rational public purpose that would be served by discouraging investment by friendly foreigners in the American telecommunications industry.

No court has addressed the FCC's misinterpretation of section 310(b). The reason why that is so is predictable and pragmatic: No experienced telecommunications lawyer could responsibly advise a foreign client not to be prepared to bear the burden of proving that its investment would advance the public interest. And no foreign firm expected to make such a showing would want to accuse its future regulator of lawlessness in its reading of section 310(b)(4). But lawlessness correctly describes the FCC's conduct: Its misinterpretation of section 310(b)(4) is "the law" only in appearance. If a court ever were to face the question of how the FCC has inverted the burden of proof and agency discretion conferred by section 310(b)(4), it would have ample grounds to take exception to the usual rule in administrative law of deferring, under *Chevron*, to the agency's interpretation of its organic statute.[103]

---

101. *Id.* at 121.

102. 5 U.S.C. § 706(2)(A).

103. Chevron, U.S.A. Inc. *v.* Natural Resources Defense Council, Inc., 467 U.S. 837 (1984).

WAIVERS OF SECTION 310(b)(4)

The FCC has made the process by which one applies for permission to exceed the foreign ownership limits equivalent to a petition for waiver of such limits. Because the Communications Act by its very language never contemplated that the FCC would enforce section 310(b)(4) through the prospective process that the agency has devised, it follows that the statute fails to instruct the FCC on how to decide when to waive the restrictions on foreign ownership and control.

The FCC's answer to that dilemma has been, once again, to claim broad discretion. When presented with a petition for waiver, the FCC has stated that it need only enforce the foreign ownership restrictions (and the pre-1996 foreign management restrictions) if it finds that "the public interest . . . would require the radio licenses to be revoked or license renewals denied."[104] As we shall see in chapter 4, the waiver process has become extremely complex and has affected in minute detail how foreigners structure the ownership and control of their investments in U.S. telecommunications firms.

RELATED PROVISIONS
IN THE COMMUNICATIONS ACT

Five other provisions of the Communications Act are pertinent to foreign involvement in U.S. telecommunications and thus must be read in conjunction with section 310(b).

Section 310(a)

The first such provision, section 310(a), is a model of brevity: "The station license required under this Act shall not be granted to or held by any foreign government or the representative thereof."[105] Unlike section 310(b), section 310(a) applies to both common and private carriers, including, for example, satellite earth stations constructed for private carriage[106] and mobile services.[107] The FCC interprets the

---

104. General Elec. Co., 5 F.C.C. Rcd. 1335 ¶ 5 (Common Carrier Bureau 1990).
105. 47 U.S.C. § 310(a).
106. Licensing Under Title III of the Communications Act of 1934, as amended, of Non-common Carrier Transmit/Receive Earth Stations Operating with the INTELSAT Global Communications Satellite System, Declaratory Ruling, 8 F.C.C.

section to prohibit either de facto or de jure control of radio licenses.[108] Foreign governments may enter into limited partnerships with licensees, so long as the licensee exercises full control over the operation of the system.[109] The FCC may consult with the executive branch in determining whether a violation of section 310(a) is threatened.[110] Section 305(d), added in 1962, permits the president to authorize foreign governments to operate a low-power radio station at the site of its embassy.[111]

Obviously, the section does not prevent a foreign government or its representative from using radio carriage facilities as a customer of a common carrier. Likewise, the section does not generally bar a company from contracting with a foreign government or its representative to operate private carriage facilities on the foreigner's behalf. For example, the FCC approved an arrangement under which Banco Nacional de Mexico (Banamex) asked Satellite Transmission and Reception Specialist (STARS) to build and operate the U.S. end of a transborder satellite service to transmit Banamex's private messages between the United States and Mexico.[112] The FCC does, however, bar specialized mobile radio base station licensees from serving any entity that would not itself be eligible for licensing under FCC rules, so foreign governments or representatives of foreign governments may not use that service.[113]

Compared with section 310(b), there has been relatively little litigation involving section 310(a). The FCC has determined that it does not violate section 310(a) to grant a radio license for personal

---

Rcd. 1387, 1387 n.6 (1993).

107. *First Mobile Services Report*, 9 F.C.C. Rcd. at 1056 n.5.

108. Orion Satellite Corporation Request for Final Authority to Construct, Launch and Operate an International Communications Satellite System, Mem. Op., Order and Authorization, 5 F.C.C. Rcd. 4937 (1990).

109. *Id*. at 4939 ¶ 20.

110. *Id*.

111. 47 U.S.C. § 305(d).

112. STARS, Application for Authority to Modify Its License for a 9.2-Meter Fixed C-Band Transmit/Receive Earth Station at Sylmar, California (Call Sign E890790) for the Provision of International Services Between the United States and Canada and Mexico via All U.S. Domestic Satellites and the Anik and Morelos Satellite Systems, Mem. Op. and Order, 5 F.C.C. Rcd. 3150 (1990).

113. Amendment of Part 90 of the Commission's Rules to Eliminate Separate Licensing of End Users of Specialized Mobile Radio Systems, Notice of Proposed Rulemaking, PR Dkt. No. 92-79, 7 F.C.C. Rcd. 2885 (1992); 47 C.F.R. § 90.603(c).

use to an honorary consul of Bolivia, who received no compensation for his services from the Bolivian government.[114] In upholding the grant of a broadcast license to Loyola University, the FCC concluded and the D.C. Circuit agreed that the pope is not a foreign sovereign for purposes of section 310(a).[115] Interesting questions might arise under 310(a) as the governments of other countries begin to privatize their telecommunications networks. It has never been decided whether a privatized telecommunications monopoly would be considered a "representative of a foreign government" despite privatization.

Section 214

Foreign carriers or their U.S. affiliates must secure authorization for international service under section 214 of the Communications Act.[116] In issuing such authority on a case-by-case basis, the FCC has sought to prevent undue discrimination by the foreign parent against unaffiliated U.S. carriers. In cases of international resale, the FCC has undertaken to determine whether the foreign country on the other end of an international circuit provides equivalent opportunities to U.S. carriers to resell interconnected private lines.[117] In cases of facilities-based carriers, the FCC generally has conditioned its grant of section 214 authorization on the existence of regulatory safeguards to prevent discrimination against U.S. carriers in the terms and conditions of access to foreign markets for the origination and termination of U.S. international traffic, and on annual reporting requirements.[118] As we discuss in chapter 6, the FCC in 1995 replaced its case-by-case analysis with a rule that would condition the grant of section 214 authorization for a foreign carrier on the agency's determination that U.S. carriers had equivalent access to the

114. Russell G. Simpson, 2 F.C.C.2d 640 (1966).
115. Noe v. FCC, 260 F.2d 739 (D.C. Cir. 1958).
116. 47 U.S.C. § 214.
117. *E.g.*, Cable & Wireless, Inc., 8 F.C.C. Rcd. 1664 (Common Carrier Bureau 1993); fONOROLA Corp., 7 F.C.C. Rcd. 7312 (1992), *order on recons.*, 9 F.C.C. Rcd. 4066 (1994).
118. Atlantic Tele-Network, Inc., 6 F.C.C. Rcd. 6529 (1991), *review denied*, 8 F.C.C. Rcd. 4776 (1993), *aff'd*, Atlantic Tele-Network, Inc. v. FCC, 59 F.3d 1384 (D.C. Cir. 1995); Telefónica Larga Distancia de Puerto Rico, 8 F.C.C. Rcd. 106 (1992); AmericaTel Corp., 9 F.C.C. Rcd. 3993 (1994); MCI Comm. Corp., 9 F.C.C. Rcd. 3960 (1994).

foreign carrier's "destination markets."[119]

Submarine Cable Landing Act

In 1921 Congress passed the Submarine Cable Landing Act, which regulates the granting of licenses for landing or operating cables connecting the United States with a foreign country.[120] Congress subsequently made that law part of the Communications Act of 1934. Unlike section 310(b), those provisions explicitly contain a congressional mandate for the president to pursue a policy of reciprocity. Given that Congress had thirteen years of experience with the Submarine Cable Landing Act when it enacted the foreign ownership restrictions in the Communications Act in 1934, it is clear that Congress knew how to draft a reciprocity standard for the provision now denominated as section 310(b), had it wanted to do so.

Section 34 of the Communications Act specifies that a person operating a submarine cable between the United States and another country must first secure a written license from the president.[121] By executive order, President Dwight D. Eisenhower delegated that licensing function to the FCC in 1954.[122] The teeth in that law are in the president's power to withhold or revoke licenses under section 35.[123] After giving "due notice and hearing," the president may withhold or revoke such a license for any of three reasons. One is that such action "will promote the security of the United States."[124] Another is that such action "will assist . . . in maintaining the rights

---

119. Market Entry and Regulation of Foreign-Affiliated Entities, Report and Order, IB Dkt. No. 95-22, 11 F.C.C. Rcd. 3873, 3881 ¶ 20, 3888 ¶ 39 (1996) [hereinafter *Market Entry and Regulation Report and Order*].

120. Act of May 27, 1921, ch. 12, § 1, 42 Stat. 8 (codified at 47 U.S.C. §§ 34–37).

121. 47 U.S.C. § 34. "No person shall land or operate in the United States any submarine cable directly or indirectly connecting the United States with any foreign country, or connecting one portion of the United States with any other portion thereof, unless a written license to land or operate such cable has been issued by the President of the United States. The conditions of this Act . . . shall not apply to cables, all of which including both terminals, lie wholly within the continental United States." *Id.*

122. Exec. Order No. 10,530, § 5(a), 19 FED. REG. 2709 (1954), *reprinted at* 3 U.S.C. § 301 note.

123. 47 U.S.C. § 35.

124. *Id.*

or interests of the United States or of its citizens in foreign countries."[125] Although that factor could justify a policy of reciprocity, the third factor is explicit—the withholding or revocation of such a license "will assist in securing rights for the landing or operation of cables in foreign countries."[126] In addition to having the power to revoke or withhold the issuance of a license, the president "may grant such license upon such terms as shall be necessary to assure just and reasonable rates and service in the operation and use of cables so licensed."[127] The president (or the FCC, acting as his delegate) can enforce violations of the Submarine Cable Landing Act through an injunction forbidding the landing or operating of a cable or compelling its removal.[128] A knowing violation of section 34 is a misdemeanor punishable by a $5,000 fine or imprisonment of not more than one year, or both.[129]

Since receiving delegated authority from the president to enforce the Submarine Cable Landing Act, the FCC has aggressively pursued reciprocal access for U.S. carriers in a number of cases. In *French Telegraph Cable Co.*, the FCC denied a French carrier a license to land and operate a submarine cable for provision of authorized international record communications services from San Francisco and Washington, D.C., in part because the agency determined that France would not grant similar rights to U.S. citizens.[130] In *Tel-Optic Ltd.*, the FCC refused to take final action on an application for a license because the applicant supplied insufficient information concerning ownership and control of the cable system and its foreign landing points for the agency to determine whether the foreign country accorded U.S. carriers the reciprocal treatment required by section 35.[131] Despite the age and relative obscurity of the Submarine Cable Landing Act, the recent growth in undersea cables has generated a series of FCC decisions interpreting landing rights.[132]

125. *Id.*
126. *Id.*
127. *Id.* Section 35 also contains the proviso that "[t]he license shall not contain terms or conditions granting to the licensee exclusive rights of landing or of operation in the United States." *Id.*
128. *Id.* § 36.
129. *Id.* § 37.
130. 71 F.C.C.2d 393 (1979).
131. FCC 85-99 (adopted Mar. 1, 1985).
132. Optel Communications, Inc., 9 F.C.C. Rcd. 6153 (Int'l Bureau 1994); IDB

Section 308(c)

An additional provision in the Communications Act permits the FCC, in limited circumstances, to consider reciprocal national treatment when issuing radio licenses. Section 308(c) provides that the FCC, "in granting any license for any station intended or used for commercial communication between the United States . . . and any foreign country, may impose any terms, conditions, or restrictions authorized to be imposed with respect to submarine-cable licenses" under the Submarine Cable Landing Act.[133] Because section 308(c) is not confined by its text to wireline communications, it enables the FCC to impose a reciprocity condition on the issuance of a radio license, but *only* in the case of an international route.

Presidential Seizure under Section 606

The fourth related provision, section 606, devotes hundreds of words to setting forth the emergency powers of the president to seize all radio stations and wireline facilities for communications.[134] As chapter 2 explains, that presidential power predates the Communications Act of 1934 and was exercised to its fullest extent by Woodrow Wilson during World War I.

Section 606(c) of the Communications Act grants the president broad powers to control radio stations in the event of war or emergency:

> Upon proclamation by the President that there exists war or a threat of war, or a state of public peril or disaster or other national emergency, or in order to preserve the neutrality of the

---

Communications Group, Inc., 8 F.C.C. Rcd. 5222 (Common Carrier Bureau 1993); Telefónica Larga Distancia de Puerto Rico, 8 F.C.C. Rcd. 106 (1992); IDB Communications Group, Inc., 7 F.C.C. Rcd. 6553 (Common Carrier Bureau 1992); Alascom, Inc., 6 F.C.C. Rcd. 2969 (Common Carrier Bureau 1991); Western Union Corp., 4 F.C.C. Rcd. 2219 (Common Carrier Bureau 1989); UNC Inc., 3 F.C.C. Rcd. 7154 (Common Carrier Bureau 1988); General Elec. Co., 3 F.C.C. Rcd. 2803 (Common Carrier Bureau 1988); FTC Communications, Inc., 2 F.C.C. Rcd. 7513 (Common Carrier Bureau 1987); FTC Communications, Inc., 75 F.C.C.2d 15 (1979); Teleprompter Cable Servs., Inc., 35 F.C.C.2d 943 (1972).

133. 47 U.S.C. § 308(c); *see also* S. REP. NO. 781, 73d Cong., 2d Sess. 7 (1934).
134. 47 U.S.C. § 606.

United States, the President, if he deems it necessary in the interest of national security or defense, may suspend or amend, for such time as he may see fit, the rules and regulations applicable to any or all stations or devices capable of emitting electromagnetic radiations within the jurisdiction of the United States as prescribed by the Commission, and may cause the closing of any station for radio communication, or any device capable of emitting electromagnetic radiations between 10 kilocycles and 100,000 megacycles, which is suitable for use as a navigational aid beyond five miles, and the removal therefrom of its apparatus and equipment, or he may authorize the use or control of any such station or device and/or its apparatus and equipment, by any department of the Government under such regulations as he may prescribe upon just compensation to the owners.[135]

After its enactment in 1934, that provision was slightly amended in 1951 to clarify the president's power to use, control, or close radio facilities that an enemy might use for navigation.[136]

Section 606(d) empowers the president to seize control of wireline telecommunications facilities in the event of war:

Upon proclamation by the President that there exists a state or threat of war involving the United States, the President, if he deems it necessary in the interest of the national security and defense, may, during a period ending not later than six months after the termination of such state or threat of war and not later than such earlier date as the Congress by concurrent resolution may designate, (1) suspend or amend the rules and regulations applicable to any or all facilities or stations for wire communication within the jurisdiction of the United States as prescribed by the Commission, (2) cause the closing of any facility or station for wire communication and the removal therefrom of its apparatus and equipment, or (3) authorize the use or control of any such facility or station and its apparatus and equipment by any department of the Government under such regulations as he may prescribe, upon just compensation to the owners.[137]

That section was not part of the Communications Act in 1934.

---

135. *Id.* § 606(c).
136. Act of Oct. 24, 1951, ch. 553, § 1, 65 Stat. 611 (1951)
137. 47 U.S.C. § 606(d).

Congress added the provision in 1942, seven weeks after the attack on Pearl Harbor.[138]

Other parts of section 606 confer related powers on the president. Section 606(a) empowers the president in time of war to direct that communications in his judgment that are essential to national security be given priority with any carrier subject to the Communications Act.[139] Section 606(b) entitles the president to use the armed forces to prevent anyone from using physical force to obstruct communications during war.[140] Section 606(e) entitles the president to set the just compensation for use of facilities under sections 606(c) or 606(d).[141] Section 606(f) preserves (with exceptions) the taxation and police powers of the states against the effect of sections 606(c) or 606(d).[142] Section 606(g) explains that sections 606(c) or 606(d) do not "authorize the President to make any amendment to the rules and regulations of the FCC which the FCC would not be authorized by law to make; and nothing in subsection (d) shall be construed to authorize the President to take any action the force and effect of which shall continue beyond the date after which taking of such action would not have been authorized."[143] Section 606(h) sets out penalties for noncompliance.[144]

RELATED PROVISIONS
OUTSIDE THE COMMUNICATIONS ACT

Two other statutes that are not part of the Communications Act are relevant to foreign investment in telecommunications and address the same national security concerns addressed by section 310(b). Those statutes are the Exon-Florio Amendment and the Foreign Agent Registration Act.

---

138. Act of Jan. 26, 1942, ch. 18, § 1, 56 Stat. 18 (1942).
139. 47 U.S.C. § 606(a).
140. *Id.* § 606(b).
141. *Id.* § 606(e).
142. *Id.* § 606(f).
143. *Id.* § 606(g).
144. *Id.* § 606(h).

Exon-Florio Amendment

The 1988 Exon-Florio Amendment to the Defense Production Act of 1950 is the primary and most controversial law governing foreign investment in the United States.[145] In implementing the Exon-Florio Amendment, President Ronald Reagan established the Committee on Foreign Investment in the United States (CFIUS)[146] and authorized it to enforce and administer the amendment.[147]

Under the Exon-Florio Amendment, the federal government may block any proposed investment that appears to threaten national security. CFIUS makes that determination upon request by a potential investor or a CFIUS agency.[148] CFIUS analyzes whether the investment will result in foreign control that will impair national security. After receiving a CFIUS report, the president has fifteen days to determine the appropriate course of action[149] and to report the reasons for his decision to Congress.[150]

Considering its mode of enforcement, the Exon-Florio Amendment has rarely been implemented. As of 1994, CFIUS had investigated fewer than 5 percent of the transactions of which it had been notified,[151] and of those the president formally blocked only one.[152] A number of other foreign investors, however, have withdrawn their investment plans when faced with possible CFIUS opposition[153] or when under heavy political pressure during a CFIUS investigation.[154]

---

145. Exon-Florio Amendment to the Omnibus Trade and Competitiveness Act of 1988, Pub. L. No. 100-418, § 5021, 102 Stat. 1107, 1425 (codified at 50 U.S.C. § 2170).

146. *See* Interim Directive Regarding Disposition of Certain Mergers, Acquisitions, and Takeovers, 53 FED. REG. 43,999 (1988), *reprinted in* 50 U.S.C. § 2170.

147. *See* 50 U.S.C. § 2170(e); 31 C.F.R. § 800 (CFIUS regulations to implement Exon-Florio Amendment).

148. 50 U.S.C. § 2170(a).

149. *Id.* § 2170(c).

150. *Id.* § 2170(g).

151. GENERAL ACCOUNTING OFFICE, FOREIGN INVESTMENT—IMPLEMENTATION OF EXON-FLORIO AND RELATED AMENDMENTS (GAO/NSIAD-96-12, 1995).

152. *Id.*

153. *See* Martin Tolchin, *Agency on Foreign Takeovers Wielding Power*, WALL ST. J., Apr. 24, 1989, at D6.

154. *See* Alan F. Holmer, Judith H. Bello & Jeremy O. Preiss, *The Final Exon-Florio Regulations on Foreign Direct Investment: The Final Word or Prelude to Tighter Controls?*, 23 LAW & POL'Y INT'L BUS. 593, 611 (1992).

Two sets of additional provisions added in 1992 as part of the Defense Authorization Bill toughened the Exon-Florio Amendment.[155] The Byrd Amendment[156] requires a mandatory investigation of acquisitions or takeovers by a foreign government or by companies controlled or "acting on behalf of" foreign governments, if such a transaction could "result in control" of a domestic company engaged in activities that "could affect national security."[157] The Bingham Amendment[158] prohibits foreign government-owned companies from purchasing U.S. defense contractors that are engaged in contracts requiring access to certain proscribed categories of information, or that are involved in contracts valued at more than $500 million with the Defense or Energy Departments.[159] The provision does not apply if the Exon-Florio Amendment is not invoked to prevent the transaction.[160]

Foreign Agent Registration Act

The Foreign Agent Registration Act (FARA) of 1938[161] requires agents of foreign principals to register with the U.S. attorney general[162] and submit to the labeling of any "political propaganda" that they wish to distribute in the United States.[163] The term "foreign principal" includes not only foreign governments and foreign political parties (broadly defined to include organizations such as the Irish Republican Army),[164] but also citizens of countries other than the United States and corporations or partnerships organized outside the United States.[165] An "agent" is anyone who engages in political acts

---

155. Pub. L. No. 102-484, 106 Stat. 2315 (1992) (codified at 50 U.S.C. § 2170).

156. 50 U.S.C. § 2170.

157. *Id.* § 2170(b).

158. *Id.* § 2170(a).

159. *Id.*

160. *Id.* § 2170(a)(b).

161. 22 U.S.C. §§ 611–21.

162. *Id.* § 612.

163. *Id.* § 614.

164. *Id.* § 611(b)(1); *see also* Attorney Gen. *v.* Irish People, Inc., 684 F.2d 928, 938, 942 (D.C. Cir. 1982), *cert. denied*, 459 U.S. 1172 (1983); Attorney Gen. *v.* Irish Nat'l Aid Comm., 346 F. Supp. 1384, 1390–91 (S.D.N.Y.), *aff'd*, 465 F.2d 1405 (2d Cir.), *cert. denied*, 409 U.S. 1080 (1972).

165. 22 U.S.C. §§ 611(b)(2), (3).

or lobbying, performs public relations services, or solicits money under the control of a foreign principal.[166] Commercial activities and diplomats, among others, are exempt.[167]

FARA defines "political propaganda" as a communication intended to influence the recipient in some way relating to the policies of a foreign principal or to instigate "riot" or the overthrow of government.[168] But the statutory definition is not exclusive. Political propaganda "includes," but evidently is not limited to, the following conduct:

> any oral, visual, graphic, written, pictorial, or other communication or expression by any person (1) which is reasonably adapted to, or which the person disseminating the same believes will, or which he intends to, prevail upon, indoctrinate, convert, induce, or in any other way influence a recipient or any section of the public within the United States with reference to the political or public interests, policies, or relations of a government of a foreign country or a foreign political party or with reference to the foreign policies of the United States or promote in the United States racial, religious, or social dissensions, or (2) which advocates, advises, instigates, or promotes any racial, social, political, or religious disorder, civil riot, or other conflict involving the use of force or violence in any other American republic or the overthrow of any government or political subdivision of any other American republic by any means involving the use of force or violence.[169]

FARA also defines the "disseminating" of political propaganda inclusively rather than exclusively. The term "includes transmitting or causing to be transmitted in the United States mails or by any means or instrumentality of interstate or foreign commerce or offering or causing to be offered in the United States mails."[170]

Section 614(a) provides that persons required to register under FARA must send the attorney general within forty-eight hours two copies of any political propaganda that the agent transmits through the

---

166. *Id.* § 611(c).
167. *Id.* § 613.
168. *Id.* § 611(j).
169. *Id.*
170. *Id.*

mails or by other means to two or more persons.[171] In addition, the agent must attach an identification statement: The "political propaganda [must be] conspicuously marked at its beginning with, or prefaced or accompanied by, a true and accurate statement, in the language or languages used in such political propaganda, setting forth the relationship or connection between the person transmitting the political propaganda or causing it to be transmitted and such propaganda."[172] In 1987 the Supreme Court upheld that statute against a First Amendment challenge.[173]

Congress enacted FARA a year before the outbreak of World War II in response to the distribution of publications sponsored by

171. "Every person within the United States who is an agent of a foreign principal and required to register under the provisions of this Act and who transmits or causes to be transmitted in the United States mails or by any means or instrumentality of interstate or foreign commerce any political propaganda for or in the interests of such foreign principal (i) in the form of prints, or (ii) in any other form which is reasonably adapted to being, or which he believes will be, or which he intends to be, disseminated or circulated among two or more persons shall, not later than forty-eight hours after the beginning of the transmittal thereof, file with the Attorney General two copies thereof and a statement, duly signed by or on behalf of such agent, setting forth full information as to the places, times, and extent of such transmittal." *Id.* § 614(a).

172. *Id.* § 614(b). The identification statement must additionally state

> that the person transmitting such political propaganda or causing it to be transmitted is registered under this Act with the Department of Justice, Washington, District of Columbia, as an agent of a foreign principal, together with the name and address of such agent of a foreign principal and of such foreign principal; that, as required by this Act, his registration statement is available for inspection at and copies of such political propaganda are being filed with the Department of Justice; and that registration of agents of foreign principals required by the Act does not indicate approval by the United States Government of the contents of their political propaganda.

Id. The attorney general, "having due regard for the national security and the public interest," is empowered to prescribe by regulation "the language or languages and the manner and form in which such statement shall be made and require the inclusion of such other information contained in the registration statement identifying such agent of a foreign principal and such political propaganda and its sources as may be appropriate." *Id.*

173. Meese *v.* Keene, 481 U.S. 1035 (1987); *see also* United States *v.* Peace Information Center, 97 F. Supp. 255 (D.D.C. 1951); United States *v.* Auhagen, 39 F. Supp. 590 (D.D.C. 1941).

various Nazi and Communist groups.[174] FARA was not intended to limit the distribution of such propaganda, but to require the distributor to disclose the source so that its American audience could make an informed decision about the material's accuracy.[175] Congress made significant amendments to FARA in 1963 to expand the statute to control the activities of foreign lobbyists.[176]

It is important to note how FARA overlaps with section 310(a) of the Communications Act. *Neither* FARA nor section 310(a) directly bars any foreigner from speaking. Section 310(a) indirectly inhibits speech by barring foreign governments from owning transmission media, but not from using others' networks. Likewise, FARA does not prevent foreign principals from speaking; rather, it labels their speech. Nevertheless, FARA is the broader measure. Section 310(a) targets foreign governments and their representatives; FARA covers almost any political activity, whether or not sponsored by a foreign sovereign. And FARA covers customers of any carriage facility, a group excluded by section 310(a). FARA also covers the print media as well as electronic transmissions.

⟨ CONCLUSION

Most of the laws intended to combat pernicious alien influences on the American public vest authority over such matters in the executive branch. FARA empowers the U.S. attorney general to label propagandistic content. The Exon-Florio Amendment empowers the president to shut off the flow of foreign capital for specific acquisitions. Section 606 of the Communications Act empowers the president to seize transmission media. The level of control that those laws authorize can simultaneously reduce individual liberty and strengthen national defense. Both the threat to liberty and the usefulness of those laws arise from the same factor—the broad discretion that the laws vest in the executive branch. However ambivalent one may be about

---

174. Act of June 8, 1938, 52 Stat. 631; *see also* Viereck v. United States, 318 U.S. 236, 244 (1943); Michael I. Spak, *America for Sale: When Well-Connected Former Federal Officials Peddle Their Influence to the Highest Foreign Bidder—A Statutory Analysis and Proposals for Reform of the Foreign Agents Registration Act and the Ethics in Government Act*, 78 Ky. L.J. 237 (1990).

175. United States v. Auhagen, 39 F. Supp. 590 (D.D.C. 1941).

176. *See* Spak, *supra* note 174, at 246–49, 276.

Congress's giving such broad powers to some instrument of the federal government, it is nevertheless probably superior from the perspective of protecting national security with the least infringement of individual liberty that such discretion resides in the executive branch and not in an unaccountable independent agency.

From that perspective it is understandable why the plain language of section 310(b) of the Communications Act did not leave the FCC with nearly so much discretion on matters of national security as the agency and the courts have presumed in their perfunctory recitation of the goals of section 310(b). Nevertheless, the FCC has claimed broad discretion to interpret section 310(b). Even if that discretion legitimately existed, the FCC has exercised it unwisely. In direct conflict with the plain language of section 310(b)(4), the FCC has created a presumption that foreign investment disserves the public interest if it exceeds 25 percent of the equity of an American radio licensee. That policy is contrary to the FCC's legal authority under the Communications Act and thus is unlawful under the Administrative Procedure Act.

An examination of the other provisions in the Communications Act and in related statutes reveals that it is unlikely that the FCC's zeal in enforcing 310(b)(4) is essential either to protect national security or to advance the "public interest" in any other respect. Despite their exemption from section 310(b), private carriers and cable television have not become tools of today's version of "the Hun." As telecommunications markets become increasingly global in scope, the main effects of the FCC's error are to obstruct the natural flow of capital, to raise transactions costs, and to stifle competition.

APPENDIX: OTHER FEDERAL STATUTES
RESTRICTING FOREIGN DIRECT INVESTMENT

Section 310(b) of the Communications Act is only one of several federal statutes that restrict foreign direct investment on an industry-specific basis. Federal statutes also restrict foreign direct investment in the air transportation, shipping, banking, mining, energy, and fishing industries. None of those statutes, however, implicates speech or the transmission of information. And, quite conspicuously, no industry-specific statute restricts the foreign ownership of newspapers, publishers, motion picture studios, software companies, Internet service providers, book stores, theaters, or video rental outlets.

Foreign participation in the air transportation industry is subject to substantial restrictions, under authority of the Federal Aviation Act of 1958,[177] as implemented by the Department of Transportation.[178] Under the Federal Aviation Act, foreign airlines are prohibited from operating in U.S. domestic air service.[179] Domestic airline operations are limited to aircraft registered in the United States, for which only a U.S. citizen[180] or a corporation organized in the United States,[181] with aircraft based and used primarily in the United States, may register. Aliens may engage in international air transportation serving U.S. and foreign destinations, but they must operate as "foreign air carriers"[182] and must acquire a permit from the Department of Transportation (DOT).[183] DOT regulations set forth the requirements for permit applications[184] and the terms, conditions, and limitations of foreign air carrier permits.[185]

Foreign companies may invest in U.S. airlines, but only to the extent that the recipient airline would remain a U.S. citizen under the Federal Aviation Act. That act requires that no more than 25 percent of the voting stock, and 49 percent of the equity, be held by foreign investors.[186] Investments below that threshold are subject to review by the DOT if the investor is an air carrier that would operate jointly with the domestic airline in which it invested.[187] Certain indirect investments by foreign-owned U.S. companies in the airline industry are not regulated. A foreign-owned corporation may own U.S. aircraft and operate domestic air service if it is organized in the United States and if its aircraft are based and used in the United States.[188]

---

177. 49 U.S.C. §§ 1301–1557.

178. 14 C.F.R. §§ 211–98.

179. 49 U.S.C. § 1301(3).

180. *Id.* § 1401(a)–(b).

181. *Id.* § 1401(b)(1)(A)(ii).

182. *See id.* §§ 1301(22), (24).

183. *Id.* § 1372(a).

184. 14 C.F.R. § 211.

185. *Id.* § 213.

186. *See* 49 U.S.C. § 1301(16).

187. *See* 14 C.F.R. § 399.88.

188. *See* GENERAL ACCOUNTING OFFICE, AIRLINE COMPETITION: IMPACT OF CHANGING FOREIGN INVESTMENT AND CONTROL LIMITS ON U.S. AIRLINES (Jan. 10, 1993).

Foreign investors also face two types of restrictions on their participation in the U.S. maritime industry: those on domestic transport and those on international transport. In addition to explicit restrictions on foreign maritime investment, federal law discourages foreign investment by curtailing the operations of foreign-owned marine transport companies and vessels.

The Merchant Marine Act of 1920[189] contains the "Jones Act"—a law providing that merchandise moving between ports in the United States must be transported on U.S.-built, owned, and registered vessels.[190] Under the Jones Act, the secretary of the Treasury has discretion to suspend those requirements with respect to countries that grant the United States reciprocal treatment, but only under extraordinary circumstances.[191]

The Shipping Act of 1916 sets forth a direct restriction on foreign investment.[192] The Shipping Act requires the approval of the secretary of the Treasury before a U.S. vessel may be transferred to an alien,[193] with narrow exceptions.[194] Other laws prevent foreign vessels from carrying certain U.S. government cargo by authorizing or directing government agencies to ship preferentially on U.S. flag vessels on noncompetitive terms. The Cargo Preference Act of 1954 requires that at least half of specified government cargo be transported on privately owned flag vessels, when they are available at reasonable rates.[195] The 1954 act requires that all cargoes covered be shipped on U.S.-flag vessels, subject to waiver.[196]

Banking is also an area in which the United States restricts foreign activity and investment. The Foreign Bank Supervision Enhancement Act of 1991 establishes stringent rules for federal supervision of foreign banks seeking to invest or operate in the United States.[197] The act requires federal review before a foreign bank's establishment of branches, agencies, or commercial lending company

---

189. 46 U.S.C. §§ 861–99.

190. *Id.* § 883.

191. *Id.*

192. *Id.* §§ 801–42.

193. *Id.* §§ 808(c), 835(b), (e).

194. *See id.* §§ 1181, 1241 (b)(1).

195. *Id.*

196. *Id.*

197. Pub. L. No. 102–242, 105 Stat. 2286 (codified in various sections of 12 U.S.C.).

subsidiaries in the United States.[198] The act also authorizes the Federal Reserve Board to terminate a foreign bank's U.S. activities and offices if it finds that the bank has violated domestic law or engaged in unsafe or unsound banking practices.[199] General investments by foreigners in U.S. banks are also regulated in certain circumstances. In particular, the act requires disclosure of any purchase of shares in a national bank with the use of loans from a foreign bank that were secured by such shares.[200] Penalties can be severe—up to $25,000 per day of a continuing violation, plus additional penalties for failure to make required reports.[201]

There are also restrictions on bank operations by foreigners at the federal level. The National Bank Act of 1863 provides that all directors of national banks be U.S. citizens.[202] Further, at least two–thirds of the directors on the board of a national bank must reside in the state where the bank is located, or within 100 miles of the bank itself.[203]

Foreign investment in various sectors of the energy industry is also restricted under U.S. law. The Mining Law of 1987 limits the right to explore for minerals and to purchase lands containing mineral deposits to "citizens of the United States and those who have declared their intention to become such."[204] The Department of the Interior enforces those prohibitions. Foreign-owned corporations are considered citizens of the United States if they are organized under U.S. law and thus may exercise federal mining claims under the law.[205]

Foreign investment in the nuclear energy industry is largely proscribed. The Atomic Energy Act of 1954[206] effectively bars foreign ownership of companies operating in the nuclear power industry. The act regulates the licensing of production facilities that use nuclear materials and prohibits the Nuclear Regulatory Commission (NRC) from issuing a license to "an alien or corporation or other entity if the

---

198. 12 U.S.C. § 3105(d).

199. *Id.* § 3105(e).

200. *See id.* § 1817(j)(9).

201. *Id.* §§ 3110(a)(1), (c).

202. *Id.* §§ 21–215.

203. *Id.* § 72.

204. 30 U.S.C. § 22.

205. *Id.* § 24.

206. Atomic Energy Act of Aug. 30, 1954, 68 Stat. 921 (codified as amended in various sections of 42 U.S.C.).

NRC knows or has reason to believe it is owned, controlled, or dominated by an alien, a foreign corporation, or a foreign government."[207] The act also authorizes the Department of Energy (DOE) to issue leases or permits for exploring or mining nuclear source material in land belonging to the United States.[208] Under that authority, DOE may restrict the mining rights of foreign entities.[209]

The Mineral Lands Leasing Act of 1920 restricts foreign procurement of leases to explore and extract deposits of coal, oil, oil shale, gas, and various nonfuel minerals on U.S.[210] government lands, with certain exceptions based on reciprocity.[211] The Outer Continental Shelf Lands Act of 1953 authorizes the leasing of oil and natural gas in the offshore area comprising the continental shelf of the United States.[212] The regulations implementing the law provide that leases may be issued only to U.S. citizens, aliens, or corporations organized under the laws of the United States.[213]

The Federal Power Act of 1920[214] authorizes the Federal Energy Regulatory Commission to issue licenses to construct and operate power plants on public lands.[215] Under the act, licenses may be granted only to U.S. citizens and domestic corporations.[216] The Geothermal Steam Act of 1970[217] restricts the issuance of leases for geothermal steam development and utilization to U.S. citizens and corporations organized under U.S. law.[218]

The Coast Guard Authorization Act of 1989 limits foreign investment in the U.S. commercial fishing industry by imposing restrictions on foreign ownership of fishing vessels.[219] Vessels documented for U.S. fisheries must be owned by U.S. citizens.[220]

---

207. 42 U.S.C. § 2133(d).
208. *See id.* § 2097.
209. *Id.*
210. 30 U.S.C. §§ 181–287.
211. *Id.* § 181; 43 C.F.R. § 3102.2.
212. 43 U.S.C. §§ 1331–56.
213. 43 C.F.R. § 3102.1 (Interior Dep't).
214. 16 U.S.C. §§ 791(a)–825.
215. *Id.* § 797(a).
216. *Id.* § 797(e).
217. 30 U.S.C. §§ 1001–27.
218. *Id.* § 1015.
219. 46 U.S.C. § 12102.
220. *Id.* § 12102(c).

4

Ownership and Control

LIKE ANY COMPLEX regulatory constraint, the foreign-ownership restrictions in the Communications Act of 1934 challenge managers, aided by their lawyers and investment bankers, to structure the ownership and control of a multinational telecommunications firm so that, on the one hand, the firm complies with the law and, on the other, it maximizes returns to shareholders by exploiting competitive opportunities while trying to counteract the agency costs engendered by the regulation. As briefly mentioned in chapter 3, the FCC enforces section 310(b)(4) through a protracted waiver process. In this chapter we examine how foreign investors have structured the ownership and control of their investments in U.S. telecommunications firms either to avoid seeking waivers or to facilitate securing waivers that the FCC can be expected to grant.

TRANSACTIONS COSTS, AGENCY COSTS, AND THE STRUCTURE OF THE FIRM

Two questions lie at the heart of economic analysis of the firm. First, What is the optimal size of the firm? That question in turn invites many others. What are the optimal scale and scope of the firm's activities? How much of the total demand for a given product should the firm attempt to supply? Should the firm produce many products or only one? Should the firm vertically integrate into supply or distribution activities? The pursuit of answers to those questions makes up much of the discipline within economics known as industrial organization.

The second fundamental question concerning the firm is, Who should own and who should govern the enterprise? That question also invites many others. What is the firm's optimal capital structure? How should executive compensation be structured? What should the firm's dividend policy be? If the firm is a corporation, in what jurisdiction should it be incorporated? What should management's policy be toward unsolicited tender offers or proxy contests? How does the structure for the ownership and governance of the firm change in the face of technological or regulatory change? Those questions are the domain of corporate finance.

Transactions Costs

Many of the recipients of the Nobel Prize in economics have contributed to understanding those questions of industrial organization and corporate finance.[1] A recurring theme in the analysis that they and other eminent economists have undertaken is that the set of activities that define the boundaries of the firm, and the set of contracts that define the ownership and control of the firm, depend critically on "transactions costs."

Why, for example, do some economic activities occur within the firm, while others are procured by contract in the marketplace? The famous answer that Ronald Coase gave was that a firm takes the place of contracts only when it offers lower transactions costs to produce a particular good or service.[2] So, for example, General Motors incurred lower transactions costs by producing car bodies internally, through its acquisition of Fisher Body, than by contractually specifying and procuring the same bodies from a separately owned and managed Fisher Body.[3] Conversely, every obstacle to cooperation that

---

1. They include Kenneth J. Arrow, Ronald H. Coase, Franco Modigliani, Harry M. Markowitz, Merton H. Miller, William F. Sharpe, and George J. Stigler.

2. Ronald H. Coase, *The Nature of the Firm*, 4 ECONOMICA (n.s.) 336, 386–405 (1937). Other pioneering papers in this area are Armen A. Alchian, *Uncertainty, Evolution, and Economic Theory*, 58 J. POL. ECON. 211 (1950); Armen A. Alchian & Harold Demsetz, *Production, Information Costs, and Economic Organization*, 62 AM. ECON. REV. 777 (1972).

3. *See* Benjamin Klein, Robert G. Crawford & Armen A. Alchian, *Vertical Integration, Appropriable Rents, and the Competitive Contracting Process*, 21 J.L. & ECON. 297 (1978).

increases a firm's costs makes it less likely that the firm will exist at all; instead, the economic functions that it would perform will be conducted by individual parties assembling factors of production on an ad hoc basis—as is the case, for example, in the production of a typical motion picture.[4] Thus, economists argue that the scale, scope, and vertical integration of the firm all reflect the transactions costs of assembling and directing factors of production within a managerial hierarchy rather than through arm's-length contractual relationships.

Transactions costs also affect the ownership and control of the firm. Through an evolutionary process, firms gravitate toward efficient governance structures. In particular, it is efficient to divide functions between investors and managers even though investors consequently must expend resources to specify and monitor the performance of managers.[5] Someone possessing capital may lack management expertise, and someone possessing management expertise may lack capital or may wish to avoid the risks of owning the productive assets that he manages. In other words, separating ownership from control facilitates risk diversification.

Risk Diversification, Moral Hazard, and Control Transactions

Financial risk, which is inherent in any economic undertaking, consists of specific risk and market risk. Specific risk is unique to a particular firm—such as the uncertainty of whether the firm will be awarded a valuable patent or win a valuable contract with a large customer. Market risk, such as the risk of war, is endemic to all firms and all industries. An investor cannot diversify away market risk because it accompanies all economic ventures to varying degrees. In an efficient capital market, only nondiversifiable risk matters for

---

4. For a synthesis of the transactions costs literature on the nature of the firm, see OLIVER E. WILLIAMSON, THE MECHANISMS OF GOVERNANCE (Oxford University Press 1996); PAUL R. MILGROM & JOHN ROBERTS, ECONOMICS, ORGANIZATION AND MANAGEMENT (Prentice-Hall 1992); OLIVER E. WILLIAMSON, THE ECONOMIC INSTITUTIONS OF CAPITALISM: FIRMS, MARKETS, RELATIONAL CONTRACTING (Free Press 1985).

5. *See* Eugene F. Fama & Michael C. Jensen, *Separation of Ownership and Control*, 26 J.L. & ECON. 301 (1983); Michael C. Jensen & William H. Meckling, *Theory of the Firm: Managerial Behavior, Agency Costs and Ownership Structure*, 3 J. FIN. ECON. 305, 308–10 (1976).

the pricing of an asset. A security that is risky in isolation, but is uncorrelated with the market, has nothing but diversifiable risk, and it will earn a return on average no higher than the return on riskless investments.

An investor can reduce the specific risk facing his portfolio by placing a relatively small share of his capital in each of a large number of investments.[6] But the more that an investor reduces the financial risk facing his portfolio by spreading his funds across many firms that seek investment capital, the smaller will be the proportion of his total wealth that depends on the performance of any given firm and the smaller, therefore, will be his incentive to oversee or participate in the management decisions facing any one of those numerous firms. "Since he holds the securities of many firms precisely to avoid having his wealth depend too much on any one firm," Eugene Fama has observed, "an individual security holder generally has no special interest in personally overseeing the detailed activities of any firm."[7] Shareholders are "residual claimants" of the firm who contract for the right to the residual net cash flows of the venture; in contrast, salaried managers have a less immediate personal stake in the firm's profits or losses.[8] Principal-agent problems can arise when investors hire professional managers to manage the firm in which those investors have passively invested.[9] The professional managers do not internalize the costs of deviating from behavior that maximizes the firm's profit. The potential for "moral hazard" thus arises.[10]

---

6. *E.g.*, RICHARD A. BREALEY & STEWART C. MYERS, PRINCIPLES OF CORPORATE FINANCE 136–39 (McGraw Hill, 4th ed. 1991).

7. Eugene F. Fama, *Agency Problems and the Theory of the Firm*, 88 J. POL. ECON. 288, 291 (1980).

8. *See* Eugene F. Fama & Michael C. Jensen, *Agency Problems and Residual Claims*, 26 J.L. & ECON. 327, 328 (1983).

9. Harold Demsetz & Kenneth M. Lehn, *The Structure of Corporate Ownership: Causes and Consequences*, 93 J. POL. ECON. 1155 (1985).

10. "Moral hazard . . . arises in agreements in which at least one party relies on the behavior of another and information about that behavior is costly. The owner of a firm hires a manager and wants the manager to maximize profits. . . . Because it is costly for the principal to know exactly what the agent did or will do, the agent has an opportunity to bias his actions more in his own interest, to some degree inconsistent with the interests of the principal." Armen A. Alchian & Susan E. Woodward, *The Firm Is Dead; Long Live the Firm*, 26 J. ECON. LIT. 65, 68 (1988); *see also* KENNETH J. ARROW, *Insurance, Risk and Resource Allocation*, in ESSAYS IN THE THEORY OF

The potential for the separation of ownership and control to cause moral hazard simply means that the firm's owners must create desirable incentives for managers through other governance mechanisms. The firm's owners, for example, can design compensation packages for managers to be an increasing function of the firm's net cash flows, an objective that could be achieved by giving those managers stock.[11] The purpose of such stock ownership by management is to align managers' incentives with those of the firm's shareholders, thereby reducing moral hazard. Another important mechanism for reducing moral hazard associated with the separation of ownership and control is the market for corporate control, in which management can be displaced by investors who acquire enough voting shares to elect a majority of the corporation's directors. An extensive literature maintains that potential competition from alternative management teams is an important means to induce managers to maximize share value.[12] In anticipation of improving share value, bidders are willing to offer premiums to the target's shareholders.

Legal Structures for Risk Bearing

Legal structures for governing a firm differ in the degree to which they allocate risk among its managers, investors, and creditors. The limited liability of a corporation permits small, risk-averse investors to invest in a firm with large capital requirements that might incur operating losses or liabilities exceeding the personal wealth of any one investor. Thus, limited liability makes ownership shares more readily alienable (and thus more liquid) because, as Judge Richard A. Posner has observed, "without limited liability a shareholder would not even be allowed to sell his shares without the other shareholders' consent,

---

RISK-BEARING 134, 142–43 (1970).

11. *See* Michael C. Jensen & Kevin M. Murphy, *Performance Pay and Top-Management Incentives*, 98 J. POL. ECON. 225 (1990).

12. The pioneering work is Henry G. Manne, *Mergers and the Market for Corporate Control*, 73 J. POL. ECON. 110 (1965). For analysis of the subsequent literature, see ROBERTA ROMANO, THE GENIUS OF AMERICAN CORPORATE LAW (AEI Press 1993); FRANK H. EASTERBROOK & DANIEL R. FISCHEL, THE ECONOMIC STRUCTURE OF CORPORATE LAW (Harvard University Press 1991); Roberta Romano, *A Guide to Takeovers: Theory, Evidence and Regulation*, 9 YALE J. ON REG. 119 (1992); Michael C. Jensen, *Takeovers: Their Causes and Consequences*, 2 J. ECON. PERSPECTIVES 21 (1988).

since if he sold them to someone poorer than he, the risk to the other shareholders would be increased."[13] Although limited partners are also insulated from vicarious liability under most circumstances,[14] a partnership extends liability to a general partner to a greater degree than the typical publicly traded corporation extends liability to a shareholder. Thus, a general partner is vicariously liable for the acts or omissions of his fellow partners.[15] And, of course, an unincorporated sole proprietor only shifts risk to third parties to the extent that personal bankruptcy would permit him to shift the burden of liabilities onto his creditors.

In short, a variety of sophisticated legal institutions has developed to accommodate the demands of the capital markets to permit ownership to be separated from control and to ensure that doing so does not create significant agency costs. Most of the scholarly understanding of the function of such legal institutions did not exist as recently as the 1960s. We shall now examine how section 310(b) imposes significant transactions costs on foreign investors and U.S. licensees that seek to affiliate with one another. Those costs are more than the obvious burden of hiring Washington communications lawyers, consultants, and publicists. The larger costs are the agency costs of inferior structures for ownership and control of the resulting international telecommunications venture. The FCC neither acknowledges nor bears those costs. Like all costs, however, those agency costs ultimately work their way into the price that consumers pay for the firm's services and the return that the firm's owners earn on their invested capital.

The Statutory Scheme

Section 310(b) of the Communications Act restricts foreign ownership or management of broadcast, common carrier, or aeronautical en route or aeronautical fixed radio station licenses. First, consider

---

13. Richard A. Posner, Economic Analysis of Law 394 (Little, Brown & Co. 4th ed. 1992).

14. Rev. Unif. Limited Partnership Act § 303(a) (1985), 6 U.L.A. 282 (1987 supp.).

15. "Traditionally, partners are said to be jointly liable for the partnership's contractual obligations and jointly and severally liable for its tort liabilities." Robert C. Clark, Corporate Law 6 (Little, Brown & Co. 1986).

application of that ban under sections 310(b)(1) and 310(b)(2). Those sections apply the ban to "any alien or the representative of any alien" and to "any corporation organized under the laws of any foreign government."[16] Here we see the first level of distortion of business behavior caused by the statute. Individual foreigners who want to hold U.S. radio licenses simply may not; they must incur the costs of finding some other, more removed corporate vehicle to hold the license on their behalf, either as remote investors or as customers.

Those foreigners who do not give up altogether in their attempt to make a direct investment in a U.S. wireless company would proceed under section 310(b)(3) or 310(b)(4), neither of which absolutely bars foreign affiliation. Rather, those provisions ban, or subject to extraordinary regulatory oversight, only certain configurations of equity interests and foreign voting power. Additionally, foreigners who would not have been interested in holding a radio license but who merely want to make an investment in a licensee will seek to operate under those sections. Thus, the FCC decisions analyzed in this chapter involve the application of sections 310(b)(3) and 310(b)(4). The interpretation of those provisions by the FCC, and to a lesser degree by the courts, determines the extent of the transactions costs that any foreign investor must bear.

Section 310(b)(3) governs the level of permissible foreign investment that may be made directly in a licensee. Following its amendment in 1996, the section applies to "any corporation of which more than one-fifth of the capital stock is owned of record or voted by foreigners or their representatives or by a foreign government or representatives thereof or by any corporation organized under the laws of a foreign country."[17] The practical effect of the provision is to ban licensees of radio spectrum from raising capital by selling more than 20 percent of their stock to foreign buyers.

Section 310(b)(4) governs the level of foreign investment in a licensee's holding company, applying to "any corporation directly or indirectly controlled by any other corporation of which more than one-fourth of the capital stock is owned of record or voted by foreigners, their representatives, or by a federal government or representative thereof, or by any corporation organized under the laws

---

16. 47 U.S.C. §§ 310(b)(1), (2).
17. *Id.* § 310(b)(3).

of a foreign country, if the Commission finds that the public interest will be served by the refusal or revocation of such license."[18] In some cases, choosing between section 310(b)(3) and 310(b)(4) is straightforward. Obviously, section 310(b)(3) would apply to a foreigner seeking to purchase stock in a radio licensee. Likewise, section 310(b)(4) would apply to a foreigner seeking to purchase stock in a holding company (that is, a company that holds a controlling interest in a radio licensee). Suppose, however, that the foreigner seeks to buy stock in a corporation that is not itself a licensee, but which owns a *non*controlling interest (less than 50 percent) in a licensee. Then, paradoxically, the FCC will apply section 310(b)(3) with its lower benchmark for foreign ownership, not section 310(b)(4).[19]

This glance at sections 310(b)(3) and 310(b)(4) reveals two ways in which those provisions might affect an investor's transactions costs. First, investors must incur costs to comply with the restrictions in each provision. Second, the FCC may decline to enforce the restrictions of section 310(b)(4) if "the public interest" requires it. Investors must therefore consider the costs of that waiver process. The extent of those compliance costs will become clearer presently when we examine the scope and nature of the FCC's enforcement of section 310(b). In addition to those direct compliance costs, investors incur agency costs that could otherwise be avoided if section 310(b) did not constrain their choice of structures for the ownership and control of a U.S. radio licensee.

FCC ENFORCEMENT PREROGATIVES

The Communications Act empowers the FCC to enforce section 310(b) in a variety of contexts. The FCC might be called on to enforce section 310(b) in an application proceeding for an initial grant of license,[20] in a license renewal proceeding,[21] in an application to

---

18. *Id.* § 310(b)(4).

19. Ronald W. Gavillet, Jill M. Foehrkolb & Simone Wu, *Structuring Foreign Investments in FCC Licensees Under Section 310(b) of the Communications Act*, 27 CAL. W. L. REV. 7, 10–11 (1990).

20. Jireh's Broadcasting, L.P., 5 F.C.C. Rcd. 3308, 3309 ¶ 4 (1990) (dismissing application because foreign partner whose U.S. citizenship was pending never filed an amendment indicating subsequent grant of citizenship); Edward F. & Pamela J. Levine, 6 F.C.C. Rcd. 4679, 4769 ¶ 3 (1991) (application will be dismissed when it does not

transfer control of a license,[22] or in a license revocation proceeding.[23] Even if the FCC has approved a foreigner's affiliation with a licensee on many occasions, that fact does not insulate the affiliation from later FCC review.[24] When section 310(b)(1), 310(b)(2), or 310(b)(3) is involved, the FCC will consider only the issue of compliance. When section 310(b)(4) is involved, the FCC may consider not only compliance with the limits on ownership and voting power, but also whether to waive those limits. The FCC asserts the power to approve a transaction only on the condition that it give prior approval to further changes in the licensee's corporate structure.[25] In that manner the agency in effect becomes a third party in the licensee's corporate governance.

The FCC has construed section 310(b)(4) to grant the agency the discretion to permit foreign ownership beyond the 25 percent benchmark if the investor and licensee can show that the investment would be in the public interest. As chapter 3 explains, however, the FCC has misread the plain language of section 310(b)(4). The statute actually requires the FCC to show that the public interest would call for the applicant's license to be revoked or denied. Nevertheless, the FCC's construction still stands—though it was explicitly disowned by the House Commerce Committee in 1995[26]—and, for the time being, the foreign investor must contend with that agency *mis*interpretation of the statute in his business planning.

The FCC's waiver process for section 310(b)(4) adds uncertainty to any investment that would come under that subsection. The

---

contain information to verify compliance with section 310(b)); Loudon Broadcasters, Inc., 3 F.C.C. Rcd. 796 ¶ 2 (1988), *recons. dismissed sub nom.* Lauderdale-McKeeham Christian Broadcasting Corp., 4 F.C.C. Rcd. 8095 (1989).

21. *See* Spanish Int'l Comm. Corp., 2 F.C.C. Rcd. 3336, 3339 ¶ 14–16 (1987) [hereinafter *SICC*], *aff'd sub nom.* Coalition for the Preservation of Spanish Broadcasting *v.* FCC, 931 F.2d 73 (D.C. Cir. 1991), *cert. denied*, 502 U.S. 907 (1991); *see also* Fox Television Stations, Inc., 77 Rad. Reg. 2d (P & F) 1043, 1046 ¶ 5 (1995) [hereinafter *Fox*].

22. 47 U.S.C. § 310(d).

23. *See* KOZN FM Stereo 99, Ltd., 59 Rad. Reg. 2d (P & F) 628, 629 ¶¶ 4, 7 (1986).

24. Seven Hills Television Co., 2 F.C.C. Rcd. 6867, 6868–74 ¶¶ 7–27 (1987), *recons. dismissed*, 3 F.C.C. Rcd. 826 (Rev. Bd. 1988), *recons. denied* 3 F.C.C. Rcd. 879 (Rev. Bd. 1988), *rev. dismissed* 4 F.C.C. Rcd. 4062 (1989).

25. Gavillet, Foehrkolb & Wu, *supra* note 19, at 49.

26. H.R. REP. No. 204, 104th Cong., 1st Sess. 120–21 (1995).

agency's inquiry into whether the grant of a waiver would serve the public interest is essentially arbitrary, as chapter 5 documents in the cases of the largest foreign investments made in U.S. telecommunications firms as of 1996. The FCC has created a laundry list of factors, many of which bear little or no relation to the statute's original purpose. The factors that do relate to the statutory purpose are applied with inexplicable zeal. Before the FCC's 1995 rulemaking on foreign entry, its decisions on waivers under section 310(b) exemplified Judge Henry Friendly's observation that the authors of FCC opinions "remain free to pull [prior authorities] out of the drawer whenever the agency wishes to reach a result supportable by the old rule but not the new."[27] Whether the consistency of those decisions will improve in light of the FCC's rulemaking remains to be seen.

In its 1995 rulemaking the FCC described the history of its inquiry in section 310(b)(4) cases as follows:

> [T]he Commission . . . has generally considered the following factors: national security, the extent of alien participation in the parent holding company, and the nature of the license, including whether the licensee exercises control over content. In addition, the Commission may consider any other public interest factors appropriate. . . . One of the Congress' principal reasons for enacting Section 310 of the Communications Act of 1934 was its concern for national security and preventing alien activities against the government during a time of war. Accordingly, the Commission has traditionally sought to ascertain whether a country with which a prospective licensee or its parent is associated enjoys "close and friendly relations with the United States" and, therefore, is not a "national security concern." . . . The Commission has also traditionally considered the extent of alien participation in the parent corporation of a Title III radio [broadcasting] licensee. More specifically, the Commission has considered where the parent corporation is incorporated (the United States or elsewhere); the citizenship of the stockholders, officers and directors of the parent corporation; and whether there are intermediate corporations between the licensee and the parent corporation that are incorporated in the United States, are owned by U.S. citizens or interests, and have U.S. officers and directors. In addition, the

---

27. Henry J. Friendly, The Federal Administrative Agencies 63 (Harvard University Press 1962).

Commission has traditionally considered the type of radio license at issue in assessing whether the public interest would be disserved by foreign ownership in a parent corporation exceeding the Section 310(b)(4) benchmarks. For example, the Commission has concluded that concern about the effect of foreign ownership on national security is lessened when common carrier radio licenses are involved because they are "passive" in nature and the licenses confer no control over the content of transmissions. Finally, the Commission may also consider other relevant factors, including the furtherance of established Commission policies such as increased competition or the wide dissemination of licenses.[28]

Among the "other relevant factors" that the FCC has considered is a concern with preserving U.S. jobs.[29] The FCC has also inquired into the licensee's need for capital and into the value of the services that the company provides, second-guessing management in the former case and customers in the latter.

The FCC acknowledged in its 1995 rulemaking that Congress enacted section 310(b) because of national security concerns.[30] The FCC's view, however, seemed to be that such concerns had become anachronistic:

> The original national security rationale for limiting foreign ownership in a parent corporation has less applicability today than it had in the 1930's. Today there is a plethora of service providers. No single licensee which is owned in part by a foreign corporation could take over the wireless or wireline services in the United States in a time of war.[31]

The FCC therefore suggested in 1995 adopting trade policy as an additional determinant of the public interest when considering requests for waivers under section 310(b)(4), as well as when processing foreign carriers' section 214 applications.[32] As chapter 7 discusses in

---

28. Market Entry and Regulation of Foreign-Affiliated Entities, Notice of Proposed Rulemaking, IB Dkt. No. 95-22, 10 F.C.C. Rcd. 5256, 5263–64 ¶¶ 15–19 (1995) (quoting GRC Cablevision, Inc. 47 F.C.C.2d at 467, 468) [hereinafter *Market Entry and Regulation NPRM*].

29. Gavillet, Foehrkolb & Wu, *supra*, note 19, at 17, 49.

30. *Market Entry and Regulation NPRM*, 10 F.C.C. Rcd. 5256, 5263 n.16.

31. *Id.*

32. *Id.* at 5269 ¶ 33.

greater detail, the FCC ultimately adopted a reciprocity test under which the determining factor would be whether the government of a foreign carrier's home market permits U.S. citizens to invest in similar licensees.[33] In congressional testimony in 1995, Reed E. Hundt, chairman of the FCC, stated that "a section 310(b) which links effective access to overseas markets to access to our own markets" would more effectively serve the public interest than the present section.[34] He urged that any legislation that enacts "an effective market access approach . . . vest the FCC with the discretion—as does the current section 310(b)(4)—to determine on a case-by-case basis whether the standard is met."[35]

The FCC has claimed broad latitude in interpreting "the public interest." In the case of waivers under section 310(b)(4), it is extraordinary how far the FCC has wandered from the concerns with national security that motivated section 310(b). Should the FCC adopt its new trade policy proposal, the FCC will have strayed even farther from the law's intended purpose, and the outcome of the agency's waiver inquiries will be even more uncertain.[36] Not only has the FCC never laid down rules as to what will be allowed and what will not, opting instead to decide each waiver on a case-by-case basis that maximizes agency control over international investment decisions, but the policies adopted in one case might not be the same policies adopted in another. Such arbitrariness has added a further element of risk to business planning under section 310(b).

In short, compliance with section 310(b) is a continuing

---

33. Market Entry and Regulation of Foreign-Affiliated Entities, Report and Order, IB Dkt. No. 95-22, 11 F.C.C. Rcd. 3873, 3890 ¶¶ 43–44 (1996) [hereinafter *Market Entry and Regulation Report and Order*].

34. *Trade Implication of Foreign Ownership Restrictions on Telecommunications Companies: Hearings on Section 310 of the Communications Act of 1934 Before the Subcomm. on Commerce, Trade, and Hazardous Materials of the House Commerce Comm.*, 104th Cong., 1st Sess. 18 (1995) (statement of Reed E. Hundt, Chairman, Federal Communications Commission).

35. *Id.* at 19.

36. Addressing its treatment of foreign carriers' section 214 applications, the FCC has acknowledged that case-by-case review leads to uncertainty: "[O]ur case-by-case review of foreign carrier applications has caused uncertainty in the market due to the lack of a clear standard for evaluating applications by foreign carriers with different degrees of market power in their home markets." *Market Entry and Regulation NPRM*, 10 F.C.C. Rcd. at 5266 ¶ 23.

obligation. Section 310(b) requires the licensee and the would-be foreign investor (or consultant, or officer, or director) to adjust all his dealings, all the time, to comply with section 310(b). The gymnastics that managers, investors, and lawyers must perform to comply with section 310(b) raise transactions costs that distort the design of the efficient scale, scope, ownership, and control of firms that seek to compete as full-service networks for international telecommunications.

WAIVERS FOR FOREIGN OFFICERS
AND DIRECTORS BEFORE 1996

One means by which an investor can monitor his investment is by becoming involved in the firm's management. Before passage of the Telecommunications Act of 1996, section 310(b)'s restrictions on foreign officers and directors frustrated that method of reducing agency costs. Although Congress repealed those restrictions in 1996, it is still useful to examine the FCC's decisions in that area, for they show how the agency made ad hoc attempts to mitigate the adverse economic consequences of that portion of section 310(b). That approach may continue to affect the FCC's application of what remains of section 310(b) after 1996.

Section 310(b)(3) formerly barred "any corporation of which any officer or director is an alien" from holding most kinds of radio licenses, including a license to broadcast radio or television programming or to serve as a common carrier.[37] The FCC had no discretion under section 310(b)(3) concerning foreign officers and directors.[38] The presence of a single foreign officer or director in a radio licensee violated the restriction.

Section 310(b)(4) formerly barred holding companies of licensees from having any foreign officer or from having more than one-fourth foreigners on the board of directors, "*if* the Commission finds that the public interest will be served by the refusal or revocation of such license."[39] The FCC interpreted that language to establish a presump-

---

37. 47 U.S.C. § 301(b)(3).

38. *See* Request for Declaratory Ruling Concerning the Citizenship Requirements of Sections 310(b)(3) and (4) of the Communications Act of 1934, as Amended, 103 F.C.C.2d 511, 524 ¶ 21 (1985) (*Wilner & Scheiner*), *recons. in part*, Reconsideration Order, 1 F.C.C. Rcd. 12 (1986) (*Reconsideration Order*).

39. 47 U.S.C. § 301(b)(4) (emphasis added).

tion that those affiliations were not in the public interest, but the agency would grant waivers of the limits. Virtually all waiver requests under section 310(b)(4) concerned restrictions on the citizenship of foreign officers and directors of the holding company, as opposed to the stock ownership limitations on the holding company. In considering whether to waive the limits on foreign officers and directors imposed by section 310(b)(4), the FCC considered the scope of its "public interest" inquiry to be broad and to encompass, among other considerations, the following factors.[40]

Was the Foreigner's Country of Citizenship Friendly?

In assessing the extent to which foreign officers or directors could pose a national security threat, the FCC considered whether the foreigner's country of citizenship had close and friendly relations with the United States.[41] Under that standard, the FCC approved a license where the licensee's parent corporation had 50 percent Canadian directors[42] and one where one officer was Canadian.[43] The FCC also approved a license where four of the parent's directors were Swedish, British, and Swiss[44] and another where one officer was British.[45]

The FCC evidently never appreciated the presumption in a telecommunications regulatory agency's making pronouncements about foreign affairs. Nor did the FCC decisions in that area recognize that, by creating such a criterion for deciding whether to grant a waiver under section 310(b)(4), the agency created the potential for international embarrassment if it had ever denied a waiver on the grounds that the foreigner's country was not friendly to the United States. For example, despite having been America's ally in two world wars, France sometimes conspicuously abstained before

---

40. *See* Regulatory Policies and International Telecommunications, 2 F.C.C. Rcd. 1022, 1032 ¶ 73 (1986) [hereinafter *Regulatory Policies*]; Gavillet, Foehrkolb & Wu, *supra* note 19, at 49.

41. *Regulatory Policies*, 2 F.C.C. Rcd. at 1032 ¶ 73 & n.126.

42. Vermont Tel. Co., Inc., 10 F.C.C. Rcd. 9337, 9337 ¶¶ 1, 3 (1995) .

43. Application for Consent to Transfer Control of Hughes Communications, Inc. to General Motors Corp., 59 Rad. Reg. 2d (P & F) 502, 502 ¶ 5 (1987).

44. Comsat Gen'l Corp., 3 F.C.C. Rcd. 4216, 4218 ¶¶ 24–26 (1988).

45. Data Gen'l Corp., 2 F.C.C. Rcd. 6060, 6060 ¶¶ 4–8 (1987).

1996 from supporting American foreign policy and on occasion denied U.S. warplanes passage through French airspace when flying from Britain to the Mediterranean to launch air strikes. That was not very friendly, but surely it would not have justified the FCC's restricting French investment in American radio licensees on national security grounds. If the FCC had ever taken it seriously, the friendly-citizenship criterion would have been a Pandora's box. Had Congress not repealed the restrictions on officers and directors, one could have expected the FCC to have invoked the criterion only when it would have granted a waiver anyway, and then only when the case would not have been a close call.

Would the Facility
Exercise Editorial Control?

The FCC also asked whether the licensed facility involved was editorially passive, as is a common carrier, or had control over the content of transmissions. Thus, in *Data General Corporation* the FCC permitted a foreign officer in the licensee's parent because the licensee lacked editorial control.[46] In *Millicom Inc.* the FCC ruled that common carrier microwave facilities are passive, and thus the agency approved the foreign officers.[47]

Did the Foreigner Offer
Valuable Expertise or Capital?

The FCC will consider the financial or professional qualifications of the applicant. In *Houston International Teleport, Inc.*, the agency approved a foreign officer who had valuable management, technical, and business expertise.[48] The FCC also considered whether foreign participation would help ensure the continued vitality of a business, thus preserving U.S. jobs and shareholder value.[49]

---

46. *Id.* at 6060 ¶ 7.
47. 4 F.C.C. Rcd. 4846, 4847 ¶ 12 (1989).
48. 2 F.C.C. Rcd. 1666 ¶¶ 5–6 (1987).
49. *Millicom*, 4 F.C.C. Rcd. at 4848 ¶¶ 15–16.

Did the Foreigner
Control Operations?

The FCC examined whether the foreign officers or directors also controlled or supervised operations of the licensee. In *Comsat General Corporation* the FCC approved foreign officers and directors in excess of one-fourth of the board on the condition that they would not exercise direct control over the licensee.[50] The FCC also considered the extent to which other officers and directors were U.S. citizens.[51] The FCC routinely permitted holding companies to have foreign officers when officers would exercise no control or supervision over the operation of the licensed facilities. For example, in *International Telephone & Telegraph Corporation*, the FCC allowed two officers, one Canadian and one British, neither of whom would be involved in the activities of ITT's communications subsidiaries.[52] And, in *American Satellite Corp.*, the FCC granted a waiver to the parent (Continental Telephone Company) of a common carrier radio station licensee to permit the parent to have a British vice president for finance who was also a director.[53] The FCC did not explain its reasoning, despite the fact that the applicant did *not* pledge that the officer would not be involved in supervising the licensee subsidiary.

---

50. 3 F.C.C. Rcd. at 4218 ¶ 26; *see also* Atlantic Tele-Network, Inc., 7 F.C.C. Rcd. 6634, 6635 ¶¶ 6–7 (1992); GCI Liquidating Trust, 7 F.C.C. Rcd. 7641 ¶5 (1992); Continental Ill. Venture Corp., 6 F.C.C. Rcd. 1944 ¶ 5 (1991); McCaw Cellular Comm., Inc., 5 F.C.C. Rcd. 6258 ¶¶ 6–7 (1990) (where foreign officers would not be involved in day-to-day activities of the licensee, excess over section 310(b)(4) limit on officers was not contrary to public interest).

51. General Elec. Co., 5 F.C.C. Rcd. 1335 ¶ 6 (1990) (twelve of thirteen officers and thirty-two of thirty-three directors to have been U.S. citizens); *Atlantic Tele-Network*, 7 F.C.C. Rcd. at 6634 ¶ 6.

52. 67 F.C.C.2d 604, 605 ¶ 8 (1978); *see also* Advanced Mobile Phone Serv., Inc., 54 Rad. Reg. 2d (P & F) 354, 357 (Common Carrier Bureau 1983) (reading *ITT* decision as promulgating general rule authorizing foreign officers of corporate parent when officers would exercise no control or supervision over the operation of licensed facilities); *accord, General Elec.*, 5 F.C.C. Rcd. at 1335 ¶ 6 (British manager would be "the only one of thirteen elected officers . . . not a United States citizen").

53. 80 F.C.C.2d 254, 271–72 ¶¶ 60–63 (1980).

THE VACUITY OF THE FOREIGN
OWNERSHIP RESTRICTIONS AFTER THE
TELECOMMUNICATIONS ACT OF 1996

The Telecommunications Act of 1996 repealed the restrictions on foreign officers and directors contained in sections 310(b)(3) and 310(b)(4). In other words, since February 1996 Congress has permitted a foreigner to control the management of any U.S. radio licensee, even a radio or television station. That deregulation of foreign control occurred notwithstanding the recurrent rhetoric in prior FCC and D.C. Circuit decisions that "'Congress intended that aliens be prohibited entirely from holding . . . high level management positions' in radio licensees."[54] By implication, therefore, the 1996 legislation overruled broad categories of FCC and appellate decisions (and portions of decisions) interpreting when foreigners were deemed to have acquired unlawful control of a U.S. radio licensee, either through impermissible participation as a managerial officer or through too many foreigners' sitting on the licensee's board of directors.

Stated in economic terms, what is the statutory purpose of the 1996 legislation with respect to the optimal structuring of ownership and control? Unfortunately, the 1996 legislation lacks any apparent goal. Granted, it will unambiguously improve the economic welfare of American consumers and producers for there to be an unrestricted market in managerial capital in the U.S. telecommunications sector. But what remains of section 310(b) after the partial deregulation of foreign investment in 1996 is a statute that can only be described as anomalous and irrational. Sections 310(b)(3) and 310(b)(4) continue to limit ownership of both stock and voting rights to the 20 and 25 percent benchmarks that the FCC has interpreted as ceilings in the absence of a waiver. Thus, foreigners may lawfully exercise control of a U.S. radio licensee in their capacities as officers and directors, but not in their capacity as shareholders exercising their voting rights through the proxy process. It is simply irrational for Congress to say as a matter of federal telecommunications policy that active foreign

---

54. Sacramento RSA Limited Partnership *v.* FCC, No. 94-1532, 1995 U.S. App. LEXIS 15190 (D.C. Cir. 1995) (unpublished opinion) (quoting Request for Declaratory Ruling Concerning the Citizenship Requirements of Sections 310(b)(3) and (4) of the Communications Act of 1934, as amended, 103 F.C.C.2d 511, 520 n.43 (1985) (*Wilner & Scheiner*)).

managers may control a U.S. radio licensee but passive foreign investors may not. As a practical matter, of course, no foreign investor with a sizable holding in a U.S. radio licensee would ever be so naïve after the 1996 legislation as to rely exclusively on the proxy process and forgo the opportunity as an officer or director to participate in the control of the licensee.

There is one statutory purpose that can be concocted after the fact to explain the peculiar shape that section 310(b) has assumed since 1996: Congress may have intended to capture for American investors a disproportionate share of the returns from potentially superior foreign managerial capital. If a team of foreign managers can run a U.S. telecommunications company more productively than a team of American managers can, then Congress will not stand in the way. But Congress *will* see that American investors get the lion's share of the gains. By retaining the stock ownership limits in sections 310(b)(3) and 310(b)(4), Congress in 1996 in effect ensured that, absent prior approval by the FCC, American investors would reap 80 and 75 percent, respectively, of the superior profitability achieved by the foreign management team. That result is the mirror image of a problem, described in chapter 6, that U.S. telecommunications firms frequently encounter when seeking to invest in overseas markets: The American company supplies essentially 100 percent of the managerial capital but can receive dividends based only on its much lower level of permissible stock ownership.

MAXIMIZING FOREIGN INVESTMENT

Section 310(b) and the FCC decisions interpreting it block the simplest means by which foreign investors could affiliate themselves with U.S. radio licensees. That impediment does not mean that foreign investment in the kinds of radio licensees covered by section 310(b) will not occur. Rather, it implies that the cost of structuring such investment will be higher, and the return on such investment lower, than if the foreign ownership restrictions did not exist. Given those regulatory constraints, how do a foreign investor and a U.S. radio licensee structure their transaction to maximize the allowable level of foreign investment in the licensee?

Although, as the rest of this chapter will show, a multitude of financial instruments is available to tailor the ownership structures to the business needs of the particular deal, there are four generic steps to structuring a transaction to maximize the level of foreign owner-ship. A foreigner may hold up to 20 percent ownership or voting interest directly in a licensee under section 310(b)(3) *and* up to a 25 percent ownership or voting interest in a company controlling the licensee under section 310(b)(4). An investor may therefore increase his stake in a licensee by arranging with the licensee to split off a holding company. For purposes of the following discussion, assume that a foreign investor, Foreign Corporation, seeks to invest in a radio licensee called American Wireless Corporation. Assume further that a waiver of sections 310(b)(3) and 310(b)(4) is unavailable.

First, American Wireless would create a separate entity, American Wireless Exempt Corporation, to hold (1) all radio licenses granted to American Wireless that are exempt from section 310(b) and (2) all assets that are used in conjunction with the broadcasting licenses or radio common carrier licenses granted to American Wireless. Foreign Corporation could acquire 100 percent of American Wireless Exempt.[55] The assets and licenses of American Wireless Exempt would potentially be worth more separately than if they were subsumed in an entity holding licenses burdened by section 310(b).

Second, American Wireless would create American Wireless Holdings Corporation, a holding company organized under American law that held 80 percent of the stock of American Wireless. Foreign Corporation would acquire 25 percent of the voting equity of American Wireless Holdings, as section 310(b)(4) permits. In addition, American Wireless Holdings could appoint foreigners to positions of senior management and to all the seats on its board of directors.

Third, Foreign Corporation would acquire a 20 percent interest directly in American Wireless, as permitted by section 310(b)(3). Like its holding company, American Wireless could appoint foreign-ers to positions of senior management and to all the seats on its board of directors. By combining its direct and indirect holdings in American Wireless, Foreign Corporation could have the right to 40 percent of the net cash flows of the radio licensee—20 percent of American Wireless and 25 percent of the remaining 80 percent of American Wireless owned by American Wireless Holdings.

---

55. *See, e.g.*, Telefónica Larga Distancia de Puerto Rico, 8 F.C.C. Rcd. 106, 107 ¶ 4 (1992).

Fourth, in a transaction akin to a sale-leaseback, American Wireless could procure from Foreign Corporation services using radio licenses that are exempt from section 310(b) and unregulated assets that American Wireless would need to use in conjunction with its broadcasting licenses or radio common carrier licenses.

COMPLIANCE WITH FOREIGN
STOCK OWNERSHIP LIMITS

Section 310(b) limits the stake that a foreign investor may hold in a radio licensee by purchasing stock, either in the licensee itself or in a corporation that in turn owns stock in the licensee. The import of those limits is that the foreign shareholders will always be in the minority. There is no assurance that the majority shareholders will protect the interests of the minority shareholders. Again, the foreign investor faces an impediment to monitoring the firm's management. If the interests of the majority shareholders do not coincide with the foreign investor's, he will have no other way to protect his investment by monitoring the performance of management.

Sections 310(b)(3) and 310(b)(4) apply to foreign holdings of "capital stock." Capital stock traditionally includes preferred stock, common stock, voting and nonvoting stock,[56] and convertible nonvoting stock.[57] Arguably, section 310(b) ought to distinguish among those different types of stock, for they differ in the extent to which they bundle ownership and control into a single financial instrument. Different types of stock will give investors different rights to the residual net cash flows in the radio licensee and different voting rights with which to control the licensee's management.

Perhaps reflecting the undeveloped state of academic research on corporate finance in 1934, section 310(b) fails to distinguish between stock that confers the right to control the corporation and stock that merely confers the right to receive the corporation's residual net cash flows. How then does the FCC determine compliance with section 310(b) in the most straightforward cases, where the agency has already determined that the equity interest involved is "capital stock"?

---

56. Spanish Int'l Comm. Corp., 4 F.C.C. Rcd. 2153, 2154 ¶ 12 (1989) ("Voting and non-voting stock interests are indistinguishable for purposes of section 310(b).").

57. *Wilner & Scheiner*, 103 F.C.C.2d at 521 n.37; Gavillet, Foehrkolb & Wu, *supra* note 19, at 24.

Section 310(b)(3)

In determining whether a licensee has complied with section 310(b)(3), the FCC will count the percentage of shares foreigners hold in the licensee corporation. The FCC will also count the foreigner's shares in a corporation that owns less than 50 percent of the shares of a licensee. "Any ownership or voting interest held by an individual other than a United States citizen or by an entity organized under the laws of a foreign government is counted in the application of the statutory benchmarks."[58]

The Multiplier. How should the FCC measure the extent of a foreigner's interest when he holds, say, a 10 percent interest in a corporation with a 20 percent interest in a radio licensee? Clearly, it would not be correct to say that the foreigner holds either a 10 percent or a 20 percent interest in the licensee. Instead, the FCC concludes that the foreigner holds 10 percent of the 20 percent interest—giving it only a 2 percent interest in the licensee. The FCC employs that simple calculation, known as the "multiplier," in seeking to measure a foreigner's interest in a licensee when the foreigner holds stock in an intervening corporation rather than in the licensee directly.

There is an additional complication. Suppose that the foreigner holds a 51 percent interest in a corporation with a 20 percent interest in a licensee? The FCC would then assume that the foreigner controlled the corporation, just as if he held a 100 percent interest in the corporation.[59] Consequently, the FCC would not generally use the multiplier when a foreigner's ownership interest in the intervening corporation exceeded 50 percent; nonetheless, the FCC will still use the multiplier if the foreigner's interest is in the form of nonvoting stock.[60]

If the FCC determines that a licensee does not comply with the 20 percent cap on foreign investment directly in a licensee, the

---

58. *Wilner & Scheiner*, 103 F.C.C.2d at 521 ¶ 17.

59. *Id.*; *see also* Corporate Ownership Reporting and Disclosure by Broadcast Licensees, Dkt. No. 20251 *et al.*, 97 F.C.C.2d 997, 1018 ¶ 41 (1984) (*Attribution Order*), *recons. granted*, 58 Rad. Reg. 2d (P & F) 604 (1985), *recons. of later order granted in part*, 1 F.C.C. Rcd. 802 (1986).

60. Gavillet, Foehrkolb & Wu, *supra* note 19, at 12, 24.

agency will require the foreigner to divest itself of its interest in the licensee.[61] Thus, the FCC forced one company to sell its television stations to unrelated buyers with no foreign affiliations, although the agency did not enforce its policy of requiring an unqualified licensee to sell its station to a qualified party for less than its market value.[62]

When Section 310(b)(4) Also Applies. Suppose that a foreign investor holds a 10 percent interest directly in a radio licensee and also holds a 20 percent interest in a holding company with a controlling interest in the same licensee. In determining whether the foreigner has caused the licensee to violate section 310(b)(3), would the FCC add the two interests together?

The FCC has determined that it will not count—or "flow through"—interests in a holding company to determine whether a foreigner has violated section 310(b)(3). In *Data Transmission Co.*, a Swiss citizen named Haefner owned debentures that could be converted into a 22 percent interest in a company that held an 85 percent interest in a microwave radio licensee called Data Transmission.[63] If the FCC had applied the multiplier in that case, it would have found a 19 percent foreign interest in the licensee—22 percent × 85.5 percent = 19 percent. In addition, Haefner held a 9.5 percent foreign interest in the licensee itself. Haefner's 22 percent interest did not by itself violate section 310(b)(4), and his 9.5 percent interest did not by itself violate section 310(b)(3). Data Transmission asked the FCC for a ruling that Haefner's potential 19 percent interest in the licensee would not be added to the 9.5 percent that he held directly. If the interests were added, Haefner's holdings in Data Transmission would have exceeded the 20 percent limit in section 310(b)(3).[64]

The FCC ruled that Congress "did not intend a 'flow through' effect whereby ownership in a parent corporation would be included with the ownership interest of the subsidiary licensee."[65] A foreigner may therefore hold up to 20 percent ownership or voting interest directly in a licensee and up to a 25 percent interest in the company

---

61. *Wilner & Scheiner,* 103 F.C.C.2d at 518 ¶ 12.

62. *SICC,* 2 F.C.C. Rcd. at 3339–40 ¶¶ 21–22.

63. 52 F.C.C.2d 439 (1975) (*Datran II*).

64. *Id.* at 439–40.

65. *Id.* at 440 (*citing* S. REP. NO. 781, 73d Cong., 2d Sess. 7 (1934)).

controlling the licensee.[66] In that manner, a foreign investor may enlarge his stake in a licensee without violating section 310(b). He will, of course, still not hold a controlling interest in either corporation.

Section 310(b)(4)

Section 310(b)(4) applies to foreign ownership interests when another corporation controls a licensee. That section allows foreign investors to own up to 25 percent of a parent or holding corporation.[67] Just as the FCC before 1996 was permitted to waive the section 310(b)(4) restrictions on foreign officers, so also the FCC may waive section 310(b)(4)'s limits on stock ownership.[68] The FCC has based its waiver decisions on the following "public interest" factors.

Broadcast. In *Banque de Paris de Pays Bas*,[69] a French bank sought to obtain a beneficial interest in 18 percent of the stock of the parent corporation (Columbia Pictures) of several broadcast licensees in addition to the 20 percent that the bank already held. The bank stated that the additional shares would be acquired in trust with a domestic bank as trustee. The FCC approved the transaction on four conditions. First, the bank could not acquire any additional shares. Second, the bank could not enter into any agreement concerning the manner in which the stock held in the bank's name would be voted. Third, the bank would have to report to the FCC annually as to all agreements made and all actions taken regarding the stock. Fourth, the bank could not "take any action looking toward assertion of control by it alone or in concert with any other person over [the parent company]."[70]

Until the FCC's 1995 decision concerning Fox Television

---

66. Gavillet, Foehrkolb & Wu, *supra* note 19, at 19–20.

67. *See, e.g.*, McCaw Cellular Comm., Inc., 4 F.C.C. Rcd. 3784, 3788 ¶ 31 (1989) (permitting British Telecom's acquisition of 22 percent of McCaw's stock); MMM Holdings Cos., Inc., 4 F.C.C. Rcd. 8243, 8247 ¶ 26 (1989) (permitting British Telecom to hold just under 25 percent of McCaw's stock).

68. 47 U.S.C. § 310(b)(4). *See also* Moving Phones Partnership L.P. *v.* F.C.C., 998 F.2d 1051, 1057–58 (D.C. Cir. 1993), *cert. denied*, 114 S. Ct. 1369 (1994); Telemundo, Inc. *v.* F.C.C., 802 F.2d 513, 516 (D.C. Cir. 1986).

69. 6 F.C.C.2d 418 (1966).

70. *Id.* at 418.

Stations, Inc.—which is described in detail in chapter 5—the decision in *Banque de Paris de Pays Bas* was the only case in which the FCC had permitted a broadcast licensee to exceed the 25 percent benchmark.[71] The FCC's 1995 notice of proposed rulemaking concerning foreign investment hinted that the agency might be more lenient in the future: "It may be appropriate now to revisit our restrictive approach to alien investment in broadcasting. In contrast to the situation that existed in 1927, there are currently a plethora of broadcast and other mass communications facilities available to the general public."[72] But the FCC's report and order in that docket failed to liberalize foreign investment in broadcasting in any respect.[73]

U.S. Owners, Officers, and Directors. In deciding whether to permit foreign ownership in excess of the 25 percent benchmark, the FCC before 1996 often considered the level of foreign ownership in light of U.S. presence in other areas—such as ownership, officers, and directors. The harsh conditions the FCC imposed in 1966 in *Banque de Paris de Pays Bas* contrast significantly with the relatively relaxed approach the agency took in 1974 in *GRC Cablevision, Inc.*,[74] where Canadians owned 50 percent of a 60 percent corporate parent of a cable relay service licensee. The FCC gave five reasons for permitting the arrangement. First, the parent corporation, though majority foreign-owned, was a U.S. corporation. Second, a majority of the parent corporation's directors were U.S. citizens.[75] Third, the foreign shareholders were from a country with a tradition of close and friendly ties with the United States. Fourth, there was no other adverse information concerning the foreign investors. Fifth, the facility in question would be used for relay of broadcast signals and was thus "largely passive in operation."[76]

In *Teleport Transmission Holdings*, the FCC in 1993 approved 65 percent foreign ownership in a licensee's parent corporation where

---

71. Ian M. Rose, Note, *Barring Foreigners from Our Airwaves: An Anachronistic Pothole on the Global Information Highway*, 95 COLUM. L. REV. 1188, 1194 (1995).

72. *Market Entry and Regulation NPRM*, 10 F.C.C. Rcd. at 5298 ¶ 102.

73. *Market Entry and Regulation Report and Order*, 11 FCC Rcd. at 3875–76 ¶¶ 2–4.

74. 47 F.C.C.2d 467, 467 ¶ 2 (1974).

75. *Accord*, Upsouth Corp., 9 F.C.C. Rcd. 2130, 2131 ¶ 12 (1994).

76. *GRC Cablevision*, 47 F.C.C.2d at 468 ¶ 5.

the parent had 75 percent U.S. directors and officers.[77] In *MCI Communications Corp.*, the FCC in 1994 approved 28 percent foreign ownership of the parent where 80 percent of its directors and 100 percent of its officers were U.S. citizens.[78] In *A Plus Communications*, the FCC approved transfer of a corporation to an entity with foreign ownership known only to be under 50 percent because a majority of its directors were U.S. citizens.[79] In *IDB Communications Group, Inc.*, the FCC approved 26.2 percent foreign ownership of the parent in which all officers and 75 percent of the directors were U.S. citizens.[80]

If the FCC thought before 1996 that high levels of U.S. citizenship among the licensee's officers and directors would justify the agency's approval of higher levels of foreign ownership, then what does that imply for the FCC's policy following repeal of the restrictions on officers and directors in sections 310(b)(3) and 310(b)(4) upon passage of the Telecommunications Act of 1996? Congress in effect made that "plus factor" on the FCC's list of considerations irrelevant by permitting the levels of U.S. involvement in the licensee's management and on its board of directors to be nonexistent. Thus, it would be inappropriate for the FCC to rely on that factor as a consideration supporting the agency's refusal under section 310(b)(4) to permit foreign ownership exceeding 25 percent.

Common Carriage. In 1976 in *Data Transmission Co.*, the FCC set for hearing the transfer to a foreigner of the parent corporation of a common carrier licensee and listed as one of the issues whether section 310 "precluded alien control of the parent corporation of a common carrier radio licensee and, if not, what criteria and policy considerations should be developed for considering an application proposing such."[81] The case was mooted, however, by the licensee's bankruptcy. The FCC has since treated a common carrier's requests for a waiver under section 310(b)(4) more generously than a

---

77. 8 F.C.C. Rcd. 3063, 3065 ¶¶ 10–11 (1993).
78. 9 F.C.C. Rcd. 3960, 3964 ¶ 22 (1994).
79. A Plus Comm. of Puerto Rico, Inc., File No. 22913-CD-TC-(2)-82 (May 13, 1982), *discussed in* Gavillet, Foehrkolb & Wu, *supra* note 19, at 15.
80. 6 F.C.C. Rcd. 4652, 4653 ¶ 10 (1991).
81. 59 F.C.C.2d 909, 912 ¶ 8 (1976).

broadcaster's.[82] The agency summarized its reasoning in 1995:

> The distinction between common carrier and broadcast licensees
> in terms of content control has been the basis for our traditionally
> disparate treatment of these licensees under Section 310(b)(4).
> While the FCC has granted applications permitting foreign
> ownership of a parent holding company of a non-broadcast
> licensee to exceed 25 percent, the FCC has consistently declined
> to do so in broadcasting because of a broadcast licensee's ability
> to control the content of its transmission.[83]

The D.C. Circuit, however, has not accepted the proposition that
common carriers are lesser national security risks than broadcasters.
Referring to the "national security policy underlying section 310(b),"
the court in 1993 explained that "the rationale is equally applicable to
common carrier radio stations, as they, also, are a part of the nation's
communications network."[84]

Capital Contributions. The FCC has ruled that using a simple "count
the shares" approach may not accurately measure the extent of foreign
ownership, especially when the corporation has issued more than one
class of stock.[85] When it suspects that stock ownership is disproportionate to equity interest, the FCC will consider the amount of foreign
capital contribution to a corporation in determining compliance with
section 310(b)'s benchmarks.[86] In applying that standard in *Fox*, the
FCC granted the petitioner's application for renewal of its license,
subject to a showing of its compliance with section 310(b)(4).[87]

Sophisticated Capital Structures

FCC rules and policies, adjudicatory decisions, and a small number
of court decisions have supplemented and clarified the statutory
restrictions of section 310(b). Still, section 310(b) leaves unanswered
a number of specific questions regarding the use of sophisticated

---

82. *Market Entry and Regulation NPRM*, 10 F.C.C. Rcd. at 5296–98 ¶¶ 99–103.
83. *Id.* at 5296–97 ¶ 100.
84. *Moving Phones*, 998 F.2d at 1055–56.
85. *Fox*, 77 Rad. Reg. 2d (P & F) at 1051 ¶ 36.
86. *Id.*
87. *Id.* ¶ 11.

equity instruments and partnership structures. Following the broad definition of "corporation"[88] in the Communications Act, the FCC has reasoned that Congress intended section 310(b) to cover all forms of business association,[89] and that an overly restrictive interpretation of "corporation" would permit circumvention of the foreign ownership limits.[90] Likewise, the FCC has interpreted the term "capital stock" to encompass "the alternative means by which equity or voting interests are held in these businesses."[91] The FCC's definitions of "corporation" and "capital stock" are now so expansive as to encompass policyholders of insurance companies,[92] as well as members of a church[93] and of a labor union.[94]

Nonvoting Stock and Preferred Stock

Section 310(b) does not distinguish between voting and nonvoting stock, or between common and preferred stock. To the contrary, the statute encompasses stock "owned . . . *or* voted." In 1986 a group of applicants requested that the FCC exclude preferred stock from the definition of "capital stock" in determining compliance with sections 310(b)(3) and 310(b)(4) "where the applicant certifies that the preferred stock contains none of the indicia normally associated with equity ownership."[95] The FCC noted that the term "capital stock" generally encompasses various classes of stock (including preferred stock) and indicated that the legislative history did not support any different interpretation of the term.[96]

That interpretation of "capital stock" differs from the FCC's distinction between voting equity and nonvoting equity under the

---

88. *See* 47 U.S.C. § 153(j) ("'Corporation' includes any corporation, joint-stock company, or association.").

89. *Attribution of Ownership Interests*, 97 F.C.C.2d at 1009 ¶ 22.

90. *Wilner & Scheiner*, 103 F.C.C.2d at 516 ¶ 7.

91. *PrimeMedia*, 3 F.C.C. Rcd. at 4295 ¶ 9 (emphasis omitted).

92. Farragut Television Corp., 4 Rad. Reg. 2d (P & F) 350, 352–53 ¶¶ 5–6 (1965).

93. Kansas City Broadcasting Co., 5 Rad. Reg. (P & F) 1057, 1094 ¶ 16 (1952).

94. Chicagoland Television Co., 4 Rad. Reg. 2d (P & F) 747, 752 ¶ 8 (1965), *application for review denied*, F.C.C. 65-367 (May 5, 1965).

95. *Reconsideration Order*, 1 F.C.C. Rcd. at 13–14 ¶ 15 (preferred stock); *Wilner & Scheiner*, 103 F.C.C.2d at 519 n.37 (nonvoting stock).

96. *Wilner & Scheiner*, 103 F.C.C.2d at 511 ¶ 9.

ownership attribution rules in cases not involving foreign ownership. To those cases the FCC has applied the following rule: "Holders of non-voting stock shall not be attributed an interest in the issuing entity."[97] The FCC has said that the attribution criteria are merely "instructive" in making section 310(b) determinations.[98] It would therefore seem to serve no purpose, for example, for an American radio licensee to recapitalize itself so that a foreign entity could acquire only 20 percent of voting control while simultaneously acquiring a larger percentage of the U.S. firm's net cash flows.

Convertible Debentures

As a general matter, the FCC does not class convertible instruments such as warrants, convertible debentures, and options under section 310(b)'s definition of "capital stock." Such interests are not considered for attribution of ownership purposes.[99] The FCC has concluded that the holder of a convertible interest has no control of the licensee if the right to convert the interest is beyond the holder's power.[100] Thus, the FCC has deemed the holder of a convertible interest to have little control over the licensee, because the holder's threat to convert would be an empty one. A foreigner's convertible interest has concerned the FCC only if he could actually control the licensee.[101] Even that caveat ceases to make sense any longer in light of Congress's repeal in 1996 of the restrictions on direct managerial control of radio licensees by foreign officers or directors.

The FCC addressed the applicability of section 310(b) to convertible debentures in three decisions involving the ultimately unsuccessful efforts of a Swiss national to revive the Data Transmission Company (Datran), a radio common carrier licensee, with increasingly large infusions of various forms of equity and debt.[102]

---

97. 47 C.F.R. § 73.3555, note 2(f); *see also Attribution Order*, 97 F.C.C.2d at 1020–21 ¶ 45–47.

98. *Wilner & Scheiner*, 103 F.C.C.2d at 520–24 ¶ 16–22.

99. *See Attribution Order*, 97 F.C.C.2d at 1021–22 ¶ 48.

100. *Id.* at 1021 ¶¶ 46–48; *see also* William S. Paley, 61 Rad. Reg. 2d (P & F) 413, 415 (1986), *recons. denied*, 62 Rad. Reg. 2d (P & F) 852 (1987); Gavillet, Foehrkolb & Wu, *supra* note 19, at 27.

101. Channel 31, Inc., 45 Rad. Reg. 2d (P & F) 420, 421–23 (1979); *Datran I*, 44 F.C.C.2d at 935; *Datran II*, 52 F.C.C.2d at 439; *Datran III*, 59 F.C.C.2d at 909; Gavillet, Foehrkolb & Wu, *supra* note 19, at 27.

102. *Datran I*, 44 F.C.C.2d at 935; *Datran II*, 52 F.C.C.2d at 439; *Datran III*, 59

The cases established that, even before the 1996 amendments to section 310(b), a radio licensee issuing a foreigner debentures that were nominally convertible to the voting stock of the licensee (or its parent) that exceeded the statutory limits did not violate section 310(b), provided that the debentures contained a restriction forbidding their conversion if conversion would create foreign-owned shares in excess of the statutory maximums.[103]

In *Datran II,* the FCC considered the Swiss national's ownership of convertible debentures, which, if exercised, would increase his ownership to over 20 percent.[104] But the debentures contained restrictions on conversions that would result in ownership exceeding the 20 percent benchmark, and the FCC approved the proposed investment.[105]

Datran III, however, stands for the proposition that the use of debt instruments such as convertible debentures does not automatically immunize a transaction from a determination that a foreign entity has obtained de facto control over the parent of a licensee in violation of section 310(b)(4). In *Datran III,* the FCC was troubled by the total configuration of foreign debt and equity interests and, in particular, by certain "equity" features of the debt. By the time of the agency's decision, the total foreign investment (both debt and equity) in Datran had risen to thirty times the equity investment of the nominally controlling party, and the foreign investor was the only conceivable source of future investment. Most important, the convertible debentures and other debt instruments carried severe restrictions on the management and operation of the licensee. Thus, the licensee and parent required the foreign investor's approval to sell the licensee's common stock or assets, merge it with another company, or permit it to guarantee loans or purchase the stock of other companies. To the FCC, that configuration of foreign-held equity and debt raised a serious question as to whether the foreign investor had obtained de facto control over the parent corporation.[106] Before 1996, the case

---

F.C.C.2d at 909 ¶¶ 4–5.

103. Thus, it is also clear that convertible debentures do not qualify for purposes of section 310(b) as "capital stock owned of record or voted."

104. *Datran II,* 52 F.C.C.2d at 439.

105. *Id.*

106. *Id.* at 910–12. The issue was never finally decided because Datran went bankrupt, mooting the case.

indicated that the FCC would take a harder look at transactions where the total foreign investment was large relative to that of Americans, where the foreign-held debt instruments gave some measure of control over the parent or licensee, or where factors suggested that debt instruments were in fact equity investments.[107] Since the enactment of the Telecommunications Act of 1996, however, the case has continued relevance only with respect to the last of those three principles.

Debt and Leases

The FCC has not construed section 310(b) to cover debt supplied by foreign creditors. In its 1985 declaratory ruling on foreign ownership, the FCC stated: "Unlike limited partners, creditors do not possess either an ownership or voting interest over the licensee and consequently the direct restrictions embodied in section 310(b) are not applicable to debt interests."[108] Nonetheless, the FCC will scrutinize debt financing if it suspects that such financing is a sham intended to conceal what in effect is equity ownership.[109] The FCC's ownership attribution rules similarly concluded that both leases and lease-backs were not cognizable ownership interests: "There is no direct influence or control which pertains to them, and any indirect influence or control, if it occurred, would be too irregular and involve too many other factors for the FCC to oversee."[110]

The FCC's tolerance of debt and lease-backs has created an attractive means for structuring acquisitions by foreigners. The typical corporate loan agreement will have not only elaborate representations and warranties, but also remedies for the creditor in the event that the debtor fails to meet certain objectively specified financial criteria. Those remedies sometimes include the right to name one or more members to the debtor's board. Similarly, commercial leases (retail leases, for example) may have a profit pass-through provision pursuant to which rental payments include a component that varies

---

107. Gavillet, Foehrkolb & Wu, *supra* note 19, at 28–31.

108. *Wilner & Scheiner*, 103 F.C.C.2d at 519 ¶ 14; Omninet Corp., 2 F.C.C. Rcd. 1734, 1735 n.15 (1987).

109. Fox Television Stations, Inc., Second Mem. Op. and Order, 11 F.C.C. Rcd. 5714, 5719 ¶ 14 (1996).

110. *Attribution Order*, 97 F.C.C.2d at 1022 ¶ 49.

with the lessee's sales or profits; such a provision causes the lease to resemble a kind of equity.

Debt also may be relevant to the availability of a waiver under section 310(b)(4). If the licensee is already heavily leveraged, then a foreign investor may have a stronger argument that its infusion of equity capital will serve the public interest, even if that investment exceeds the statutory benchmark. Thus, the FCC has stated that it "has the discretion to consider . . . debt transactions . . . in evaluating whether to grant an exemption from a strict application of the statutory benchmarks contained in section 310(b)(4) in a specific factual situation where such an exemption would further the public interest."[111] The FCC takes an especially close look at debt financing when the debt is a substantial part of the licensee's capitalization. In *Pan Pacific Television, Inc.*, for example, the FCC designated for an evidentiary hearing the question of whether unlawful foreign control existed because a Taiwanese citizen interested in providing Chinese-language programming to an applicant for a television license near San Francisco had helped to secure financing and equipment and otherwise had helped with the station's construction and operation.[112]

Partnerships

Section 310(b) does not mention partnerships. The FCC nonetheless stated in 1984 that each limitation contained in section 310(b) "applies equally to all financial interests in all business forms of licensees."[113] No general partner may be a foreigner, the representative of a foreigner, a foreign government or the representative of a foreign government, or any corporation organized under the laws of a foreign country.[114]

The FCC treats limited partnerships the same as corporations for purposes of the foreign ownership restrictions.[115] Excluding limited partnership interests would allow easy circumvention of the restrictions.[116] Also, although limited partners may lack control of the

---

111. *Wilner & Scheiner*, 103 F.C.C.2d at 519 n.38.

112. 3 F.C.C. Rcd. 6629, 6636 ¶ 37 (1988).

113. *Attribution Order*, 97 F.C.C.2d at 1009 ¶ 22.

114. Algreg Cellular Engineering, 7 F.C.C. Rcd. 8686 (1992); Great Western Cellular Partners, 8 F.C.C. Rcd. 3222 ¶ 2 (1993).

115. *Wilner & Scheiner*, 103 F.C.C.2d at 516 ¶ 9.

116. *Id.* at 515 ¶ 7.

partnership business, the FCC has held that the foreign ownership restrictions were designed to protect the United States from foreign *influence* as well as foreign control in the field of broadcasting.[117] (Again, the repeal in 1996 of the restrictions on foreign officers and directors calls that reasoning into question.) A limited partner may own up to 20 percent of the partnership's total equity.

Partners as Officers or Directors. The FCC has held that, for purposes of sections 310(b)(3) and (4), partners have sufficient indicia of control to be treated as officers or directors of a corporation.[118] Interpreting section 310(b)(3), the FCC in *Jireh's Broadcasting* denied a radio license to a company with one Canadian partner.[119] The FCC has applied that analysis to limited partnerships as well as general partnerships[120] because it reasoned that limited partners have the same authority to bind a company as do general partners.[121] After all, a limited partner is "limited" only in the sense that he is protected from total liability in the event the enterprise becomes insolvent. If the limited partner has been sufficiently insulated from partnership affairs, however, the FCC has applied the multiplier to his interest in the parent corporation of the licensee.[122] In at least one instance, the FCC denied a license where a parent company's limited partners lacked such insulation.[123]

The effect of the 1996 legislation has been to overrule all those agency decisions to the extent that they implicate direct involvement of the foreign limited partner in the licensee's management (as opposed to the foreigner's exercise of voting rights). It is debatable, however, whether that change in law has much practical effect for reasons that have nothing to do with telecommunications regulation per se: Presumably the limited partner would risk losing his limited liability if he were to participate actively in the management of the licensee.

---

117. *Id.* at 517 ¶ 11.

118. Delta Cellular Partners, 5 F.C.C. Rcd. 5525 ¶ 4 (1990); Addison Broadcasting Co., 2 F.C.C. Rcd. 6357 ¶ 3 (1987).

119. 5 F.C.C. Rcd. at 3308 ¶¶ 3–4.

120. *See, e.g.,* Cellwave Tel. Servs. L.P. *v.* FCC, 30 F.3d 1533 (D.C. Cir. 1994).

121. *Reconsideration Order,* 1 F.C.C. Rcd. at 14 ¶ 19.

122. *Wilner & Scheiner,* 103 F.C.C.2d at 522 ¶ 20. For a full analysis and examples, see Gavillet, Foehrkolb & Wu, *supra* note 19, at 32–34.

123. *See* Catherine L. Waddill, 8 F.C.C. Rcd. 2169, 2169 ¶¶ 4–5 (1993).

Partners as Stockholders. For purposes of section 310(b) the FCC has considered limited partners to be "stockholders" whose ownership interests are determined by equity contribution.[124] In calculating those limited partner ownership interests, the FCC employs the same multiplier that it uses for attribution purposes. The FCC has explained its process as follows:

> [A]ssume that (1) Company A, a domestically organized limited partnership, holds 22 percent ownership interest in the licensee; (2) [AI], a natural person who is not a citizen of the United States, is a limited partner with a 25 percent ownership interest in Company A; and (3) all other direct or indirect interests in the licensee are held by United States citizens. If [AI]'s interest is adequately insulated, under the "multiplier" approach, [AI] would be attributed with 5.5 percent ownership interest in the licensee. If [AI] is not insulated from active participation in the business, the multiplier would not be used, and [AI] would be attributed with 22 percent ownership interest in the licensee, thereby violating the ownership benchmark established in section 310(b)(3).[125]

"Where applicable," the FCC has noted, "the 'multiplier' is utilized in any link on the ownership chain to determine the amount of alien ownership or voting interests in the licensee."[126] The FCC takes the same general approach to partnership interests under section 310(b)(4).[127]

Irrevocable Trusts

The FCC has ruled that section 310(b) covers beneficiaries of irrevocable trusts.[128] In *PrimeMedia Broadcasting, Inc.*, the FCC considered an irrevocable trust of which a foreigner was the benefi-

124. *Wilner & Scheiner*, 103 F.C.C.2d at 520 ¶ 16; Continental Cellular, 5 F.C.C. Rcd. 691, 692 n.8 (1990).

125. *Wilner & Scheiner*, 103 F.C.C.2d at 523 n.51.

126. *Id.*

127. Gavillet, Foehrkolb & Wu, *supra* note 19, at 33.

128. PrimeMedia Broadcasting, Inc, 3 F.C.C. Rcd. 4293, 4295 ¶ 11 (1988); Teleport Transmission Holdings, 8 F.C.C. Rcd. 3063, 3064 ¶ 8 (1993).

ciary.[129] The trustee was to exercise legal title and control over the stock in the licensee corporation for the foreigner's benefit, and the foreigner was to retain the equitable interest in the stock itself. The FCC noted that, although *PrimeMedia* presented a case of first impression as to section 310(b)(3), the agency had previously considered the insulating nature of trusts in other areas of its rules.[130] The FCC then stated that Congress did not intend to exclude equitable ownership interests that do not confer actual control. Notwithstanding that a trust divorces control and the indicia of ownership from the residual benefit of receiving profits, the FCC ruled that irrevocable trusts are not exempt from section 310(b)(3).[131] That odd result would seem to survive the 1996 repeal of the restrictions on foreign officers and directors, because that legislation did not liberalize the restrictions on a foreigner's ability to be a passive recipient of the profits of a radio licensee.

FOREIGN CONTROL

The Communications Act does not define "control" for purposes of section 310(b). The FCC has written its own broad definition. In *Powell Crosley, Jr.*, the FCC stated in 1945 that a "realistic definition of the word 'control' includes any act which vests in a new entity or individual the right to determine the manner or means of operating the licensee and determining the policy that the licensee will pursue."[132] The FCC has conveniently asserted that "corporate control varies from case to case and cannot be precisely defined," for it involves facts that are case-specific.[133] "Control" is thus a phantom that can take many forms: actual (de facto) or legal (de jure), direct or indirect, negative or affirmative.[134] The paucity of standards and

---

129. 3 F.C.C. Rcd. at 4292 ¶ 7.

130. *Id.* (citing Rust Craft Broadcasting Co., 68 F.C.C.2d 1013, 1017–19 ¶¶ 9–11 (1978); *Attribution Order*, 97 F.C.C.2d at 1023–24 ¶¶ 53–56).

131. *Id.* at 4295 ¶ 11.

132. 11 F.C.C. 3, 20 (1945); *see also* Metromedia, Inc., 98 F.C.C.2d 300, 304 ¶ 8 (interpreting section 309), *recons. denied*, 56 Rad. Reg. 2d (P & F) 1198 (1984), *appeal dismissed sub nom.* California Ass'n of the Physically Handicapped *v.* FCC, 778 F.2d 823 (D.C. Cir. 1985); WHDH, Inc., 17 F.C.C.2d 856, 863 ¶ 17 (1969).

133. Storer Comm., Inc., 101 F.C.C.2d 434, 441 ¶ 22 (1985); *see also Datran III*, 59 F.C.C.2d at 910 ¶ 6; *Datran I*, 4 F.C.C.2d at 935.

134. *Fox,* 77 Rad. Reg. 2d (P & F) at 1070 ¶ 150; *Seven Hills*, 2 F.C.C. Rcd. at

the breadth of the FCC's assertion of authority to find "control" has produced uncertainty and litigation.

The FCC will examine any affiliation to determine whether it gives a foreigner de facto or de jure control of a licensee.[135] The repeal in 1996 of the restrictions on foreign officers and directors has implicitly repealed many of the FCC's prior rulings on foreign control, but not all of them. It is therefore necessary to survey the range of decisions to determine which remain good law. De jure control is typically determined by whether a shareholder owns more than 50 percent of the voting shares of a corporation.[136] De facto control has not been so precisely defined.[137] A minority shareholder, for example, "controls" a corporation only if he has the power to "dominate the management of corporate affairs."[138] The FCC has acknowledged that influence and control are different.[139] The agency defends its de facto control standard as being necessary because operational reality may diverge from legal technicality.[140] As a result, "even in instances in which the technical statutory requirements are met, the FCC may still find that aliens exercise an effective control over the operations of a station that is contrary to statutory policy."[141]

Foreign Consultants

Concern over de facto foreign control has arisen in broadcast cases in which the licensee has hired a foreigner to consult on management or programming. In *Spanish International Communications Corporation*, the FCC designated seven television licenses for a hearing on the issue of whether the licensees were under the de facto control of Mexican interests.[142] In such cases the FCC has been satisfied that de

---

6878 ¶ 38; WWIZ, Inc., 2 Rad. Reg. 2d (P & F) 169, 191 ¶ 3 (1964), *aff'd sub nom.* Lorain Journal Co. v. FCC, 351 F.2d 824 (D.C. Cir. 1965), *cert. denied*, 383 U.S. 967 (1966).

135. *Fox*, 77 Rad. Reg. 2d (P & F) at 1070–73 ¶¶ 149–65.

136. *Id.* at 1070 ¶ 151.

137. Univision Holdings, Inc., 7 F.C.C. Rcd. 6672, 6675 ¶ 15 (1992).

138. *MCI*, 9 F.C.C. Rcd. at 3961 ¶ 11.

139. WWOR-TV, 6 F.C.C. Rcd. 193, 199–200 ¶¶ 13–14 (1990).

140. *Telemundo*, 802 F.2d at 513.

141. *Id.*; Channel 31, Inc., Debtor-in-Possession, 45 Rad. Reg. 2d (P & F) 420, 421 (1979).

142. *SICC*, 2 F.C.C. Rcd. at 3336 ¶¶ 1–2.

facto foreign control has been corrected when the licensee has cancelled the consulting contracts. That case law cannot survive the 1996
amendments to section 310(b). If foreign officers and directors may
lawfully control the management and programming of a radio
licensee, then there could be no rational basis for the FCC to restrict
that licensee's ability to hire a foreign consultant.

Special Covenants

In *McCaw Cellular*, the FCC found that covenants giving a minority
shareholder the power to block certain major transactions do not, by
themselves, constitute corporate control under the Communications
Act.[143] Restrictions that limit the otherwise normal financial preroga-
tives of a board of directors need not represent control.[144] A minority
shareholder's right to prevent any change in a company's bylaws or
charter does not constitute control.[145] Neither does requiring a minority shareholder's consent before the corporation may amend its bylaws
or articles of incorporation. The FCC has deemed such requirements
to be protective of the minority shareholder's investment, as they prevent the dilution of its stock holdings.[146] The FCC has taken the same
view of steps taken by creditors to force the sale of a licensee.[147]
Simply entering into an affiliation agreement with a network that is
also an equity partner does not establish control either.[148]

The FCC nonetheless has found that special covenants do give
de facto control under numerous circumstances. In *Stereo Broadcasters, Inc.*,[149] the FCC found that control also included the ability to
direct the applicant's finances, personnel, and programming. Where
a foreign minority stockholder has full veto power on all or most
issues (including financial plans, business plans, and other daily

---

143. 4 F.C.C. Rcd. at 3789 ¶¶ 37–38; *see also* News Int'l, plc, 97 F.C.C.2d 349
(1984).

144. Flathead Valley Broadcasters, 5 Rad. Reg. 2d (P & F) 74, 76 ¶ 6 (Rev. Bd.
1965).

145. *News Int'l*, at 357–58 ¶¶ 19–22.

146. *McCaw Cellular*, 4 F.C.C. Rcd. at 3789 ¶ 39.

147. Turner Comm. Corp., 68 F.C.C.2d 559, 562–63 ¶¶ 9–11 (1978).

148. BBC License Subsidiary L.P., 10 F.C.C. Rcd. 7926, 7931–32 ¶¶ 35–36
(1995); NBC, Inc., 6 F.C.C. Rcd. 4882, 4883 ¶ 4 (1991).

149. 87 F.C.C.2d 87, 88–89 ¶¶ 3–33 (1981) (interpreting section 310(d)), *recons.
denied*, 50 Rad. Reg. 2d (P & F) 1346 (1982).

operations of the licensee), the FCC would likely find the foreigner to have de facto control.[150] After the 1996 amendments to section 310(b), the creation of de facto control through special covenants would be problematic only if the foreign shareholder were attempting to exercise control through the proxy process. Because section 310(b) allows the foreign investor to exercise control directly as an officer or director of the radio licensee, it would be naïve for a foreign investor who was inclined to participate in the licensee's management to fail to exercise de facto control in the statutorily permitted manner rather than the manner that has raised the FCC's concerns before 1996.

Limited Partnerships

Before 1996 the FCC indicated that foreign limited partners could possess control of a licensee in violation of section 310(b). In *Sacramento RSA Ltd. Partnership*, the FCC reconsidered whether a limited partnership's application for a cellular license violated section 310(b)(3).[151] The FCC held that it would have to examine whether a foreign limited partner had control comparable to that of an officer or director of a corporation. In particular, the FCC indicated that it would scrutinize the partnership agreement to ascertain the degree to which the foreign limited partner was insulated from management.[152] Given that the 1996 legislation removed all restrictions on foreigners participating as officers or directors of the radio licensee, Congress implicitly repealed that body of agency law. Furthermore, as noted earlier, the law of limited partnerships imposes the binding constraint on a foreign limited partner's participation in the management of the radio licensee in such a situation.

---

150. Satellite Transmission & Reception Specialist Co. and Transmission Operator Provided Systems, Inc., DA 90–927 (July 13, 1990) (*STARS/TOPS*) (FCC found de facto control by foreign-controlled company where CEO of parent was also CEO of licensee and contract provisions between the parties provided for exclusive use of licensed earth stations by the parent), *discussed in* Gavillet, Foehrkolb & Wu, *supra* note 19, at 45; *Pan Pacific*, 3 F.C.C. Rcd. at 6636 ¶ 37 (substantial involvement by Taiwanese minority shareholder in financial affairs of applicant for television license held to indicate de facto control).

151. 9 F.C.C. Rcd. 3182, 3183 ¶ 7 (1994), *aff'd*, Sacramento RSA L.P. *v.* FCC, 1995 U.S. App. LEXIS 15190 (D.C. Cir. 1995) (unpublished decision).

152. 9 F.C.C. Rcd. at 3183 ¶ 7.

CONCLUSION

Every strategy that a foreign investor must adopt to avoid contravening section 310(b) forces him to run a gauntlet of transactions costs. The strategies that an investor may adopt to increase its investment in a radio licensee subject to section 310(b) require extensive planning by high-priced lawyers. The agreements must have the FCC's prior approval and sometimes entail hearings and appeals. And all the investor's potential competitors will be alerted to come have their say in the matter, too. Ultimately, no strategy that an investor may adopt in maximizing its interest in a radio licensee enables it to minimize the agency costs of monitoring the licensee. Control of the licensee itself remains separated from ownership. For every investor who decides to spend the resources necessary to invest in a radio licensee subject to section 310(b), many other investors surely decide not to bother.

A defender of section 310(b) might argue that, while the statute does impose those costs, it provides the desired benefit for the public interest—protecting national security by ensuring that foreign investors do not control radio licensees. Like the costs, our hypothetical defender might argue, the magnitude of that benefit is inestimable. Congress has simply decided that an inestimable benefit outweighs an inestimable cost.

There are several problems with that analysis. As an instrument of national security, section 310(b) is ineffectual. Chapter 2 documented the historical evidence on that score. The point is further supported by the gaping loopholes in section 310(b)'s coverage, as chapter 3 showed. Section 310(b) allows foreigners to use common carrier and most private carrier radio networks as the customer of a licensee. It does not prevent foreigners from buying cable television systems or wireline telephone companies. Yet propagandists, spies, and saboteurs have not inflicted any known damage by their unrestricted use of those media. Indeed, one must wonder whether a true enemy of the United States would even bother to get a radio license from the FCC before using the airwaves. The vast majority of foreign investors surely pose no danger to the national security whatsoever. If a tiny minority of them do, the president possesses ample means, detailed in chapter 3, to deal with the problem at much lower cost to the nation. In that respect, then, section 310(b) imposes costs without deriving any real benefit whatsoever.

Despite such a burdensome statute, some foreigners do make substantial investments in U.S. radio licensees. And, just as an infinite number of monkeys typing on an infinite number of typewriters would eventually produce *War and Peace*, enough lawyers could invent ways around section 310(b) to keep the FCC busy for generations to come. The statute's principal effect, then, is to prevent foreigners from investing in American telecommunications companies in the manner that minimizes transactions costs and agency costs.

5

Foreign Direct Investment in the United States

SUBSTANTIAL BENEFITS flow from foreign direct investment. Foreign direct investment in U.S. telecommunications service providers can reduce their cost of capital and increase their access to new technologies and management techniques. Those two factors in turn are likely to heighten competition in the U.S. market and thus benefit all consumers of telecommunications services. With greater domestic competition and the increased likelihood of access to foreign markets that would result from a policy of permitting more foreign investment in U.S. telecommunications, U.S. firms will be better conditioned to compete in the delivery of telecommunications services on a global scale. The few significant foreign investments that have been made in U.S. telecommunications service providers illustrate the benefits of foreign direct investment and suggest the potential gains from U.S. policies that would be more hospitable to such investment.

THE CAUSES OF FOREIGN DIRECT INVESTMENT

Firms invest abroad to obtain competitive advantages stemming from technological knowledge, management skills, and vertical integration of suppliers.[1] The need to control the activities of firms operating in other countries is, according to the influential theory of Steven Hymer, the driving force behind foreign direct investment.[2] Hymer theorized that the advantage of foreign direct investment may arise from imperfect competition. A firm might have special access to information about production and a means of capturing increasing returns to scale.[3] For example, a foreign firm with exclusive access to valuable information might outbid a domestic firm for land and plant in an industry.

Extending that analysis, Edward Graham and Paul Krugman have suggested that a foreign firm might possess some firm-specific knowledge or assets that enable it to manage the U.S. firm more ably than its American managers.[4] In a related vein, Robert Lipsey has claimed that overseas investment enables a firm to raise the value of its firm-specific assets—for example, its technologies, patents, or unique skills—by extending the range of markets that it can serve.[5] Even modern game theory supports the competitive advantage framework. Taking the reactions of a firm's competitors into account, Graham has developed a model to explain the strategic decision making of multinational enterprises.[6] Under the assumptions of two monopolistic firms facing constant marginal costs and possessing complete information, he found, not surprisingly, that the lower a firm's relative marginal costs are to its foreign rival's and the smaller the relative size of a firm's market is to the foreign market, the greater the probability that the firm will choose to enter the rival's market.

Other explanations exist for foreign direct investment that are

1. *See* James R. Markusen, *The Boundaries of Multinational Enterprises and the Theory of International Trade*, 9 J. ECON. PERSPECTIVES 169 (1995).

2. STEVEN H. HYMER, THE INTERNATIONAL OPERATIONS OF NATIONAL FIRMS: A STUDY OF DIRECT FOREIGN INVESTMENT (MIT Press 1976).

3. *Id.* at 25–30.

4. *See* EDWARD M. GRAHAM & PAUL R. KRUGMAN, FOREIGN DIRECT INVESTMENT IN THE UNITED STATES 35–36 (Institute for International Economics, 3d ed. 1995).

5. ROBERT LIPSEY, OUTBOUND DIRECT INVESTMENT AND THE U.S. ECONOMY 1 (National Bureau of Econ. Research Working Paper No. 4691, 1994).

6. Edward M. Graham, *Strategic Management and Transnational Firm Behavior: A Formal Approach, in* THE NATURE OF THE TRANSNATIONAL FIRM 1 (C. N. Pitelis & R. Sugden eds., Routledge 1991).

unrelated to Hymer's theory. For example, foreign direct investment may allow multinational enterprises to retain or increase world market share in the face of fluctuating exchange rates. Diversification of a multinational enterprise's portfolio reduces risks such that if one locale suffers a productivity shock, the firm can shift resources to a country where productivity is higher. Firms may also invest abroad to circumvent trade barriers or to gain proximity to foreign consumers.

FOREIGN DIRECT INVESTMENT
IN THE UNITED STATES

By the end of 1993, foreign direct investment in the United States (FDIUS) was $445.2 billion.[7] Despite that high level, foreign investment slowed between the 1980s and the early 1990s, and there has been a trend of U.S. affiliates' incurring losses or paying dividends to foreign parents in excess of current earnings. In 1995, however, there was evidence of a resurgence in FDIUS.[8]

What caused the surge in FDIUS during the 1980s? A popular explanation is the increased competitiveness of companies in Europe, Japan, and Canada relative to their U.S. rivals.[9] There are at least four other explanations. One is that the share of production capacity located abroad increases as exchange rate volatility rises.[10] A second theory links the surge in FDIUS to the value of the dollar: As the dollar falls, exports to the United States slow, and U.S. pro-

7. U.S. DEP'T OF COMMERCE, BUREAU OF ECONOMIC ANALYSIS, 74 SURVEY OF CURRENT BUSINESS, no. 8, at 98 (Aug. 1994). The Bureau of Economic Analysis measurement records historical costs rather than market value (estimated to be $745.6 billion at the end of 1993) and includes earnings retained by subsidiaries in the United States and transfers of funds from parent firms. The figure omits subsidiaries' investments financed from borrowed funds within the United States or from a third country.

8. Bernard Wysocki, *Foreigners Find US a Good Place to Invest*, WALL ST. J., Aug. 7, 1995, at A1.

9. EDWARD M. GRAHAM & PAUL R. KRUGMAN, FOREIGN DIRECT INVESTMENT IN THE UNITED STATES 44 (Institute for International Economics, 2d ed. 1991) [hereinafter GRAHAM & KRUGMAN 2D ED.].

10. LEE GOLDBERG & CHARLES KOLSTAD, FOREIGN DIRECT INVESTMENT, EXCHANGE RATE VARIABILITY AND DEMAND UNCERTAINTY 15 (National Bureau of Econ. Research Working Paper No. 4815, 1994).

ductive assets and U.S. labor become cheaper.[11] A third theory attributes the rise in FDIUS to foreign firms' lower cost of equity: When a foreign stock exchange discounts future earnings at a lower rate than the New York Stock Exchange, foreign firms can offer higher bids than their U.S. rivals.[12] A fourth theory is that FDIUS is a means to evade actual or potential U.S. tariffs and other trade barriers, as may be the case with the production of Japanese automobiles and color televisions in the United States.[13]

Foreign-owned firms currently represent a substantial share of total U.S. manufacturing production. The four major industry groups in the manufacturing sector that absorbed the greatest amount of FDIUS are chemicals, industrial machinery and equipment, electronics, and transportation equipment.[14] FDIUS has been mainly characterized by acquisitions of existing plants (88 percent in 1990) rather than "greenfield" investment in the construction of new sites.[15] The leading source countries of FDIUS in 1993 were Japan (21 percent), the United Kingdom (21 percent), and the Netherlands (15 percent).[16]

Job Creation

By 1988 FDIUS provided nearly 9 percent of all U.S. manufacturing jobs.[17] It is difficult, however, to credit foreign firms with the actual creation of jobs in the U.S. economy. FDIUS has little effect on the number of local jobs but instead represents the transfer of ownership from the U.S. firm to a foreign firm.[18] Graham and Krugman have argued that the increased demand for labor as a

---

11. Cletus Coughlin, *Foreign-Owned Companies in the United States: Malign or Benign?*, 74 FED. RESERVE BANK ST. LOUIS BULL. 17, 24 (1992).

12. Robert Laster & Martin McCauley, *Making Sense of the Profits of Foreign Firms in the U.S.*, 19 FED. RESERVE BANK N.Y. Q. REV. 44, 47 (1994).

13. Coughlin, *supra* note 11.

14. GRAHAM & KRUGMAN 2D ED., *supra* note 9, at 42.

15. *Id.* at 24.

16. *Id.* at 22.

17. Coughlin, *supra* note 11, at 19.

18. Norman Glickman & Dennis Woodward, *Industry Location and Public Policy, in* REGIONAL AND LOCAL DETERMINANTS OF FOREIGN FIRM LOCATION IN THE UNITED STATES 190, 191 (Henry W. Herzog, Jr. & Alan M. Schlottman eds., University of Tennessee Press 1991).

result of foreign direct investment only influences employment levels in the short run, and they have concluded that the net impact of FDIUS on the number of U.S. jobs is negligible in the long run.[19] Rachel McCulloch has argued that when jobs abroad are sacrificed as a result of FDIUS, global demand falls enough to offset the gains in U.S. jobs in the targeted industry with job losses in other sectors of the domestic economy, and that those losses are accelerated by foreign firms' propensity to source from abroad.[20]

If there is little evidence that FDIUS creates jobs in the United States, it is even more questionable that FDIUS *destroys* jobs in the United States. Nonetheless, Clyde Prestowitz, founder of the Economic Strategy Institute, has envisioned that massive job losses in the U.S. airline industry would result from British investors' being allowed access.[21] He has reasoned as follows. British Air's partial acquisition of USAir would result in the loss of 3,500 jobs to the U.S. economy.[22] Without corresponding access to the British market, U.S. carriers could not provide one-stop flights from midsize markets in the United States to London. As British Air gained more of the transatlantic market, it might transfer U.S. jobs to Britain in large numbers. Prestowitz, however, neglected to address, among other countervailing factors, the extent to which British investors would reinvest in their American subsidiaries the profits earned in the United States.

Another concern that FDIUS arouses is the "headquarter effects," or the extent to which foreign owners will shift R&D activities outside the United States. To test that claim, Cletus Coughlin compared R&D expenditure per worker in the manufacturing sector for U.S.-based and foreign-based multinationals operating in the United States and found that U.S. firms spent only slightly more per worker on R&D ($4,640 versus $3,780) than their foreign-based counterparts.[23] If foreign owners had actually been shipping R&D jobs overseas, then one would expect to have seen a larger divide

---

19. GRAHAM & KRUGMAN, *supra* note 4, at 60–62.

20. Rachel McCulloch, *Foreign Investment in the U.S.*, 30 FIN. & DEV. 13, 15 (1993).

21. CLYDE V. PRESTOWITZ, JR., THE FUTURE OF THE AIRLINE INDUSTRY (Economic Strategy Institute 1993).

22. *Id.* at 34.

23. Coughlin, *supra* note 11, at 27.

between what foreign-owned and American-owned manufacturing firms had spent in the United States. Coughlin's result suggests that R&D jobs do *not* go overseas as foreigners increase their ownership of U.S. assets.

Wages

A major concern to policymakers is how foreign affiliates treat U.S. workers. Statistics show similar value added per worker and compensation between foreign affiliates and U.S. firms within the same industry.[24] There is even some evidence that foreign firms pay their U.S. workers more handsomely than do American-owned firms. One study found that workers of foreign affiliates in chemicals and transportation equipment earned 20 percent more per hour than the average U.S. worker in manufacturing in 1992,[25] although it must be noted that workers employed by U.S.-based firms in those industries also earn higher wages than the average manufacturing workers.

By measuring pay in terms of compensation per employee including employee benefits, another study found that workers of foreign-owned affiliates earned $5,300 more than their counterparts employed by U.S.-owned firms ($38,300 compared with $33,000).[26] Sixty percent of the difference was due to the mix of industries—if U.S.-based firms invested in the same industries as U.S. affiliates of foreign firms, the wage disparity would be greatly reduced. In an effort to explain the wage disparity within an industry, the study used regression analysis to control for plant size and capital intensity. Foreign ownership could not explain the change in wages, and it should therefore be *associated*, but not credited, with higher wages.

Productivity

Foreign investment provides both direct and indirect benefits to the host countries with regard to productivity. Foreign direct investment

---

24. GRAHAM & KRUGMAN, *supra* note 4, at 71–72.

25. JAN ONDRICH & MICHAEL WASYLENKO, FOREIGN DIRECT INVESTMENT IN THE UNITED STATES 162 (Upjohn Institute 1993).

26. Ned Howenstice & William J. Zeile, *Characteristics of U.S. Manufacturing Establishments*, 74 SURVEY OF CURRENT BUSINESS, no. 8, at 34, 45 (Aug. 1994).

directly increases productivity by providing host countries with access to modern technology that they cannot provide themselves.[27] Indirectly, foreign direct investment boosts productivity through "intraindustry" and "interindustry" spillovers. Intraindustry spillovers involve effects that influence the efficiency of the host country's existing producers. Interindustry spillovers benefit local suppliers and customers. Such spillovers include: increased competition that forces existing inefficient firms to raise investment in physical and human capital; advanced training techniques for labor and management that diffuse throughout the general economy; and sophisticated techniques of intermediate supply in areas such as quality control, reliability, and speed of delivery. Magnus Blomström has defended the "spillover benefit hypothesis" with evidence that productivity levels of domestic firms increase with the foreign subsidiaries' share of the market.[28]

When capital stocks increase, productivity and wages should rise. Because foreign firms are credited with increasing the host country's capital stock, should they also be credited with increasing productivity and wages for American workers? Commerce Department data reveal that foreign firms in the United States had higher levels of productivity than their domestic counterparts in 1990. Labor productivity (as measured by the value added per production hour) was $22 higher in foreign-owned manufacturing firms than in their U.S.-owned counterparts.[29] Similar to the breakdown for wage differentials, 70 percent of the difference was due to the effects of industry mix and 20 percent was due to the effects within industry. After controlling for plant size, capital intensity, and employee skill level, the study concluded that the difference in productivity due to foreign ownership was insignificant.

Output

Some proponents of FDIUS oppose states offering incentive packages to lure foreign direct investment. They argue that such packages transfer wealth from state taxpayers to the foreign firms and that

---

27. MAGNUS BLOMSTRÖM, HOST COUNTRY BENEFITS OF FOREIGN INVESTMENT, (National Bureau of Econ. Research Working Paper No. 3615, 1991).

28. *Id.*

29. Howenstice & Zeile, *supra* note 26, at 42.

such transfers entail a loss in U.S. output because of the inefficiency of redistributing income through the political process.[30]

Apart from the question of the possible inefficiency of such incentive packages, one would expect FDIUS to expand U.S. output by augmenting capital shortfalls and raising the productivity of U.S. workers. Modern capital stocks elicit technological advancements and make the United States more competitive in the global economy. One study tested the proposition that foreign firms establish themselves in the United States to copy ideas and export them back to the parent company. The method of the study was to examine the receipts of royalties and licenses transferred between foreign parents and their U.S. affiliates. If foreign firms were copying U.S. ideas, then one would expect to see a net flow of funds from the foreign parents to the domestic affiliates. To the contrary, the study found that in 1990 U.S. affiliates paid six times as much on royalties as their foreign parents.[31] The study concluded that the transfer of technology moves from the parent to the U.S. affiliate.

Trade

The trade deficit for U.S. affiliates of foreign firms peaked at $95.4 billion in 1987 and has since declined to $90.6 billion in 1990 and $81.5 billion in 1992.[32] The common explanations for that deficit are the natural import bias of the types of firms targeted by foreigners (for example, wholesalers) and the tendency of affiliates to source, at least initially, from their foreign parents.

Graham and Krugman have argued that the latter factor is a small contributor to the trade deficit.[33] They examined exports per worker in both U.S. foreign affiliates and parent companies of U.S.-based multinational enterprises. In 1990 foreign multinational manufacturing firms imported approximately $21,000 of materials per worker versus only $12,000 per worker for domestically owned

---

30. Glickman & Woodward, *supra* note 18, at 201.

31. Coughlin, *supra* note 11, at 23.

32. William J. Zeile, *Foreign Direct Investment in the United States: 1992 Benchmark Survey Results,* 74 SURVEY OF CURRENT BUSINESS, no. 7, at 154 (July 1994); U.S. DEP'T OF COMMERCE, FOREIGN DIRECT INVESTMENT IN THE UNITED STATES: AN UPDATE 78 (Government Printing Office 1993).

33. GRAHAM & KRUGMAN, *supra* note 4, at 70.

firms.[34] Nonetheless, that pattern is understandable. New assembly initially requires imported inputs while foreign affiliates familiarize themselves with local suppliers. Over time, that dependency on imports diminishes. For example, the domestic content of Honda automobiles manufactured in the United States rose from an initial 30 percent to 60 percent by 1987.[35]

THE POTENTIAL BENEFITS OF FOREIGN DIRECT INVESTMENT IN U.S. TELECOMMUNICATIONS SERVICES

In addition to capturing some of the general benefits of foreign direct investment described above, the United States can gain in at least three specific ways from opening its telecommunications industry to greater foreign direct investment.

Increased Competition in U.S. Telecommunications Services

Foreign investment can increase competition in the market for telecommunications services in the United States and thus improve quality and decrease prices for American consumers. There can be little doubt, for example, that AT&T will face greater competition in its integration of its wireline long-distance facilities and its McCaw Cellular wireless facilities because of BT's investment in (and proposed merger with) MCI and because of Deutsche Telekom's and France Télécom's investment in Sprint (which, in turn, is a substantial holder of personal communications services (PCS) licenses through its WirelessCo venture with TCI and other cable television firms). Such competition would be one means to drive down the high price-cost margins, net of access charges, that Paul MacAvoy has argued exist in multiple segments of the long-distance market.[36] Likewise, one could imagine another foreign

---

34. *Id.*

35. *Id.* at 79 (citing GENERAL ACCOUNTING OFFICE, FOREIGN INVESTMENT: GROWING JAPANESE PRESENCE IN U.S. AUTO INDUSTRY (Government Printing Office 1988)).

36. PAUL W. MACAVOY, THE FAILURE OF ANTITRUST AND REGULATION TO ESTABLISH COMPETITION IN LONG-DISTANCE TELEPHONE SERVICES (MIT Press &

carrier's (such as Canada's Stentor, which already has an extensive fiber network in North America) investing in AirTouch or the spin-off of Sprint's cellular operations or the possible spinoff of the combined cellular operations of NYNEX and Bell Atlantic.

Also, as we discuss in greater detail later in this chapter, it took the managerial and financial backing of Australia's News Corporation to produce a viable fourth television network in the United States, the Fox Network. The incumbent networks sought to impede that competitive entry by alleging that Fox violated section 310(b).

Reduced Cost of Capital

Foreign direct investment increases the supply of capital in the United States. That influx decreases the cost of capital for U.S. tele-communications firms and thus enables them to fund greater levels of expansion than would be possible in the presence of a higher cost of capital.

Some argue that foreign direct investment is not needed in the U.S. telecommunications market because U.S. capital markets can accommodate all the debt or equity offerings that U.S. telecom-munications companies could possibly want to undertake. That argu-ment is familiar and ironic. When U.S. cable television firms sought in 1976 to have the FCC apply section 310(b) to their indus-try, they curiously argued, in the FCC's words, that the FCC's "failure to restrict alien ownership now may in fact encourage for-eign participation due to the industry's present financial plight."[37] In other words, foreigners who could engage in direct investment in U.S. cable systems might be willing to assume financial risks that American investors would refuse to bear at the same return on capital.

Four years later, when U.S. cable companies again tried and

AEI Press 1996); Paul W. MacAvoy, *Tacit Collusion Under Regulation in the Pricing of Interstate Long-Distance Telephone Services*, 4 J. ECON. & MGMT. STRATEGY 147 (1995).

37. Amendment of Parts 76 and 78 of the Commission's Rules to Adopt General Citizenship Requirements for Operation of Cable Television Systems and for Grant of Station Licenses in the Cable Television Relay Service, Report and Order, Dkt. No. 20621, 59 F.C.C.2d 723, 725 ¶ 6 (1976).

failed to impose foreign ownership restrictions on their industry, they made just the opposite argument: Foreign capital was unnecessary to fund their industry's growth.[38] Yet, by the late 1980s, the industry was highly leveraged and opposed the reimposition of rate regulation in 1992 in part because it would impair the ability of cable systems to service their massive debt. In other words, even *with* the unrestricted foreign direct investment that it so much wanted to prevent, the U.S. cable industry faced oppressively high capital costs.

Indeed, the unique institution of regulation in the United States may be a factor that, if not reformed over time, will increasingly constitute a risk for investment in U.S. telecommunications that requires a premium in the cost of capital relative to the cost of capital for telecommunications firms in other capitalist democracies (such as New Zealand, the United Kingdom, or Chile) that privatized their postal, telegraph, and telephone administrations but did not then emulate America's burdensome regulatory apparatus. Given the voluminous rules that the FCC issued to implement the Telecommunications Act of 1996, and given the fuel that the new statute and its regulations provided for litigation, it would be premature to conclude either that the U.S. telecommunication industry is now "deregulated" or that a risk premium for regulatory risk is no longer required in the cost of capital for U.S. telecommunications firms.

Positive Externalities in
Technology and Management

Foreign direct investment may generate beneficial spillovers for U.S. telecommunications firms. Those benefits consist of the transfer of new technology and management practices to U.S. firms and their workers. Americans may be accustomed to thinking that U.S. firms consistently are in the vanguard of new technologies. But that view assumes that the current pace of innovation in the U.S. market is independent of the threat of future foreign competition. As the

---

38. Amendment of Parts 76 and 78 of the Commission's Rules to Adopt General Citizenship Requirements for Operation of Cable Television Systems and for Grant of Station Licenses in the Cable Television Relay Service, Mem. Op., 77 F.C.C.2d 73, 75 ¶ 6 (1980).

technology developed by U.S. firms is diffused and used throughout the world, the gap in competitiveness between U.S. and foreign telecommunications firms will lessen.

Moreover, something that is unique to the U.S. telecommunications market—namely, the heavy hand of decades of FCC regulation and antitrust decrees—denies U.S. firms many of the economies of scope from research and development across multiple product lines that foreign firms can exploit. A principal goal of the Modification of Final Judgment, after all, was to limit collaboration between manufacturers of telecommunications equipment and the Bell operating companies, which build and manage most of the nation's public telecommunications networks.[39] The repeal of the manufacturing ban in the Telecommunications Act of 1996[40] creates the opportunity for foreign direct investment in the regional Bell operating companies by overseas equipment manufacturers (such as Siemens or Ericsson or Fujitsu), which could produce valuable technology transfers. If the market for knowledge is imperfect because of free riding, firms may be less inclined to sell or license their latest technology to firms in another country than to transfer it to them through foreign direct investment. Policymakers may overlook that benefit from foreign direct investment because spillovers are difficult to measure; but spillovers do matter, as strategic trade theorists have appropriately emphasized.

The Extent of Foreign Direct Investment in U.S. Telecommunications Services

The significant foreign direct investments in U.S. telecommunications companies have had to accommodate the constraints of section 310(b). When the structure of a particular deal required the FCC to consider waiving the 25 percent ownership benchmark in section

---

39. *See* Affidavit of Robert W. Lucky, Motion of Bell Atlantic Corporation, BellSouth Corporation, NYNEX Corporation, and Southwestern Bell Corporation to Vacate the Decree, United States *v.* Western Elec. Co., No. 82-0192 (D.D.C., filed July 6, 1994) (affidavit by Vice President of Applied Research, Bell Communications Research (Bellcore), describing how the manufacturing restriction in the MFJ has impeded Bellcore's ability to ensure network reliability and to participate in the development of new products).

40. 47 U.S.C. § 273.

310(b)(4), the agency generally decided that a waiver would serve the public interest if the benchmark would be only slightly exceeded or if the investment would facilitate inconsequential entry into the U.S. market. We examine now the seven largest foreign investments made in U.S. telecommunications.

THE BT-MCI MERGER

On November 3, 1996, British Telecommunications plc (BT) and MCI Communications Corporation (MCI) announced that they had signed a definitive merger agreement.[41] Valued at $21.61 billion,[42] the merger represented the largest takeover of a U.S. firm by a foreign concern, the largest telecommunications merger, and the third largest takeover in U.S. corporate history.[43] The combined company, to be called Concert, plc,[44] was touted as being "a new, high-growth, global communications powerhouse."[45]

BT is the United Kingdom's largest provider of communication services with more than twenty-two million customers.[46] BT provides a full range of services in the United Kingdom to business and residential customers, including local, long-distance, international,

---

41. Julia Flynn, Stanley Reed & Amy Barrett, *WORLDPHONE INC.*, Bus. Wk., Nov. 18, 1996, at 54; Sarah Cunningham & Carl Mortished, *BT Seals $20Bn Takeover of MCI to Become Global Force*, TIMES, Nov. 4, 1996; *MCI and BT Announce Largest International Merger in History*, MCI, Press Release (Nov. 3, 1996) <http://www.mci.com/aboutmci/news/content/bt.shtml> [hereinafter *Largest International Merger*].

42. Jon Van, *MCI–British Telecom Deal Sets Global Tone; The Largest-Ever Telecommunications Merger Reflects a "Superplayer" Trend*, CHI. TRIBUNE, Nov. 5, 1996, at B1. The actual value of the merger varies with the price of BT stock, because MCI shareholders will be paid a combination of stock and cash.

43. *Id.* Those statistics are based on stock values when the merger was announced.

44. Concert, plc is the new holding company that will serve as parent to both MCI and BT. Concert Communications Services is the joint venture, formed in 1993 by MCI and BT, that provides voice, data, and networking services to multinational customers.

45. Cunningham & Mortished, *supra* note 41 (quoting Sir Peter Bonfield, Chief Executive of BT).

46. *See* BRITISH TELECOMMUNICATIONS PLC, 1994 SEC FORM 20-F, at 10 (1994); *BT-MCI Fact Sheet*, CONCERT WEB SITE, <http://www.concert.com/deal/merger.htm.>.

wireless, paging, Internet, conferencing, and multimedia services.[47] The world's fourth largest telecommunications company, BT has annual revenue of more than $23 billion.[48] The company was formerly a government-controlled telecommunications monopoly, and it owned 97 percent of the country's local access lines as of 1994.[49] BT is the United Kingdom's principal provider of international facilities-based services. BT also offers a range of other telecommunications products and services, including private-line circuits, mobile communications products, and paging.[50] BT is now a public limited company in which the British government holds no more than 1.5 percent.[51]

MCI is a publicly traded U.S. corporation headquartered in Washington, D.C.[52] It is the second largest long-distance carrier in the United States and provides a variety of domestic and international voice and data communications services to twenty million customers.[53] MCI conducts most of its business through subsidiaries, which hold domestic common carrier microwave licenses, international facility authorizations, cable landing licenses, and other FCC licenses and authorizations.[54] With quarterly annualized revenue of more than $18 billion in 1996, MCI was at the time of the merger announcement the fourth largest and fastest growing major carrier of international traffic in the world.[55]

BT and MCI shareholders will respectively own 66 and 34 percent of Concert, plc. Concert, plc, will be incorporated in the United Kingdom with headquarters in London and Washington.[56] To preserve brand recognition, BT and MCI will operate under their original names in their home markets.[57] Under the terms of the

---

47. *Id.*

48. *Id.*

49. *Id.*

50. *Id.*

51. *Id.*

52. *See* MCI Communications Corp., 1995 SEC FORM 10-K, at 4 (1996); *BT-MCI Fact Sheet, supra* note 46.

53. *BT-MCI Fact Sheet, supra* note 46.

54. *Id.*

55. *Id.*

56. Cunningham & Mortished, *supra* note 41; *Largest International Merger, supra* note 41.

57. *A Telecom Titan on the Line*, FIN. TIMES, Nov. 4, 1996, at 23.

agreement, at closing, each MCI share will be converted into .54 of a Concert American Depository Share (ADS), equivalent to .54 of a BT ADS, plus $6.00 in cash. As part of the BT shareholder vote approving the merger, Concert will be authorized to purchase up to 10 percent of its outstanding shares after the closing.

Concert, plc, will have a fifteen-member board of directors comprising nine outside directors (five selected by BT and four selected by MCI) and six officers of the company, three each from BT and MCI.[58] The board will be jointly chaired by Sir Iain Vallance and Bert C. Roberts, the current chairmen of BT and MCI, respectively.[59] Sir Peter Bonfield of BT will be Concert's chief executive officer while Gerald H. Taylor, current MCI president and chief operating officer, will be its president and chief operating officer.[60] Operating as a holding company, Concert will consist of five primary business units: BT, MCI, Systems Integration, Operating Alliances, and International. It will offer an integrated set of products and services including local calling, long-distance, wireless, Internet/intranet, global communications, conferencing, systems integration/consulting, call center services, multimedia, and trading systems.[61]

Concert Communications Services will become part of BT-MCI's international unit. That entity is BT and MCI's joint venture, launched in July 1994, to provide global telecommunications services to multinational corporations at the time of BT's initial 20 percent investment in MCI. By the end of 1996, Concert Communications Services had $1.5 billion worth of contracts providing a range of advanced voice, data, and networking services to some 3,000 multinational customers.[62]

---

58. *Largest International Merger, supra* note 41.
59. *Id.*
60. *Id.*
61. *Id.*
62. Even with its initial success, Concert's revenues are still small. MCI reported $160 million in overall sales for the venture in 1995. Moreover, it is still an unprofitable venture. Nonetheless, it has an effective global reach owing to operating alliances it has forged with foreign carriers. Those alliances primarily consist of joint venture companies of MCI and BT throughout the world. MCI and BT have ventures in sixteen countries in the communications market. Those include ventures in Europe, North America, Latin America, and Asia Pacific. It is expected that Concert, plc, will focus on growing those ventures in markets opening for communications compe-

With the 1996 merger announcement, executives of both BT and MCI predicted savings in operating costs and capital expenditures. The companies estimated that the cumulative synergy benefits arising from full integration would approximate $2.5 billion within five years following the closing of the merger.[63] Annual pretax synergy benefits were estimated at approximately $850 million by the fifth year following the merger.[64] Both companies predicted significant synergies from combining their information technology operations.[65] Concert's Systems Integration division is slated to combine the resources of MCI Systemhouse and BT's Syntegra unit. The new information technology company will have annual revenues of more than $1.8 billion and employ more than 10,000 people, making it one of the world's five largest suppliers of systems integration and network outsourcing.[66] It will be well positioned to compete with the biggest competitors in the system integration business, EDS and Andersen Consulting.

The merger seemed to affect the faith of MCI's interest in ASkyB, its direct broadcasting satellite (DBS) joint venture with Rupert Murdoch's News Corp. By January 1996, when MCI put up an additional $682.5 million to buy the required licenses for DBS frequencies auctioned by the FCC, it had invested approximately $2 billion in News Corp. and ASkyB. Reportedly, Bert Roberts sought to reduce MCI's stake in ASkyB from 50 percent to 20 percent,[67] which would reduce MCI's stake by at least $545 million, to just under $1.5 billion.[68] On February 24, 1997, MCI's 50 percent stake in ASkyB was effectively reduced to a 10 percent interest in the newly formed 50–50 joint venture between Rupert Murdoch and

---

tition. *A Marriage of Convenience*, ECONOMIST, Nov. 9, 1996, at 72.

63. Flynn, Reed & Barrett, *supra* note 41.

64. *Largest International Merger, supra* note 41 .

65. The services included under the information technology banner are: client/server computing, systems integration, network and project management and outsourcing, year 2000 conversion, and practice areas in finance, high tech, transportation, public safety, entertainment, and other industries and sectors.

66. *Concert: The World's First Global Communications Company, Concert Global Highlights*, CONCERT WEB SITE <http://www.concert.com/deal/globalco.htm>.

67. Cunningham & Mortished, *supra* note 45.

68. Steve McClellan, *MCI/British Telecom Reducing ASkyB Stake; TCI Said to Be Talking About DBS Venture*, BROADCASTING & CABLE, Nov. 11, 1996, at 54.

EchoStar called Sky.[69] Under the terms of the joint venture agreement, News Corp. will contribute approximately $1 billion in the form of ASkyB assets such as four satellites, twenty-eight DBS frequencies, and cash for a 50 percent interest in Sky. MCI will receive a 20 percent stake in New Corps.'s interest in Sky (10 percent of the joint venture as a whole) in return for its 50 percent interest in ASkyB.[70]

On the basis of financial statistics at the time of the merger announcement in 1996, BT and MCI combined would have revenues in excess of $40.6 billion, profits nearing $5 billion, a cash flow of $12 billion, and a market valuation of approximately $54.2 billion.[71] Collectively, the company would have 183,000 employees and forty-three million business and residential customers in seventy-two countries.[72] On the basis of the most recent data available at the time of the merger announcement, the combined company would be the sixth most profitable in the world.[73]

The merger's architects described it in predictable hyperbole. MCI's chairman said:

> This merger creates the first telecommunications company of the new century. Financial muscle, global customers and brands, and customer-driven innovation will trump the competition as we open up communications markets both domestically and around the world. Concert's scale will allow it to pursue major opportunities in new markets while maintaining the financial stability that comes from strong core businesses in the developed markets of the U.S. and U.K.[74]

BT's chairman added:

> Concert will be exceptionally well placed to play a leading role in the major growth areas of the changing global communica-

---

69. *See* Raymond Snoddy, *News Corp in Big Digital TV Move*, FIN. TIMES, Feb. 25, 1997; Mark Landler, *Deal by Murdoch for Satellite TV Startles Industry*, N.Y. TIMES, Feb. 26, 1997, at 135.

70. *Id.*

71. *BT-MCI Fact Sheet*, *supra* note 46.

72. *Id.*

73. *Largest International Merger*, *supra* note 41.

74. *Id.*

tions marketplace. The complementary strengths and skills of
BT and MCI will enable Concert to take full advantage of the
tremendous opportunities provided by the forthcoming liberalization of telecommunications markets in the U.S. and Europe. We
believe this merger will provide major benefits for the shareholders, customers and employees of both BT and MCI.[75]

Those statements shed little light on the strategic rationale for, and
the timing of, the merger. The *Economist* speculated that the merger
was a defensive move by BT to prevent MCI from aligning with
another carrier.[76] That rationale is not persuasive, however, given
the influence that BT already could exert over MCI as its largest
shareholder. It seems more plausible that the merger, by creating
the world's first global communications company capable of offering a full range of seamless end-to-end services, would cause multinational firms to favor Concert, plc, over a loose alliance of other
telecommunications carriers that might disintegrate, to the inconve-
nience of its customers.

REGULATORY HURDLES

The BT-MCI merger must clear regulatory hurdles at the Federal
Communications Commission, the U.S. Department of Justice, the
European Commission,[77] and the United Kingdom's Monopolies and
Mergers Commission. The FCC's approval is expected to be the
most demanding. In comments filed with the FCC, AT&T sought
the imposition of "conditions that [would] minimize the ability of
BT to use improperly its market power to discriminate in favor of

---

75. *Id.*

76. "As weddings go, it lacked romance. The bride and groom had been living
together for years, and already had a child. They got along pretty well, but he had
been worried that her head might be turned by another man. Better, he reckoned, to
wed her now and avoid the prospect of fighting for her heart later." *A Marriage of
Convenience, supra* note 62, at 71.

77. On January 31, 1997, the EU Commission announced that it would fully
investigate the merger. Reportedly, the issues to be considered by the commission are
"whether the merger would impair the competitive position of rivals on the British-
U.S. route, would unfairly divert American-European traffic through Britain or
would restrict the availability of trans-Atlantic cable capacity." *MCI Merger Under
Inquiry in Europe*, N.Y. TIMES, Jan. 31, 1997, at D2.

MCI and distort competition in the U.S."[78] In addition to those safeguards, AT&T called for greater liberalization of the U.K. market. Specifically, it cited a lack of equal access presubscription and dialing parity as significant remaining barriers in the United Kingdom constraining the ability of new entrants to become viable competitors of BT.[79] Moreover, AT&T sought access to BT cable installations and networks at cost and the opportunity to put its own switching facilities in the BT cable installations. Furthermore, AT&T demanded the removal of the three-digit code that customers in the United Kingdom must dial before their calls can be routed over a network other than BT's.[80] A spokesman for AT&T said that "competition can only come when we are not paying BT to lease its lines and boost its profits."[81] AT&T wanted the conditions on regulatory approval of the BT-MCI deal to be "the model for the rest of Europe."[82]

In their December 2, 1996, application requesting FCC approval of the transfer of control to BT of licenses and authorizations held by MCI subsidiaries, MCI and BT stated that the United Kingdom satisfied every element of the effective competitive opportunities standard applied by the FCC to a transfer of control made pursuant to sections 214 and 310(b) of the Communications Act of 1934.[83] MCI and BT had asserted that U.S. carriers soon would have the legal ability to control international facilities-based carriers in the United Kingdom and provide basic telephony services to and from the United Kingdom.[84]

Two weeks after the BT-MCI filing, on December 19, 1996, the British government "threw open its telecoms market a year ahead of its European Union neighbours"[85] and became "the first

---

78. Merger of MCI Communications Corp. and British Telecommunications plc, Dkt. No. GN 96-245 (Jan. 24, 1997).

79. *Id.* at 19.

80. *Id.* at 22–26.

81. Richard Halstead, *AT&T Demands More UK Access; Telecoms Giant Wants US Regulator to Insist on Concessions in Exchange for BT Merger with MCI*, INDE-PENDENT, Nov. 10, 1996, at B2 (quoting Niall Hickey).

82. *Id.*

83. *BT and MCI File Merger Application with the FCC*, BT News Release (Dec. 3, 1996).

84. *Id.*

85. Nicholas Denton, *UK Opens Telecoms Market a Year Before Rest of EU*,

country in the world to liberalise its entire international telephone traffic in this way."[86] The Department of Trade and Industry (DTI) granted forty-four operators (among them twenty American firms, including AT&T and Sprint's Global One alliance) international facilities licenses allowing the new entrants to build and operate their own networks rather than to rely on leasing lines from BT and Cable & Wireless.[87] The DTI also authorized those licensees to provide international services through resold private lines between the United Kingdom and any other country.

In their FCC filing, BT and MCI asserted that the U.K. domestic telecommunications market provided fair interconnection terms and was fully open to U.S. carriers. They noted that twenty U.S.-owned and controlled companies had been licensed to provide various services in the United Kingdom, including SBC, U S WEST, Sprint, MFS, WorldCom, AT&T, and NYNEX.[88] BT and MCI argued that those competitors were able to offer telecommunications services to almost all of BT's customer base directly over their own facilities, through interconnection with BT's network or through resale of BT's and other competitors' services. They added that BT's competitors had captured close to 28 percent of the U.K. long-distance market for business services.[89] Furthermore, BT and MCI noted that American cable companies were among the companies granted 130 franchises for the provision of combined cable TV and telephony services in the United Kingdom.[90] The companies noted that for the first quarter of 1996 NYNEX's CableComm, the second largest cable operator in the United Kingdom, had shown a 25.9 percent increase in its share of residential service in its franchise areas.[91]

FIN. TIMES, Dec. 20, 1996, at 20.

86. *The Proposed MCI Merger with British Telecom*, BRITISH INFORMATION SERVICES, Press and Public Affairs of the British Embassy, Washington, D.C., Press Release (Nov. 12, 1996) <http://britain.nyc.ny.us/bistext/misc/pr111296.htm> [hereinafter *Proposed MCI Merger*].

87. Denton, *supra* note 85.

88. Ray Allieri, *MCI and BT File Merger Application with the FCC: MCI and BT Merger Application Summary*, MCI, Investor Relations (Dec. 3, 1996) (fax at 5).

89. *The Development and Growth of the U.S. and U.K. Telecommunications Markets*, CONCERT WEB SITE <http://www.concert.com/deal/usuk.htm>.

90. *Proposed MCI Merger*, *supra* note 86.

91. Allieri, *supra* note 88.

Finally, BT and MCI argued that the United Kingdom had an effective regulatory framework and that the FCC had already found that the U.K. regulatory authority was independent of BT.[92] In the United Kingdom the Office of Telecommunications (OFTEL), an independent body, conducts regulatory oversight of the industry. OFTEL also advises the Department of Trade and Industry (DTI). The regulatory framework, BT and MCI argued, employed fair and transparent procedures that U.S. carriers could use in the event of anticompetitive conduct on the part of U.K. telecommunications companies.[93] BT and MCI also emphasized that there were no restrictions on foreign ownership in the British telecommunications market.

BT'S INITIAL ACQUISITION OF
20 PERCENT INTEREST IN MCI

In June 1994 the Department of Justice filed suit and a proposed consent decree in response to BT's proposal to purchase 20 percent of MCI and to create an international telecommunications venture jointly owned by the two companies.[94] In July 1994 the FCC approved the alliance.[95] Approximately two weeks after the FCC's approval, the European Commission also approved the alliance.[96] The Justice Department's consent decree required MCI, BT, and the joint venture to disclose certain information about arrangements between the companies and to fulfill certain conditions to ensure that no discrimination would occur.[97] The FCC imposed similar nondiscrimination obligations but nonetheless concluded that the transac-

---

92. *Id.*

93. *Id.*

94. United States *v.* MCI Comm. Corp., 1994-2 Trade Cas. (CCH) ¶ 70,730 (D.D.C. 1994).

95. MCI Comm. Corp., 9 F.C.C. Rcd. 3960 (1994).

96. *Europe Clears MCI–British Telecom Alliance*, N.Y. TIMES, July 29, 1994, at D3.

97. Department of Justice, Antitrust Division, United States *v.* MCI Comm. Corp. and BT Forty-Eight Co. ("Newco"), Public Comments and Response on Proposed Final Judgment, 59 FED. REG. 48642 (1994); *see also* AMERICAN BAR ASSOCIATION, SECTION OF PUBLIC UTILITY, COMMUNICATIONS AND TRANSPORTATION LAW, 1995 ANNUAL REPORT: INFRASTRUCTURE IN TRANSITION 140 (1995).

tion was in the public interest; the agency therefore waived section 310(b)(4)'s restrictions on foreign investment.[98]

Under the terms of the transaction, BT paid $4.3 billion for one-fifth of MCI's outstanding shares, comprising one-fifth of the voting interest.[99] BT gained the right to nominate three of MCI's fifteen directors and to veto certain actions that MCI might take that could harm BT's interest in the company.[100]

Under the agreement the two companies formed Concert Communications Services, a joint venture to provide international enhanced voice and data services and the "global platform" (transmission, switching, and other facilities) from which those services may be offered.[101] The envisioned services were international network services, frame relay, flexible bandwidth, outsourcing, and MCI's virtual private network service, as well as other products and services that the two companies might jointly develop.[102] After the transaction, BT owned 75.1 percent of Concert, and MCI owned the remaining 24.9 percent.[103] Concert presumably would not hold any U.S. radio licenses.

Given the size of BT's investment in MCI, the percentage of MCI's outstanding shares already held in foreign hands, and the possibility for the extent of foreign investment to fluctuate over time, MCI and BT sought a waiver from the FCC to allow the foreign ownership of MCI to exceed the 25 percent benchmark under section 310(b)(4). On August 23, 1993, BT and MCI petitioned the FCC for a declaratory ruling that BT's proposed 20 percent interest, which could raise the level of foreign investment in the American carrier as high as 28 percent at any given time, would be consistent with and permissible under section 310(b)(4).[104] The FCC granted the waiver, allowing foreign ownership in MCI to exceed the statutory threshold by three percentage points on the basis of the

---

98. *MCI*, 9 F.C.C. Rcd. at 3964 ¶ 23.

99. *Alliance with British Telecom Wins F.C.C. Approval*, N.Y. TIMES, July 15, 1994, at D3.

100. BRITISH TELECOMMUNICATIONS PLC, *supra* note 46.

101. Mary Lu Carnevale, *FCC Approves Purchase by BT of MCI Stake*, WALL ST. J., July 15, 1994, at B3.

102. BRITISH TELECOMMUNICATIONS PLC, *supra* note 46.

103. MCI COMMUNICATIONS CORP., *supra* note 52.

104. *MCI*, 9 F.C.C. Rcd. at 3964 ¶ 22.

agency's conclusion that it would not serve the public interest by withholding its approval.[105]

At the time of the MCI transaction, it was the FCC's practice to decide on a case-by-case basis whether foreign investment exceeding the 25 percent benchmark was in the public interest by considering the extent of American presence in other areas of the licensee, such as ownership, officers, or directors.[106] In deciding to approve BT's purchase and to permit the 3 percent waiver of the section 310(b)(4) limitation to account for the possible fluctuation in foreign ownership, the FCC noted three factors concerning the transaction between MCI and BT. First, 80 percent of MCI's directors and 100 percent of its officers were U.S. citizens. Second, because the Title III licenses in question were for common carriage, MCI exercised no control over the content of the transmissions so that no concern of foreign control of radio transmissions would arise. Finally, BT's large cash infusion would enable MCI to improve its networks and services, and thus the American consumer, economy, and work force would all benefit.[107] The FCC concluded "that the proposed 3% fluctuation in non-BT foreign ownership above the 25% statutory benchmark is not inconsistent with the public interest."[108] As a condition of the approval of BT's investment in MCI, the FCC ordered MCI to make periodic surveys of its public shareholders to monitor the level of foreign ownership.[109]

MCI quickly used the BT capital to fuel its ambition to become a full-service provider. In May 1995 MCI entered the cellular telephony business by purchasing America's largest cellular reseller, Nationwide Cellular, for $190 million in cash.[110] That same month, MCI secured a source of content to deliver over its expanding infor-

---

105. *Id.* ¶ 23.

106. *E.g.*, GRC Cablevision, Inc., 47 F.C.C. 2d 467 (1974) (allowing slightly more than 50 percent foreign ownership of the parent corporation of a radio licensee where the parent was a U.S. corporation with a majority of the board comprising U.S. citizens, the foreigners were from a nation traditionally friendly with the United States, and the nature of the radio service was "largely passive").

107. *MCI*, 9 F.C.C. Rcd. at 3964 ¶ 23.

108. *Id.*

109. *Id.* at 3973 ¶ 61.

110. Richard Waters, *MCI Pays Dollars 190M for Wireless Business*, FIN. TIMES, May 23, 1995, at 25; *MCI Buys Nationwide Cellular for $190 Million*, N.Y. TIMES, May 23, 1995, at D4.

mation networks by committing to acquire 13.5 percent of News Corp. for $2 billion.[111] As part of the News Corp. agreement, MCI agreed to contribute $200 million to a joint venture between the two companies.[112] MCI committed another $2 billion to MCI Metro,[113] MCI's project to enter the local access market by deploying interactive broadband networks in urban areas.[114] In June 1995 the FCC granted approval for MCI's purchase of a controlling interest in Goeken Group Corporation, the owner of In-Flight Phone Corporation.[115] As a result of that transaction, MCI gained control of In-Flight's air-to-ground license and its section 214 certificate to provide international switched services by resale, thus extending MCI's wireless service capabilities. If not for the cash infusion from BT's direct investment in MCI, it is questionable whether MCI could have undertaken those multiple initiatives.

In 1995 MCI's surveys of foreign ownership levels required under the BT investment approval indicated that MCI would soon exceed the 28 percent level of authorized foreign ownership.[116] In August 1995 MCI was forced to return to the FCC to petition for an increased foreign ownership limit for MCI's capital stock.[117] Specifically, MCI sought to increase its maximum level of approved foreign ownership to 35 percent. Although a previous public interest analysis led the FCC to authorize MCI to exceed the 310(b)(4) benchmark, the request to raise the maximum limit required a new 310(b)(4) public interest analysis.

MCI argued that its request for up to 35 percent foreign ownership should be granted for the same reasons that the FCC had previously approved the 28 percent level of foreign investment. MCI further asserted that the increased foreign ownership would be

---

111. Alan Cane, *World Alliance Formed by MCI and News Corp*, FIN. TIMES, May 11, 1995, at 1.

112. Edmund L. Andrews & Geraldine Fabrikant, *MCI and Murdoch to Join in Venture for Global Media*, N.Y. TIMES, May 11, 1995, at A1.

113. Mike Mills, *MCI Poised for More Changes; Cellular Phone Firm Alliance Expected*, WASH. POST, July 28, 1995, at C1.

114. MCI COMMUNICATIONS CORP., 1993 ANNUAL REPORT 19 (1994).

115. Application of In-Flight Phone Corp. for Transfer of Control to MCI Telecommunications Corp., Declaratory Ruling and Order, 10 F.C.C. Rcd. 10448 (1995).

116. MCI Comm. Corp., 10 F.C.C. Rcd. 8697, 8697 ¶ 5 (1995).

117. *Id.* at 8697 ¶ 1.

"widely dispersed" and "passive" and that no single foreign investor, other than BT, would own more than 1 percent of MCI.[118] The FCC agreed that MCI's additional foreign investors would have "neither the interest nor the ability to control MCI."[119] The FCC also emphasized (in an order that preceded Congress's repeal of the officer and director restrictions in section 310(b)(4) in 1996) that American citizens dominated MCI's control and management.[120] Finally, the agency noted, the common carrier nature of the licenses involved weighed in favor of approval because investors would have no control over the content of transmissions.[121]

The FCC again concluded that increased foreign ownership of MCI would be consistent with the public interest.[122] In its declaratory ruling, the agency reiterated its position, first stated in its approval of BT's investment in MCI, that increased foreign ownership would "enhance MCI's ability to expand and improve network services and products for American consumers."[123] The FCC also found that, in general, competition and overall economic growth would increase as a result of approval in that proceeding.[124] Nonetheless, the FCC's approval of MCI's request carried one new condition: MCI would be required to notify the FCC if any one foreign investor, other than BT, acquired and accumulated more than 1 percent of MCI.[125]

SPRINT, FRANCE TÉLÉCOM, AND DEUTSCHE TELEKOM

On December 15, 1995, the FCC approved the sale of a 20 percent equity interest in Sprint, the third largest U.S. long-distance provider, to France Télécom (FT) and Deutsche Telekom (DT), the state-owned telecommunications monopolies of France and Germany.[126]

---

118. *Id.* ¶ 5.
119. *Id.* at 8698 ¶ 9.
120. *Id.*
121. *Id.*
122. *Id.*
123. *Id.* at 8697 ¶ 1.
124. *Id.* at 8698 ¶ 10.
125. *Id.* ¶ 11.
126. Sprint Corp., 11 F.C.C. Rcd. 1850 (1996).

The previous year the three companies had announced plans to form a global telecommunications alliance.[127] By June 1995 they had signed a definitive agreement to form their proposed joint venture. Finally, on January 31, 1996, the three telecommunications firms launched their global telecommunications joint venture, Global One, with the intent of providing seamless, end-to-end international telecommunications services to business, consumer, and carrier markets worldwide.[128] Global One is a combination of several businesses (including Atlas, the joint venture between FT and DT) spun together from the three parent companies.[129] The joint venture is directed by a "Global Venture Board" in which each party has an equal vote. The venture's operating group serving Europe (excluding France and Germany) is owned one-third by Sprint and two-thirds by FT and DT. The unit for the worldwide activities outside the United States and Europe is owned 50 percent by Sprint and 50 percent by FT and DT through Atlas.[130] On April 26, 1996, less than three months after the launch of Global One, FT and DT, pursuant to the terms of their agreement, completed their $3.66 billion purchase of 86.4 million shares of a new Class A common stock for a 10 percent interest each in Sprint.[131]

Sprint, FT, and DT needed the approval of the European Union Commission, the U.S. Justice Department, and the FCC for their joint venture. In addition, the FCC had to approve the two foreign companies' investment levels in Sprint. In July 1995 the De-

---

127. Andrew Adonis, *US Telecoms Alliance for France and Germany: Dollars 4Bn Stake in Sprint*, FIN. TIMES, June 15, 1994, at 1; Tom Redburn, *Sprint Forms European Alliance*, N.Y. TIMES, June 15, 1995, at D3.

128. The business activities of the venture were initially to be in three areas: (1) global data, voice, and video services for multinational corporations, (2) international card-based services for travelers, and (3) international transport services for other carriers. *See* FEDERAL COMMUNICATIONS COMMISSION, INTERNATIONAL BUREAU, GLOBAL COMMUNICATIONS ALLIANCES, FORMS AND CHARACTERISTICS OF EMERGING ORGANIZATIONS (Feb. 1996); *With Variations, Sprint Announces European Pact*, N.Y. TIMES, June 23, 1995, at B2.

129. RICHARD CRANSTON, LIBERALISING TELECOMMUNICATIONS IN WESTERN EUROPE 111 (Financial Times Business Information 1995); Gautam Naik, *Sprint Signs $4.1 Billion Agreement with French, German Phone Carriers*, WALL ST. J., June 23, 1995.

130. *Sprint*, 11 F.C.C. Rcd. at 1851–52 ¶ 11.

131. Tony Jackson & Alan Cane, *Sprint Signs Deal with European Partners*, FIN. TIMES, June 23, 1995, at 15.

partment of Justice gave Sprint approval to proceed with the alliance.[132] As a condition of Justice Department approval, Sprint, FT, and DT entered into a consent decree stipulating that they would not give preferential treatment to one another until the French and German telecommunications markets were opened to U.S. carriers.[133] In July 1996, two years after the parties had signed a memorandum of understanding announcing their intent to form a global telecommunications joint venture, the European Union Commission granted its final approval, subject to certain conditions, to the joint ventures for Global One and Atlas, thereby removing the last regulatory hurdle facing the venture.[134] The European Union Commission's conditions prohibit DT and FT from cross-subsidizing their joint venture and discriminating against other market players.[135]

The key regulatory battle, however, was fought and won some six months before the EU's approval, on December 15, 1995, when the FCC approved both the global joint venture, Global One, among the three firms and the sale of a 20 percent equity interest in Sprint to FT and DT. The FCC's approval capped a struggle that had begun in October of 1994, when Sprint petitioned the FCC for a declaratory ruling regarding the proposed equity investments by FT and DT in Sprint and the proposed joint venture among the three carriers.[136] In its petition Sprint had sought three rulings. First, foreign ownership in Sprint of up to 28 percent, as part of the proposed transaction, would be consistent with section 310(b)(4) of the act. Second, the investments by FT and DT would not result in a transfer of control of Sprint, and prior FCC approval thus would not be required under section 310(d). Third, the proposed transaction would otherwise be consistent with public interest.[137]

On January 11, 1996, three weeks after approving the joint

---

132. Leslie Cauley, *Sprint Alliance in Europe Receives Approval in U.S.*, WALL ST. J., July 14, 1995, at B3.

133. *Id.*

134. DT, *Global One Finally Obtains EU Approval*, Press Release No. 108/96 (July 17, 1996).

135. *Id.*

136. *Sprint*, 11 F.C.C. Rcd. at 1851 ¶ 6.

137. *Id.*

venture, the FCC issued its declaratory ruling and order.[138] The FCC issued all the rulings Sprint had sought in its petition. The Sprint order holds special significance because it was the first time that the FCC evaluated foreign carrier entry issues by using rules and policies adopted in its November 1995 *Market Entry and Regulation Report and Order.*[139] The ostensible purpose of the FCC's foreign carrier order was to implement a more transparent process for conducting its public interest analysis under sections 214 and 310(b).

*Application of the Effective
Competitive Opportunities Test*

The *Market Entry and Regulation Report and Order* specified that, as part of its overall public interest analysis under section 214, the FCC would determine whether effective competitive opportunities exist for U.S. carriers in the destination markets of a foreign carrier seeking to enter the U.S. international services market either directly or through affiliation with an existing U.S. carrier.[140] Similarly, in deciding whether it would be in the public interest to permit a foreign carrier to invest in a U.S. licensee of common carrier radio facilities in excess of section 310(b)(4) benchmark, the FCC would examine whether the home market of the foreign carrier offers effective competitive opportunities for U.S. firms in similar radio-based services.[141]

Section 214 Public Interest Analysis. In analyzing whether Sprint's proposed transaction was in the public interest under section 214, the FCC, following its new *Market Entry and Regulation Report and Order*, determined whether effective competitive opportunities existed for U.S. carriers to provide international facilities-based services in France and Germany.[142] The FCC began its inquiry by

---

138. *Sprint,* 11 F.C.C. Rcd. at 1850.

139. Market Entry and Regulation of Foreign-Affiliated Entities, Report and Order, IB Dkt. No. 95-22, 11 F.C.C. Rcd. 3873 (1995) [hereinafter *Market Entry and Regulation Report and Order*].

140. *Sprint,* 11 F.C.C. Rcd. at 1855 ¶ 32.

141. *Id.*

142. *Id.* Sprint had not sought prior FCC approval of the transaction under

ascertaining whether the DT and FT investments in Sprint created an "affiliation" between the parties.[143] Under the *Market Entry and Regulation Report and Order*, a 25 percent foreign equity investment in a U.S. carrier triggers the application of the effective competitive opportunities test.[144] Even though the two foreign companies each were to hold only a 10 percent interest in Sprint, the FCC chose to aggregate their interests to 20 percent, because the carriers had also joined forces through the joint venture.[145] The FCC further reasoned that the investment in Sprint was an important part of a global alliance strategy among Sprint, FT, and DT that affected the provision of basic telecommunication services.[146] Even with the aggregation, the two foreign companies' proposed investments in Sprint did not equal the 25 percent affiliation threshold for application of the effective competitive opportunities test. Nevertheless, the FCC held that the test should be applied because of the large size of French and German markets and the significant potential impact of the transaction on competition in the U.S. basic international telecommunications services market.[147]

In the second step of its inquiry, the FCC found that France and Germany were the relevant destination markets for purposes of applying the test.[148] The FCC reasoned that the record demonstrated that both foreign companies could exercise market power in those markets because they were state-owned monopolies.[149] In citing to the definition of market power in the *Market Entry and Regulation Report and Order*, the FCC stated that, as state monopolies, FT and

---

section 214. Sprint stated in its petition that such approval was not required because there was no change in ownership of section 214 certificates held by Sprint's subsidiaries. Nonetheless, the FCC chose to subject the transaction to a review under section 214. In support of its decision, the FCC noted that Sprint was a carrier authorized to provide telecommunications services and facilities under section 214, and through its subsidiaries, held many section 214 authorizations for the provision of U.S. international facilities-based services between the United States and France and between the United States and Germany. *Id.* at 1851 n.3, 1855 ¶ 33.

143. *Id.* at 1856 ¶ 37.
144. *Id.*
145. *Id.* ¶ 38.
146. *Id.*
147. *Id.* ¶ 39.
148. *Id.* at 1856–57 ¶ 40.
149. *Id.*

DT had the ability to act anticompetitively against unaffiliated U.S. carriers through their control of bottleneck services or facilities in France and Germany.[150]

In examining the French and German telecommunications markets under the test for effective competitive opportunities, the FCC found that both markets did not permit competition in the provision of international facilities-based services.[151] Thus, the agency concluded that there was no need for it to examine whether de facto competitive conditions of entry existed.[152] The FCC also found that the implementation of international facilities competition in 1998 in France and Germany was too distant to be considered competition in the near future under the FCC's analysis of effective competitive opportunities.[153] Thus, the FCC found that effective competitive opportunities for U.S. carriers did not currently exist in France and Germany.[154]

Section 310(b)(4) Public Interest Analysis. Because of its finding that no effective competitive opportunities existed for U.S. carriers in either the French or German facilities-based telecommunication services market, and because of its ultimate conclusion that the public interest weighed in favor of granting Sprint's petition, the FCC saw no reason to conduct an analysis of effective competitive opportunities under section 310(b)(4).[155] The FCC reached that conclusion despite its finding that cumulative foreign ownership in Sprint would have exceeded the 25 percent statutory benchmark

---

150. *Id.*

151. *Id.* at 1857 ¶¶ 41–43. Although Sprint held resale authorizations to serve France and Germany, the FCC saw no need to subject those authorizations to a test for effective competitive opportunities. The FCC reasoned that such analysis was not necessary, given its finding that France and Germany did not offer effective competitive opportunities to provide international facilities-based services and its ultimate conclusion that the public interest weighed in favor of granting Sprint's petition. *Id.* ¶ 46.

152. *Id.* ¶ 41. The FCC, in both the *Market Entry and Regulation Report and Order* and *Sprint,* has identified the following as de facto factors that it will consider: (1) terms and conditions of interconnection; (2) competitive safeguards; and (3) the regulatory framework. *Id.* ¶ 41 (citing *Market Entry and Regulation Report and Order,* 11 F.C.C. Rcd. at 3890–94 ¶¶ 42–55).

153. *Sprint,* 11 F.C.C. Rcd. ¶ 44.

154. *Id.* ¶ 45.

155. *Id.* at 1858 ¶ 48.

under section 310(b)(4).[156] The FCC stated that because there was overwhelming U.S. participation among Sprint's officers and directors, approval of the 28 percent foreign ownership sought by Sprint was consistent with the previous exercises of statutory discretion in which the agency had authorized levels of indirect foreign ownership of common carrier licensees to exceed the statutory benchmark.[157]

Countervailing Public Interest Factors

Despite its determination that France and Germany did not provide effective competitive opportunities for U.S. carriers in their facilities-based markets, the FCC approved the Sprint transaction. The FCC held that, on balance, the public interest weighed in favor of granting Sprint's petition, subject to certain conditions and safeguards, because of two important countervailing public interest factors. The first factor, and in the FCC's view the most critical factor, was the commitment of the French and German governments to full competition in their telecommunications markets.[158] The FCC observed that since it had issued its notice of proposed rulemaking on foreign carrier entry in February 1995, both the French and German governments had announced and begun to implement wide-ranging liberalization plans.[159] Moreover, the FCC noted that the French and German governments had committed, in letters from senior government representatives filed with the FCC, that they would open their telecommunications services and infrastructure markets by January 1, 1998.[160] Although timely implementation and development of effective interconnection regimes remained to be accomplished, the FCC nonetheless found that those government commitments weighed in favor of granting Sprint's petition.

The second countervailing public interest factor that the FCC recognized was the procompetitive effect, both domestically and

---

156. *Id.* at 1857–58 ¶ 47. In its petition Sprint said that its latest ownership survey showed that the 25 percent benchmark would possibly be exceeded by three percentage points after the FT and DT investments.

157. *Id.* at 1865 ¶ 94.

158. *Id.* at 1860 ¶ 63.

159. *Id.* at 1860 ¶ 64, 1862 ¶ 75.

160. *Id.* at 1861 ¶¶ 65–69.

internationally, of the two foreign companies' investment in Sprint.[161] Domestically, the FCC argued, the funds flowing to Sprint would enable it to expand and upgrade its existing network and to develop new applications and services.[162] The capital would also enable it to participate fully in the development of broadband PCS.[163] In light of Sprint's $4.5 debt burden, the capital infusion from FT and DT is essential for Sprint's ambitious plans for domestic expansion.[164] One example of such expansion plans is Sprint's 40 percent interest in a partnership with Tele-Communications, Inc. (TCI), Comcast Corporation, and Cox Communications called WirelessCo, L.P.[165] By 1999 Sprint must contribute $1.67 billion to WirelessCo.[166] WirelessCo is the mechanism by which those four companies intend to compete in the U.S. telecommunications market as full-service, end-to-end providers. The companies are pursuing both wireless and wireline strategies. With regard to wireless, WirelessCo plans to provide nationwide PCS under the Sprint brand name. WirelessCo was the high bidder in the PCS auctions, paying $2.1 billion for twenty-nine licenses, including New York, San Francisco, Detroit, Dallas, and Boston; in comparison, AT&T/McCaw bid $1.7 billion for twenty-one licenses.[167] WirelessCo and its PCS affiliates will cover an area in the United States with a population of 182.4 million. With regard to their wireline strategy, the four companies plan to enter the local exchange business in areas not already covered by Sprint's local exchange carriers (LECs). The WirelessCo partnership expects to use the cable plant of the partners and other affiliates as the primary vehicle for wireline competition with the LEC incumbent.[168] WirelessCo will supplement the existing wire-based infrastructure with an extensive broadband network buildout. Infrastructure

---

161. *Id.* at 1863 ¶ 78.

162. *Id.* at 1865 ¶ 88.

163. *Id.* at 1863 ¶¶ 81–82.

164. SPRINT CORP., 1994 SEC FORM 10-K, at F-2 (1995).

165. TCI owns 30 percent of the partnership, and Comcast and Cox each owns 15 percent. *Id.* at 5.

166. Mark Berniker, *Sprint, Cable Partners Plan Phone Service*, BROADCASTING & CABLE, Apr. 3, 1995, at 39.

167. Ronald Grover, *TCI's Endless Morning After*, BUS. WK., Apr. 10, 1995, at 60.

168. SPRINT CORP., *supra* note 164, at 5.

buildout, including wireline network development and interconnection, will cost at least another $2.3 billion.[169]

Internationally, the investment would permit Sprint to participate fully in its global seamless services joint venture (Global One) with FT and DT, and would thus establish a new competitor in that area.[170] On the whole, the FCC expected that additional competition to lower prices, increase the choice of services, and encourage more technical innovation, all to the benefit of U.S. consumers.

Section 310(d) Ruling. As noted, Sprint also sought a ruling that the investments by the two foreign companies in Sprint did not constitute a transfer of control under section 310(d) of the Communications Act.[171] The FCC held that their acquisition of a 20 percent aggregate equity interest in Sprint and their right to occupy 20 percent of the seats on Sprint's board did not constitute a transfer of control.[172] The agency also found that the voting and consent rights in the agreements among the three companies did not enable FT and DT to determine Sprint's corporate policy, nor did those rights indicate that the two foreign companies would dominate the management of Sprint's corporate affairs.[173] In addition, the FCC held that Sprint's granting the joint venture use of its international facilities did not constitute its relinquishing operational control.[174]

Anticompetitive Factors. The FCC concluded that, on balance, the public interest would be served by its approval of the transaction but qualified that conclusion with two concerns regarding anticompetitive risks. First, the FCC recognized the potential for the two foreign companies to use their current de jure and de facto monopoly power to engage in anticompetitive conduct because of their financial interests in Sprint and the carriers' joint venture. Second, the agency noted the possibility that the telecommunications liberalization to which France and Germany were committed might not

---

169. Berniker, *supra* note 166.
170. *Sprint*, 11 F.C.C. Rcd. at 1864 ¶¶ 84–86.
171. *Id.* at 1853 ¶ 20.
172. *Id.* at 1854 ¶ 23.
173. *Id.* at 1855 ¶ 30.
174. *Id.* at 1854–55 ¶ 29.

occur as scheduled.[175] The FCC's key concern was that the proposed transaction would give FT and DT each a substantial financial stake in the success of Sprint and the joint venture and, in the agency's view, would therefore give each an incentive to engage in anticompetitive strategies to maximize the return on their investment.[176] Such strategic behavior, the FCC reasoned, could yield Sprint more customers, calls, and revenues than would otherwise be the case.[177] At the same time, the FCC believed, the costs of Sprint's rivals would be raised above competitive levels, which would tend to reduce competition in the market as a whole.[178] Less competition would ultimately raise prices, lower quality, and slow innovation.[179]

Conditions and Safeguards. Because of those concerns, the FCC saddled its approval with five conditions and safeguards. According to the FCC, its conditions would "prevent potential anticompetitive conduct and minimize the unfair competitive advantages accruing to Sprint from its affiliation with FT and DT."[180] The first condition was that the FCC would deem Sprint to be a "dominant carrier for U.S. international services on the U.S.-France and U.S.-Germany routes" under the new foreign carrier entry rules.[181] As a dominant carrier, Sprint would be required to (1) make tariff filings effective on fourteen days' notice; (2) obtain prior approval of circuit additions or discontinuances; (3) file quarterly reports of revenues, number of messages, and minutes; and (4) maintain records on the provisioning and maintenance of facilities and services that FT or DT supplied to Sprint, including services and facilities procured for joint venture offerings, in France and Germany.[182]

Under the second condition, Sprint would not be allowed to operate newly acquired circuits on the U.S.-France and U.S.-Germany routes until France and Germany had implemented alter-

---

175. *Id.* at 1859–60 ¶¶ 55–56.
176. *Id.* ¶ 56.
177. *Id.* at 1860 ¶ 57.
178. *Id.*
179. *Id.*
180. *Id.* at 1867 ¶ 102.
181. *Id.* ¶ 103.
182. *Id.*

native infrastructure competition for already liberalized services and basic switched-voice resale.[183] The third condition precludes Sprint from agreeing to accept special concessions from any foreign carrier or administration with respect to traffic or settlement flows between the United States and any foreign country. Sprint must also obtain written commitments from FT and DT not to offer or provide any special concessions to the joint venture relating to the provision of basic services.[184] Under the third condition, Sprint must file monthly circuit status reports for the U.S.-France and U.S.-Germany routes and make those reports publicly available on a quarterly basis. In addition, Sprint must file annual reports regarding how France and Germany are progressing in meeting their liberalization commitments.[185]

The FCC found that the Sprint transaction served the public interest only if Sprint obtained a written commitment from FT to lower the accounting rate between the United States and France to the same range as the U.S.-U.K. and U.S.-Germany accounting rates.[186] Therefore, under the FCC's fourth condition Sprint was required to obtain a written commitment from FT that it would implement those reductions no later than two years from the adoption date of the FCC's order.[187]

As a fifth condition, Sprint was required to file a report with the FCC no later than March 31, 1998, that detailed how France and Germany had implemented effective competitive opportunities.[188] If the anticipated liberalization did not occur by that date, and effective competitive opportunities still were not available in France and Germany, the FCC would take further action, including designating for hearing the issue of whether the public interest continued to be served by Sprint's holding section 214 facilities authorizations on the U.S.-France and U.S.-Germany routes.[189]

---

183. *Id.* at 1868 ¶ 109.
184. *Id.* at 1869–70 ¶¶ 116–18.
185. *Id.* at 1870–71 ¶¶ 119, 128.
186. *Id.* at 1865 ¶ 92.
187. *Id.* at 1872 ¶ 131.
188. *Id.* ¶ 132.
189. *Id.*

Telefónica de España and
Telefónica Larga Distancia de Puerto Rico

In 1992 the FCC approved the purchase of the state-owned Puerto Rican long-distance telephone company, Telefónica Larga Distancia de Puerto Rico (TLD), by a subsidiary of Telefónica de España (Telefónica).[190] To gain FCC approval Telefónica structured its purchase to avoid section 310(b)'s restrictions on Title III radio licenses.

To appreciate the limited implications of the FCC's approval of the TLD acquisition, one must understand the history of telecommunications services in Puerto Rico. The Puerto Rican telephony market is divided between local-access and long-distance services. Historically, the Puerto Rico Telephone Company (PRTC) had a monopoly on local exchange services.[191] The Puerto Rico Telephone Authority (PRTA), a public agency of the Commonwealth of Puerto Rico, purchased PRTC from All America Cables & Radio (AAC&R) in the early 1970s.[192] Until 1987 AAC&R had a monopoly over the long-distance market; in 1987 AT&T purchased that monopoly.[193] In the late 1980s the PRTA created the predecessor of TLD to compete with AAC&R (and subsequently AT&T) in the provision of long-distance service, particularly outbound off-island service originating in Puerto Rico. In 1988, after years of petitioning by TLD, the FCC authorized TLD to compete in the provision of long-distance telephone service.[194] The company commenced service in 1989.[195] TLD provides international services both as a reseller and as a facilities-based provider.[196] The PRTA owned 100

---

190. Telefónica Larga Distancia de Puerto Rico, 8 F.C.C. Rcd. 106 (1992) [hereinafter *TLD*].

191. *Id.* at 106 n.2.

192. Statement of Telefónica Larga Distancia de Puerto Rico, Inc. in Opposition to AT&T Petition for Rulemaking, Market Entry and Regulation of International Common Carriers with Foreign Carrier Affiliations, RM-8355, at 3 (filed Nov. 1, 1993).

193. Donald L. Martin, Economic Benefits to Puerto Rico from Vigorous Telecommunications Competition 11 (Glassman-Oliver Economic Consultants, Inc. June 27, 1994).

194. Telefónica Larga Distancia de Puerto Rico, 3 F.C.C. Rcd. 5937 (1988).

195. Martin, *supra* note 193, at 6.

196. *TLD*, 8 F.C.C. Rcd. at 106–7 ¶ 2.

percent of TLD when Telefónica filed its petition to purchase a majority share of the company.[197]

Four carriers provide long-distance services that are available to most of Puerto Rico's 1.1 million access lines.[198] As of 1994, TLD was the principal competitor to AT&T, the former monopoly provider of outgoing off-island telecommunications services from Puerto Rico. Measured in total minutes, AT&T had 59.9 percent of the market; TLD had 21.6 percent; and Sprint and MCI respectively had 7.0 and 6.5 percent.[199] In a region where years of monopoly control left only 65.8 percent of households with telephone service, the competition provided by TLD and the others has yielded a significant gain in consumer welfare.[200] Economists estimate that competition in off-island service has saved consumers $578 million since it was introduced.[201]

To purchase TLD, Telefónica formed a Puerto Rican corporation, called LD, to act as the acquisition company. Under the terms of the acquisition agreement, Telefónica International Holding, B.V. (TI Holding), a Netherlands corporation, owned 79 percent of LD upon closing.[202] Telefónica Internacional de España, S.A. (TISA), a Spanish corporation, owned 100 percent of TI Holding.[203] Telefónica de España, Spain's government-controlled telecommunications operator, in turn, owned 76.22 percent of TISA;[204] the Spanish government owned the remainder.[205] Under the terms of the transaction, the PRTA retained 19 percent of TLD.[206] TISA, through TI Holding, owned 79 percent of TLD,[207] and the remaining 2 percent was held in an employee stock ownership plan.[208]

---

197. *Id.*

198. MARTIN, *supra* note 193, at 3.

199. *Id.* at 4.

200. *Id.* at 10.

201. *Id.* at i.

202. *TLD*, 8 F.C.C. Rcd. at 107 ¶ 3.

203. *Id.*

204. TELEFÓNICA DE ESPAÑA, S.A., 1993 SEC FORM 20-F, at 13 (1993).

205. Comments of Telefónica Larga Distancia de Puerto Rico, Inc., Market Entry and Regulation of Foreign-Affiliated Entities, Notice of Proposed Rulemaking, IB Dkt. No. 95-22, at 3 n.3 (filed Apr. 11, 1995).

206. *TLD*, 8 F.C.C. Rcd. at 107 ¶ 3.

207. TELEFÓNICA DE ESPAÑA, S.A., 1993 SEC FORM 20-F, at 13 (1994).

208. *TLD*, 8 F.C.C. Rcd. at 107 ¶ 3.

To avoid the foreign ownership restrictions, Telefónica acquired only those assets of TLD that did not require Title III licenses, including the current and pending section 214 licenses to provide international services and to own, operate, and maintain the necessary facilities.[209] Assets of TLD that required Title III licenses and consequently were subject to sections 310(a) and 310(b) were transferred to a newly formed Puerto Rican corporation, Telecomunicaciones Ultramarinas de Puerto Rico (TUPR).[210] The PRTA owned 85.1 percent of TUPR; TISA, through TI Holding, owned 14.9 percent.[211] TLD leased access to TUPR's Title III assets.[212]

Although Telefónica structured its investment in TLD to avoid problems under section 310(b), the FCC still scrutinized the transaction under a similar public interest test pursuant to section 214. Telefónica suggested, and the FCC concurred, that significant benefits would flow from the infusion of capital and the privatization of the government-owned Puerto Rican long-distance operator. AT&T, along with other opponents to the transaction, argued that allowing Telefónica de España to assume control of a U.S. facilities-based telecommunications operator would: (1) allow Telefónica to leverage its domestic market power in Spain to injure competition in the U.S. marketplace; (2) perpetuate high, non-cost-based accounting rates; and (3) raise troubling trade policy issues by permitting foreign access to U.S. telecommunications markets while the corresponding foreign market remained closed.[213] Although the FCC noted that "the long-term solution to foreign market power, which can be abused in the United States with or without a U.S. carrier-affiliate, is greater liberalization in foreign markets," the agency, over AT&T's objections, declined to apply a strict policy of reciprocal market entry.[214] The FCC found that its nondiscrimination safeguards were sufficient to protect U.S. carriers from discrimination that might occur as a result of the imbalance between relative market access. In addition, the facilities-based authorizations that the agency had granted to Spain and other countries were very

---

209. *Id.* at 107 ¶ 4.
210. *Id.*
211. *Id.*
212. *Id.*
213. *Id.* at 108 n.13.
214. *Id.* ¶ 9.

limited and thus unlikely to cause competitive abuse.[215] It is possible that the FCC's public interest determination also was influenced by the fact that TLD was an upstart challenging the former monopolist, although the agency did not explicitly mention that consideration as a factor justifying its conclusion that it was in the public interest to approve Telefónica's acquisition of control of TLD.

BT AND McCAW

In 1989 the FCC approved a purchase by British Telecom's U.S. subsidiary of a significant equity stake in McCaw Cellular Communications, America's largest cellular operator.[216] BT acquired just over a 22 percent interest, and as of 1993, the company owned 17 percent of McCaw's total equity (constituting 21 percent of the voting interest).[217] BT is no longer a McCaw shareholder because in September 1994 McCaw became a wholly owned subsidiary of AT&T.[218] As a result of AT&T's purchase of McCaw, BT became a significant shareholder in AT&T.[219] In January 1995, BT sold its 35.9 million shares of AT&T, a 2.3 percent stake, for approximately $1.76 billion.[220]

In other words, several years before its investment in MCI, BT had already made the most substantial foreign direct investment in a U.S. wireless company since the days of British Marconi. Although BT's brief investment is now a relic, the story remains important because it shows how the foreign ownership restrictions of section 310(b) significantly shaped an investment by one of the world's largest telecommunications operators in the then-emergent U.S. cellular market.

Through subsidiaries, McCaw holds cellular, paging, and other radio licenses.[221] At the time of the transaction with BT, the owners

---

215. *Id.*

216. McCaw Cellular Comm., Inc., 4 F.C.C. Rcd. 3784 (1989).

217. BRITISH TELECOMMUNICATIONS PLC, 1994 SEC FORM 20-F, at 6 (1994).

218. AT&T CORP., 1994 SEC FORM 10-K, at 4 (1995).

219. *British Telecom Sale of AT&T Stock to Proceed*, N.Y. TIMES, Jan. 14, 1995, at 41.

220. *AT&T Says British Telecom Will Sell Stock Holdings*, N.Y. TIMES, Jan. 27, 1995, at D3.

221. *McCaw*, 4 F.C.C. Rcd. at 3784 ¶ 2.

of McCaw stock were as follows: members of the McCaw family held 41 percent of the equity, comprising 46.24 percent of the voting interest; Affiliated Publications, Inc. (Affiliated) held 11.9 percent of the equity, comprising 52.02 percent of the voting interest; and public shareholders held the remaining 47.1 percent of the equity, comprising 1.74 percent of the voting interest.[222] Affiliated was a widely held, publicly traded corporation, with its principal business in newspaper publishing.[223] Pursuant to a shareholders' agreement, Affiliated was obligated to vote a sufficient number of its shares in accordance with instructions from Craig McCaw so that he could always command a majority vote.[224]

At the time of the BT-McCaw transaction, McCaw had initiated a merger with Affiliated whereby the latter company would spin off its McCaw interest to its own shareholders. The outcome of the proposed reorganization had little impact on the FCC's opinion of the BT-McCaw transaction, however, because Craig McCaw would control over 80 percent of the voting interest once the restructuring transpired, and he would control a majority of the voting interest if it did not.[225]

BT purchased its McCaw interest through its wholly owned American subsidiary, British Telecom USA, a Delaware corporation.[226] At the time of the McCaw transaction, the British government owned 49 percent of BT (then called British Telecom).[227] BT initially acquired 22.4 percent of the total equity and 22.4 percent of the voting interest in McCaw but later reduced its holding to the levels stated above.[228] Under the agreement, McCaw expanded its board of directors from thirteen to nineteen. McCaw had the right to nominate ten directors (all of whom had to be U.S. citizens); Affiliated had the right to nominate two directors (both of whom had to be U.S. citizens); Craig McCaw had the right to nominate three independent directors (all of whom had to be U.S. citizens);

---

222. *Id.*
223. *Id.* at 3784 ¶ 5.
224. *Id.* ¶ 3.
225. *Id.* at 3791 ¶ 47.
226. *Id.* at 3784 ¶ 6.
227. *Id.*
228. *Id.* at 3785 ¶ 7.

and BT had the right to nominate four directors.[229] Thus, under the agreement, BT had the opportunity to appoint foreign nationals as just over one-fifth of the directors, or slightly less than the maximum one-fourth then specified in section 310(b)(4). In addition to its board representation, BT had the power to veto certain corporate actions that could injure BT's interests.[230]

The agreement between McCaw and BT stipulated three conditions to ensure that BT's interest in the U.S. cellular operator would never violate section 310(b). First, both parties agreed that cumulative foreign ownership (including non-BT interests) in McCaw would never exceed the 25 percent benchmark in section 310(b)(4). Second, under its articles of incorporation and the agreement between the two parties, McCaw had the right to redeem stock held by foreigners to ensure compliance with section 310(b). Third, the terms of the agreement required periodic surveying of foreign ownership.[231]

The FCC noted that the relevant foreign ownership provision was section 310(b)(4), because BT was purchasing a stake in the U.S. parent corporation of a number of radio licensees.[232] The agency then quickly dismissed any foreign ownership concerns because BT's interest in McCaw was less than 25 percent, foreign directors constituted less than one-fourth of McCaw's board, and all McCaw officers were U.S. citizens.[233] The FCC further noted that the British government's 49 percent ownership interest in BT at the time of the transaction made no difference in the outcome because section 310(b)(4) allows foreign governments, along with foreign individuals and corporations, to invest in the parent companies of American radio licensees, so long as the ownership interests fall within the confines of the ownership and control restrictions.[234]

Although the FCC did not conduct its public interest analysis of the transaction to resolve the question of permissible foreign ownership, it did so to determine whether BT's stock purchase complied with the requirements of section 310(d) concerning trans-

---

229. *Id.* at 3785 ¶ 8.
230. *Id.* ¶ 10.
231. *Id.* ¶ 12.
232. *Id.* at 3788 ¶ 31.
233. *Id.*
234. *Id.* at 3790 ¶ 44.

fers of control.[235] The FCC defined "control" in *McCaw* as the ability "to determine the manner or means of operating the licensee and determining the policy that the licensee will pursue."[236] After stating that a minority shareholder would not be deemed to have control unless that shareholder had such significant influence over the licensee as to be able to dominate corporate affairs, the FCC concluded that BT, as a 22.4 percent shareholder, did not receive control through its acquisition because Craig McCaw still had the majority voting interest and BT's ability to veto certain corporate actions conferred no ability to compel McCaw to select a particular course of action.[237] And, although the terms of the agreement precluded McCaw from entering the U.K. cellular market, the FCC concluded that this prohibition did not affect the determination of control because the restraint was limited in scope and did not reduce the ability of other U.S. cellular providers to enter the U.K. cellular market.[238]

Even though it concluded that BT's purchase of McCaw stock did not constitute transfer of control subject to section 310(d), the FCC went on to say that the transaction was in the public interest for three reasons. First, BT's large cash infusion would strengthen McCaw's financial resources. Second, BT would bring to McCaw "substantial experience, expertise and knowledge" in telecommunications and, more specifically, in cellular telephony. Third, because no horizontal integration would result from BT's investment, American consumers would benefit from enhanced competition in the U.S. cellular market.[239]

BT's investment helped fuel McCaw's rapid expansion. Shortly after BT paid $1.5 billion to McCaw,[240] McCaw used the cash and the additional debt made available by the cash infusion to purchase LIN Broadcasting, thereby establishing a nationwide footprint for its cellular operation.[241] In March 1990 McCaw made a $3.38 billion

---

235. *Id.* at 3788–90 ¶¶ 32–43.

236. *Id.* at 3789 ¶ 34.

237. *Id.* ¶ 35.

238. *Id.* at 3790 ¶ 42.

239. *Id.*

240. Jeremy Warner, *BT in Costly Pursuit of the Holy Grail*, INDEPENDENT, Nov. 8, 1992, at 2.

241. Geraldine Fabrikant, *LIN Backs Takeover by McCaw*, N.Y. TIMES, Dec. 5,

cash tender offer for 21.9 million shares of LIN Broadcasting, a large cellular operator, raising McCaw's share in LIN to 51.9 percent.[242] At the time of the transaction, McCaw and LIN together operated in 119 markets, including five of the ten largest U.S. cities, with licenses covering seventy million people.[243] Later, AT&T, in search of an expeditious way to enter cellular telephony, purchased McCaw for approximately $12.6 billion in stock and the assumption of nearly $7 billion more in existing McCaw debt.[244]

CABLE & WIRELESS

Cable & Wireless plc provides telecommunications services in the United States through its American subsidiary, Cable & Wireless, Inc. (CWI). Section 310(b) has significantly shaped this large British telecommunications operator's participation in the U.S. market. CWI is a facilities-based provider and a reseller of domestic long-distance telecommunications services in the United States; additionally, CWI provides international long-distance service, primarily on a resale basis.[245] The company had over $600 million in revenue during its 1994 fiscal year and more than 2,400 employees.[246] But with only 1 percent of the U.S. long-distance market, the company ranks a distant fifth behind the big three U.S. long-distance companies and LDDS WorldCom.[247] CWI also provides intrastate telecommunications services in several states that allowed competitive entry into the intraLATA market even before passage of the Telecommunications Act of 1996.[248] CWI appears to hold no interest in any U.S. radio licenses.

1989, at D1.

242. Geraldine Fabrikant, *McCaw Completes Bid for Rest of LIN's Stock*, N.Y. TIMES, Mar. 6, 1990, at D6.

243. Roderick Oram, *LIN Accepts 7Bn Dollar McCaw Bid*, FIN. TIMES, Dec. 5, 1994, at 21.

244. Edmund L. Andrews, *AT&T Paying $12.6 Billion for McCaw Cellular*, N.Y. TIMES, Aug. 17, 1994, at A1.

245. CABLE & WIRELESS PLC, 1994 SEC FORM 20-F, at 73 (1994).

246. Jackie Spinner, *An All-Around Player at Cable & Wireless; Gabriel Battista to Head Long-Distance Carrier That Focuses on Smaller Businesses*, WASH. POST, July 3, 1995, at F9.

247. Eric Reguly, *AT&T Still in Talks on Mercury*, TIMES, May 4, 1995.

248. CABLE & WIRELESS PLC, 1994 SEC FORM 20-F, at 74 (1994).

CWI focuses on providing telecommunications services to small to medium-sized businesses. Its strategy is to identify and market customized telecommunications services to specialized consumer groups. In January 1995 CWI announced the formation of Omnes, a joint venture with Schlumberger, the oil field services and measurement company, that was intended to provide customized global telecommunications services to the oil and gas industry.[249] Omnes was to manage and operate SINet, Schlumberger's private network, which operates in fifty-three countries.[250]

CWI has also tried to enter the emerging American PCS market but has been limited to a nonequity role. In 1994 Cable & Wireless petitioned the FCC for a general waiver of the 25 percent limit in section 310(b)(4) for all U.K. citizens and corporations to enable CWI to purchase PCS spectrum rights that the FCC would shortly auction.[251] Rebuffed by the FCC, CWI instead sought to participate in the PCS market by supplying long-distance services to a partnership of small PCS licensees.[252] CWI joined with AT&T to support a venture called North American Wireless, that was intended to weave together a consortium of independent, small PCS operators to offer a nationwide, branded PCS service.[253]

In October 1995 the FCC approved CWI's acquisition of several very small aperture terminal (VSAT) licensees, as well as CWI's operation of "U.S. international VSAT facilities to provide, as a nondominant carrier, noninterconnected private line services to Mexico and various points in Central and South America."[254] The

---

249. Alan Cane, *C&W Goes for US Telecom Link*, FIN. TIMES, Jan. 6, 1995, at 4.

250. *Id.*

251. Cable & Wireless, Inc., 1994 FCC LEXIS 155 (Jan. 12, 1994).

252. Gautam Naik, *Alliance Planned for National Wireless System*, WALL ST. J., Nov. 7, 1994, at A3.

253. Paul Wiseman, *Wireless Deal to Aid Minorities, Small Firms*, USA TODAY, Nov. 7, 1994, at 1B.

254. Petition of Cable & Wireless, Inc. for a Declaratory Ruling Regarding Control of Licensees for Very Small Aperture Terminals Used to Provide Common Carrier Services, Application of Cable & Wireless, Inc. for Authority Pursuant to Section 214 of the Communications Act of 1934, as Amended, to Provide Private Line Services to Foreign Points Using Licensed VSAT Terminals, Declaratory Ruling and Mem. Op., Order, Authorization and Certificate, 10 F.C.C. Rcd. 13177, 13177 ¶ 1 (1995).

agency concluded that approval of CWI's investment would increase competition in the market for global network services.[255] The status of CWI's parent company, Cable & Wireless plc (C&W plc), a U.K. corporation, triggered the application of section 310(b)(4). CWI's proposed investment exceeded the 310(b)(4) foreign ownership benchmark.

The FCC approved the transaction after it considered four factors that have traditionally been found relevant to the public interest.[256] First, the FCC considered the extent of foreign participation in the parent corporation. Even though CWI's ultimate parent company was owned and controlled primarily by foreigners, the FCC noted that there was a significant buffer between the radio licensee and the parent company. That buffer consisted of an intermediate holding company that was a U.S. corporation, a majority of whose directors, officers, and owners were U.S. citizens.[257] The second factor in the FCC's analysis was the nature of the license.[258] CWI argued that the FCC had traditionally found that foreign ownership of common carrier radio licensees (such as VSAT licensees) raised fewer concerns than did foreign ownership of broadcasting licensees.[259] The FCC agreed and found that factor weighed in favor of approval; it stated that control of radio licenses that were used for common carrier services was "passive" in that the foreign interests would not control the content of the transmissions.[260] The third factor in the FCC's analysis was its finding that the transaction raised no national security concerns.[261] The final factor was the transaction's effect on competition and competitors. CWI claimed that its VSAT service would increase satellite-based competition in the market for common carrier global network services.[262] The FCC agreed and further determined that the U.K. telecommunications services market in general, and the VSAT market in particular, were quite liberalized and presented no significant barriers to entry

---

255. *Id.*
256. *Id.* at 13179 ¶ 13.
257. *Id.* ¶ 16.
258. *Id.* ¶ 13.
259. *Id.* at 13178 ¶ 7.
260. *Id.* at 13179 ¶ 18.
261. *Id.* at 13180 ¶ 19.
262. *Id.* at 13178 ¶ 7.

for U.S. firms.[263] In addition, the FCC found that U.S. consumers would benefit from the increased competition in the market as well as from CWI's increased ability to provide seamless global network services.[264]

The FCC declined to delay its decision on CWI's application pending the outcome of the agency's rulemaking to formulate a new test to evaluate foreign entry. The agency nonetheless warned that CWI would "be subject to any rules of general applicability which may be adopted in the Foreign Carrier Entry NPRM proceeding."[265]

BCE AND JONES INTERCABLE

In 1994 BCE, the Canadian telecommunications conglomerate, purchased 30 percent of Jones Intercable, America's seventh largest multiple systems operator (MSO).[266] Operating fifty-five cable television systems in twenty-three states, Jones Intercable serves over 1.3 million basic cable subscribers.[267] Through its international subsidiary, Bell Canada International (BCI), BCE agreed to invest $400 million in the cable operator for nearly one-third of the total equity and an option to purchase a controlling share of Jones Intercable, exercisable in 2002.[268] With its present equity stake, BCE has the ability to nominate six of Jones Intercable's thirteen directors.[269] Should BCE choose to exercise the option, it will have the power to elect 75 percent of the Jones Intercable board.[270] BCE and Jones Intercable are also equity partners in Bell Cablemedia, a cable television and telephony venture in the United Kingdom.[271]

BCE's investment in Jones Intercable apparently required no U.S. regulatory approval. The FCC's only influence over the transaction was to cause the parties to renegotiate the purchase price after the agency implemented rate reductions under the Cable Tele-

---

263. *Id.* at 13180 ¶ 21.
264. *Id.* ¶ 22.
265. *Id.* ¶ 21.
266. JONES INTERCABLE, INC., 1994 ANNUAL REPORT 30 (1995).
267. BCE INC., 1994 SEC FORM 20-F, at 27 (1994).
268. *Id.*
269. *New Bell Canada–Jones Deal*, N.Y. TIMES, June 6, 1994, at D2.
270. JONES INTERCABLE, INC., 1994 ANNUAL REPORT 30 (1995).
271. Dinah Zeiger, *Jones Completes 30 Percent Sale to Bell Canada*, DENVER POST, Dec. 20, 1994, at C1.

vision Consumer Protection and Competition Act of 1992.[272]

Jones Intercable identified three reasons for the equity sale to BCE. First, Jones Intercable valued BCE's telephony expertise as the cable company prepares for the joint offering of video and telephony services in the United States. Second, BCE offered Jones Intercable access to BCE's sophisticated telecommunications technology. Third, Jones Intercable considered the capital infusion critical to further Jones's strategic development.[273]

NEWS CORP. AND FOX
TELEVISION STATIONS

In 1995 the FCC granted the license renewal application of Fox Television Stations, Inc. (FTS) for television station WNYW in New York City, despite the agency's finding that foreign ownership in FTS's parent exceeded the section 310(b)(4) statutory benchmark.[274] That decision marked the first time that the FCC had permitted a broadcast licensee to exceed the benchmark.[275] Before the decision, FCC rulings had "effectively created the presumption in the broadcast area that, absent special considerations that outweigh[ed] the statutory concerns, the public interest [would] be served by denying licenses to entities with alien ownership above 25 percent."[276]

FTS was incorporated in 1985 to purchase six television broadcasting stations, including WNYW, from Metromedia.[277] Although FTS was controlled by Rupert Murdoch of Australia, who became

---

272. Dinah Zeiger, *Jones, Bell Canada Amend Deal in Wake of FCC Rate Rollbacks, Firms Agree to Alter Timetable*, DENVER POST, Mar. 29, 1994, at C1.

273. JONES INTERCABLE, INC., *supra* note 266, at 10–12.

274. Fox Television Stations, Inc., Second Mem. Op. and Order, 11 F.C.C. Rcd. 5714 (1996) [hereinafter *Fox*].

275. The FCC disputed the assertion that in Banque de Paris et des Bas Pays, 6 F.C.C.2d 418 (1966), the agency had allowed foreign ownership of a broadcast licensee to exceed the benchmark. First, the FCC asserted, in *Banque de Paris* the agency did not actually recognize foreign ownership above the benchmark. Second, the ownership interest in *Banque de Paris* was held in trust by a foreign entity, which was not attributable to the beneficiary of the trust at that time. *Fox*, 11 F.C.C. Rcd. at 5722 n.8.

276. *Id.* ¶ 21.

277. *Id.* at 5728 n.15.

an American citizen in 1985, FTS's parent company, Twentieth Holdings Corp. (THC) was almost entirely owned by an Australian company, News Corp., through FTS. When FTS sought renewal of its license for WNYW-TV in 1995, the Metropolitan Council of NAACP Branches (Metro NAACP) petitioned to deny the application, in part because it alleged that News Corp.'s ownership of more than 99 percent of THC violated section 310(b)(4).[278] The FCC faced a difficult choice: It could refuse to renew a license that it had initially granted under basically the same ownership structure almost ten years earlier, or for the first time it could countermand its presumption that foreign ownership of a broadcaster in excess of the benchmark was not in the public interest. The FCC chose the latter.

In an earlier order, *FTS I*, the FCC had conditionally granted FTS's renewal application for WNYW-TV after determining that foreign ownership of FTS's parent corporation exceeded the 25 percent benchmark in section 310(b)(4) for foreign ownership of a licensee's holding company.[279] The agency gave FTS "the opportunity either to advise the FCC how it would bring itself into compliance with the benchmark, or to demonstrate that its present ownership structure (or some other structure above the benchmark) [was] consistent with the public interest."[280] FTS elected not only to restructure its paid-in capital as debt, but also to make a showing that renewal of its license would serve the public interest.[281] Even though the FCC found that "FTS's recapitalization [did] not bring FTS into compliance with the statutory benchmark," it removed the condition on renewal of FTS's license for WNYW-TV anyway and even removed similar conditions on the renewal of other FTS licenses.[282] The FCC said that "the unique facts of this case" justified the level of foreign investment.[283]

---

278. *Id.* at 5715 ¶ 4.

279. Fox Television Stations, Inc., Mem. Op. and Order, 10 F.C.C. Rcd. 8452, 8466–77 ¶¶ 31–55 (1995) [hereinafter *FTS I*].

280. *Fox*, 11 F.C.C. Rcd. at 5716 ¶ 5 (citing *FTS I*, 10 F.C.C. Rcd. at 8524 ¶ 183).

281. *Id.* at 5714 ¶ 2.

282. *Id.* at 5715 ¶ 3.

283. *Id.* The "unique facts" language is a familiar FCC ruse by which the agency attempts to prevent certain of its decisions from having precedential effect. That artifice is silly because *every* case has unique facts. That is what distinguishes

Debt Contributions versus
Equity Contributions

In response to *FTS I*, FTS revised its capital structure, bringing News Corp.'s share of its paid-in capital below 25 percent. Nonetheless, the FCC found FTS's recapitalization unsatisfactory because the restructuring did not affect the underlying ownership structure. Under the restructuring plan THC distributed to News Corp., through Fox, Inc., cash equal to virtually all of THC's paid-in capital and in turn received from Fox, Inc., a substantial portion of that money in the form of debt secured by a promissory note payable by THC to Fox, Inc. Through that exchange of debt for equity, FTS argued that New Corp.'s share of THC's paid-in capital (and therefore its ownership interest) was reduced from 99 to less than 25 percent.[284]

In its subsequent *Fox* order, the FCC sought to clarify the distinction between debt ownership interests and equity ownership interests in assessing compliance with section 310(b)(4). The agency stated that it "must examine the economic realities of the transactions under review and not simply the labels attached by the parties to their corporate incidents."[285] In examining the restructuring plan, the FCC looked for guidance in the body of law distinguishing debt contributions from capital contributions under the Internal Revenue Code.[286] To keep its options open, however, the agency did not deem that law controlling for the purposes of its analysis under the Communications Act. The FCC did note that in the context of tax law, Congress had specified five factors to be relevant in distinguishing debt contributions from capital contributions: (1) whether there was a written, unconditional promise to prepay the money on demand and to pay a fixed rate of interest; (2) whether there was subordination to, or preference over, any indebtedness of the company; (3) the company's debt-equity ratio; (4) whether the alleged debt was convertible to stock; and (5) the relationship between holdings of stock in the corporation and holdings of the interest in

---

adjudication from legislation or rulemaking.
284. *Id.* at 5716 ¶ 6.
285. *Id.* at 5719 ¶ 14.
286. *Id.* at 5720 ¶ 16.

question.[287] After applying the five factors to THC's recapitalization plan, the FCC concluded that "the underlying structure of the News Corp./FTS relationship [would] not be affected by the revised capitalization."[288] The FCC therefore considered the intercompany loan between News. Corp. and THC a capital contribution and not a debt.[289] The FCC emphasized that the resultant debt-equity ratio was 1,400 to 1,[290] and that after recapitalization News Corp.'s equity capital contribution was reduced to $24,000, for which it was nonetheless entitled to 24 percent of FTS's voting rights and virtually all its profits.[291]

Public Interest Analysis

Having found that after the recapitalization News Corp.'s ownership of THC and FTS still exceeded the section 310(b)(4) benchmark, the FCC proceeded to determine whether it would nonetheless serve the public interest to grant FTS's license renewal petition.

Section 310(b)'s Purpose. The agency began its public interest inquiry by positing that section 310(b)'s purpose was to safeguard domestic station licensees from undue foreign influence and control and to "insure [*sic*] the American character" of licensees.[292] Without pausing to explain what the American character was, the FCC said that "that intent"—presumably the preservation of the American character and the prevention of undue foreign influence—"was particularly strong when combined with concerns for national security."[293] The agency saw the statutory benchmark as reflecting "Congress's judgment concerning the point at which foreign ownership and voting may conflict with the national interest."[294] The FCC acknowledged that it "had effectively created the presumption in the

---

287. *Id.* (citing 26 U.S.C. § 385(b)).

288. *Id.* ¶ 17.

289. *Id.*

290. *Id.*

291. *Id.* at 5721 ¶ 18.

292. *Id.* at 5722 ¶ 21 (citing S. REP. NO. 781, 73d Cong., 2d Sess. 7 (1934)).

293. *Id.*

294. *Id.* (citing Request for Declaratory Ruling Concerning the Citizenship Requirements of Sections 310(b)(3) and (4) of the Communications Act of 1934, as Amended, 103 F.C.C. 2d 511, 517 (1985) (*Wilner & Scheiner*)).

broadcast area that, absent special considerations that outweigh the statutory concerns, the public interest would be served by denying licenses to entities with alien ownership above 25 percent" and that it "had never allowed alien ownership of a broadcast licensee to exceed that level."[295]

FTS overcame that virtually insuperable presumption. The FCC balanced the equities favoring a decision for approval with the statutory interests in protecting broadcasting from foreign influence. The FCC found it "unjust to require the costly restructuring of a company that ha[d] been in continuous operation as a licensee for almost a decade and that structured its affairs in good faith reliance on [the FCC's] published rulings."[296] "The combination of inequity and expense," the FCC stated, "outweigh[ed] the statutory interest in limiting alien participation in broadcasting" and overcame "the presumption that foreign investment above the benchmark disserves the public interest."[297] The FCC could not find "that in the absence of restructuring 'the public interest [would] be served by the refusal or revocation' of FTS's license for WNYW-TV."[298]

Reasonable and Extensive Reliance. The FCC determined that FTS relied on its good-faith understanding of section 310(b)(4) in creating its corporate structure.[299] The evidence showed that FTS's counsel designed FTS's structure in 1985 in express reliance on its understanding of the statute—an understanding that the agency had itself deemed reasonable in light of the its published decisions at that time.[300] The FCC noted that in reliance on its approval of FTS's acquisition of the stations, FTS had devoted considerable resources over the previous decade to establishing and developing its broadcast interests.[301]

The FCC concluded that "FTS's reasonable and extensive reliance present[ed] an equitable reason to allow FTS to retain its present ownership and not to overturn, ten years later, FTS's settled

---

295. *Id.*
296. *Id.* at 5725 ¶ 28.
297. *Id.*
298. *Id.* (quoting 47 U.S.C. § 310(b)(4)).
299. *Id.* at 5723 ¶ 23.
300. *Id.*
301. *Id.*

expectations."[302] Nevertheless, the FCC did not find FTS's reliance to be dispositive in its public interest inquiry.[303] Rather, the agency examined as well the extent of the costs that it would impose on FTS if the agency required FTS to restructure its ownership. Ultimately, the FCC weighed the costs of restructuring against the putative benefits of limiting foreign investment in broadcasting.

The Recapitalization Plan as a Relevant Equitable Factor. Although the FCC had held that FTS's recapitalization plan did not comply with the foreign ownership benchmark, it deemed the recapitalization plan relevant to the equities that the agency would weigh in its public interest determination.[304] FTS had asserted that THC's recapitalization would produce the same structure that would have existed had News Corp. been aware in 1985 of the FCC's distinction between debt and equity in determining compliance with the statutory benchmark. Furthermore, FTS claimed, the FCC would have approved such a structure in 1985. The FCC agreed and concluded that, in light of FTS's reliance, it would be inequitable to require FTS to change a structure that FTS could reasonably have concluded the FCC would have approved ten years earlier.[305]

Cost of Restructuring. The FCC's public interest inquiry weighed the costs that it would impose on FTS's investors if it required the company to restructure, especially in light of FTS's good-faith reliance on the FCC's earlier rulings. FTS told the FCC that, if it were to restructure by selling a large part of the company to investors other than News Corp., "the transaction would result in a capital gains tax ranging from $540 million to $720 million."[306] In rejecting Metro NAACP's assertion that the tax liability was closer to $200 million, the FCC wrote: "Whether the true magnitude of FTS/News Corp.'s costs is $700 million or $200 million, . . . the precise number is not as important as the apparent fact that a restructuring would cost a substantial amount of money."[307] The agency added

---

302. *Id.*
303. *Id.*
304. *Id.* ¶ 24.
305. *Id.* at 5723–24 ¶ 24.
306. *Id.* at 5724 ¶ 25.
307. *Id.* ¶ 26.

that the public interest would not be served by imposing those substantial restructuring costs on FTS to correct a situation that occurred despite FTS's good-faith reliance on the FCC's earlier published rulings.[308] The FCC rejected Metro NAACP's argument that allowing FTS to retain its present structure would result in a $540 million "tax break."[309] That argument, the agency reasoned, was "based on the assumptions that FTS knew in 1985 that it would be required to restructure and face the tax consequences, and that FTS ha[d] 'deliberate[ly] conceal[ed]' its true ownership for the last ten years."[310] The FCC concluded that both those assumptions were unsupported.[311]

Statutory Interest. In its public interest analysis the FCC also considered the statutory interest in protecting broadcasting from the effects of foreign participation above the benchmark.[312] It found that the equities favoring FTS's existing ownership structure outweighed the statutory interest because of "the unique circumstances" of the case.[313] The FCC reasoned that the statute gave it the authority to permit foreign ownership levels exceeding 25 percent. Acting on that authority, the agency had permitted such higher levels of foreign ownership in the common carrier area, although not so high a level as sought in *Fox.*[314] Moreover, "to the extent that the statute's purpose [was] to protect the country's airwaves from undue foreign influence," the agency reasoned, "Rupert Murdoch, an American citizen, exercised de jure and de facto control over FTS and had substantial influence at News Corp. as well."[315] The FCC found no evidence of foreign influence at FTS during its ten years as a licensee and no reason to believe that national security would suffer by its permitting Australian investments that exceeded the statutory benchmark.[316]

---

308. *Id.* at 5725 ¶ 28.
309. *Id.* at 5724 n.10.
310. *Id.*
311. *Id.*
312. *Id.* at 5725 ¶ 27.
313. *Id.*
314. *Id.*
315. *Id.*
316. *Id.*

The agency concluded that its decision in *FTS I* would not open the door to routine findings that foreign ownership in excess of the benchmark was in the public interest.[317] In turn, the agency rejected Metro NAACP's argument that allowing FTS to retain its foreign ownership would result in "a huge influx of foreign capital entering the U.S. market, crowd[ing] out minority broadcasters."[318] The FCC reiterated that *Fox* was based upon unique equitable factors that the agency did not expect to face again.[319]

Retroactive Agency Adjudication. The FCC found additional support for its ruling in the body of administrative law concerning retroactive agency adjudication. Even though the FCC stated that it did not need to reach the issue of retroactivity in *Fox*, it conceded that when FTS was originally issued its licenses, the agency's published rulings had not included the interpretation of section 310(b) that the FCC later applied in *FTS I*.[320] The agency in turn outlined the D.C. Circuit's five-factor balancing test for determining whether a new rule should apply retroactively. The factors were: (1) whether the case was one of first impression; (2) whether the new rule was an "abrupt departure" from past practices or just an attempt to "fill in a void" in the law; (3) the extent of reliance on the former rule; (4) the burden that retroactivity would impose; and (5) the statutory interest in applying the new rule despite reliance on the old one.[321] The FCC stated that it had weighed those five factors in its decision and that its conclusion therefore comported with the D.C. Circuit's retroactivity test.[322]

FTS's Contributions to Broadcasting. In its public interest analysis, the FCC refused to consider a number of FTS's arguments premised

---

317. *Id.*
318. *Id.* at 5725 n.12.
319. *Id.* ¶ 27.
320. *Id.* at 5726 ¶ 29.
321. Retail, Wholesale & Department Store Union *v.* NLRB, 466 F.2d 380 (D.C. Cir. 1972); *see also* Consolidated Freightways *v.* NLRB, 892 F.2d 1052, 1058 (D.C. Cir. 1989), *cert. denied*, 498 U.S. 817 (1990).
322. *Fox*, 11 F.C.C. at 5726 ¶ 29.

on its positive contribution to the quality of broadcast programming. First, FTS cited "the benefits of having created a fourth television broadcast network, a policy objective that no one before had been able and willing to pursue," and an accomplishment that had "injected new life and vigor into free over-the-air television" and had created many jobs.[323] Second, FTS pointed to the Fox Network's positive effect on UHF stations, many of which, FTS argued, were in dire financial straits before Fox came to their rescue.[324] Third, FTS asserted that "the public interest was served by the content of its programming."[325] FTS claimed that it had provided high-quality children's programming, extensive local news coverage, and programming that had "particular appeal" to ethnic minorities.[326] FTS further argued that its programming had helped minorities own broadcast properties and produce programming.[327]

The FCC stated that it did not need to consider FTS's programming.[328] Nor did the agency find it necessary to address FTS's arguments concerning the benefits to consumers from creating a fourth television network and strengthening UHF stations.[329] The FCC reasoned that the equitable factors that it had already considered furnished an adequate basis for its decision.[330] Thus, it refrained from making "sensitive content-related judgments that would inevitably be involved" in considering FTS's contributions to broadcasting.[331]

The FCC's logic here is unpersuasive. FTS's creation of a fourth viable network clearly increased consumer welfare. Moreover, for the FCC to have taken into account the public interest implications of new competition from a fourth network in no way would have required the agency to make the "sensitive content-related judgments" that it sought to avoid. FTS's contribution to competition in broadcasting should have been the *first* factor that the

---

323. *Id.* at ¶ 30.

324. *Id.*

325. *Id.*

326. *Id.* at 5727 ¶ 30.

327. *Id.* Metro NAACP challenged nearly all of FTS's assertions regarding its contributions to broadcasting. *Id.* ¶ 31.

328. *Id.* at 5727 ¶ 32.

329. *Id.*

330. *Id.*

331. *Id.*

FCC weighed in its balance of equities. Why the FCC would have considered competition so incidental to the public interest is baffling.

The Scope of the FCC's Determination

The FCC determined the scope of its finding in *Fox* to include pending and future FTS activities. The agency reasoned that to confine its decision to stations that FTS already owned would have "unnecessarily hindered the company's ability to expand and frustrate[d] its reasonable expectations of doing so."[332] Consequently, the FCC found it consistent with public interest for FTS to acquire additional broadcast stations.[333] Subject to certain qualifications, FTS was allowed to invest in other broadcast properties.[334] If the investment sought by FTS triggered the application of section 310(b)(4), the FCC would attribute FTS's foreign ownership to the holding company as if FTS's foreign ownership were 24 percent.[335] If, on the other hand, FTS's investment triggered section 310(b)(3), the full percentage of FTS's foreign ownership would be considered for attribution purposes.[336] The agency noted that it had no discretion to permit levels of direct foreign investment that exceeded section 310(b)(3)'s 20 percent benchmark.[337] Thus, the FCC reasoned, section 310(b)(3) did not empower it to resort to the equitable remedy that the agency had crafted under section 310(b)(4).[338]

CONCLUSION

There are few cases of significant foreign direct investment in U.S. telecommunications under the present regulatory structure. Of the seven major ones, only three are old enough for their effects to be

---

332. *Id.* at 5728 ¶ 34.

333. *Id.* The FCC allowed FTS to acquire additional broadcast stations up to the allowable maximum then set forth in the FCC's ownership rules.

334. *Id.* at 5728 ¶ 35.

335. *Id.* at 5729 ¶ 35. The 24 percent represented the postrecapitalization percentage of News Corp.'s ownership (that is, corporate vote) of THC, the holding company of FTS.

336. *Id.* at 5729 ¶ 36.

337. *Id.*

338. *Id.* (citing *Wilner & Scheiner*, 103 F.C.C.2d at 522 n.51).

assessed anecdotally. In all cases, foreign investment led to greater consumer welfare. Telefónica de España's investment in TLD stimulated competition in the Puerto Rican long-distance market, BT's investment in McCaw led to the first nationwide cellular network, and News Corp.'s investment in Fox produced a viable fourth television network.

The U.S. recipients of other foreign direct investments have used or plan to use the capital in ways that can be expected to benefit American consumers. MCI has already allocated its $4.3 billion from BT to pay for three different projects that will position the company to compete in local telephony, information services, and video programming. Sprint has committed huge sums to its domestic wireless and broadband initiatives, even before receiving a single dollar from its proposed foreign investors. And Jones Intercable will use BCE's investment to fund the development of an interactive broadband network. We should expect a policy of greater hospitality toward foreign direct investment in the U.S. telecommunications industry to produce more of those kinds of benefits for American consumers.

6

Trade Policy

NO EVIDENCE EXISTS in the text of section 310(b) or in the legislative history of the Communications Act of 1934 that Congress intended that statutory provision to be a tool of international trade policy. Nonetheless, the FCC has increasingly used the foreign ownership restrictions for that purpose. By the end of 1995, the FCC, anticipating federal legislation that would give the agency an explicit international trade mandate, issued rules that interpreted section 310(b) to create a bilateral test by which a foreigner's ability to invest in a U.S. radio licensee would be made conditional on the treatment given U.S. investors by the foreigner's government. As a legal matter, the FCC climbed out on a limb that Congress proceeded to saw off: Given that Congress explicitly considered but failed to enact a reciprocity standard for section 310(b) when passing the Telecommunications Act of 1996, it became much clearer that the FCC lacked the authority to adopt its own reciprocity test through regulation. And, as an economic matter, it is doubtful that such a test would truly advance the public interest.

Neither classical trade theory nor strategic trade theory suggests that the FCC's reciprocity model will benefit the United States. It would be better for Congress simply to repeal section 310(b). If repeal is politically infeasible, however, Congress would best advance the public interest by amending section 310(b) in a manner that neither denies American consumers the benefits of foreign telecommunications carriers' competitive entry into the U.S. markets nor provokes other nations to retaliate against the United States by closing their markets to direct investment by U.S. telecommunications firms. An amendment to section 310(b) that passed the House of Representatives in August 1995 could, if modified in certain respects, provide the means to accomplish those twin objectives in a second-best world.

TRADE IN CAPITAL VERSUS
TRADE IN GOODS

Most trade theory addresses the movement of goods and services across international boundaries. That theory has far less to say about international flows of factors of production—labor, capital, managerial expertise, and technological know-how. Some caution is therefore necessary before imputing to trade in factor inputs the same analytical conclusions that emerge from the theory of trade in goods and services.

The foreign investment restrictions in section 310(b) of the Communications Act limit the free flow across national boundaries of factors of production, not final products. Foreign direct investment in telecommunications involves the international movement of financial capital, technology, and human capital in the form of managerial expertise.[1] The outward flow of managerial expertise from the American telecommunications firms to foreign markets is likely to be especially valuable, because the domestic managers of formerly state-owned postal, telegraph, and telephone administrations may lack the experience of American managers in running a firm in a competitive market or in dealing with an independent regulatory body.

Despite that important distinction between goods and factors of production, the policy debate concerning section 310(b) has proceeded without any recognition of the possibly limited relevance of analogies to classical trade theory or strategic trade theory. "Beyond the simple issue of fairness," observed economist Steven Globerman in a rare and incisive article on foreign ownership restrictions in telecommuni-

---

1. *See* EDWARD M. GRAHAM & PAUL R. KRUGMAN, FOREIGN DIRECT INVESTMENT IN THE UNITED STATES 36–37 (Institute for International Economics, 3d ed. 1995); PAUL R. KRUGMAN & MAURICE OBSTFELD, INTERNATIONAL ECONOMICS: THEORY AND PRACTICE 149, 160–61 (HarperCollins, 3d ed. 1994).

cations, "there has been little critical analysis of the merits of a policy of reciprocity in the area of foreign ownership restrictions."[2] Of course, it may be unduly naïve to assume that *any* economic theory truly influences the FCC or Congress in interpreting or amending section 310(b). For example, the FCC's 1995 notice of proposed rulemaking concerning international competition and foreign investment was conspicuously devoid of *any* discussion of international trade theory.[3] Nonetheless, because at least some arguments offered in support of reinterpreting or amending section 310(b) allude to strategic policies toward trade in goods, it is necessary to start our analysis by examining classical trade theory and strategic trade theory.

CLASSICAL TRADE THEORY

The classical theory of free trade, articulated by Adam Smith and David Ricardo and subsequently refined in the twentieth century by Gottfried Haberler, argues that the unrestricted movement of goods and services across international boundaries leads to specialization and the exploitation of comparative advantage and thus increases the wealth of both exporting and importing countries.[4] Trade enables countries to benefit from their different endowments of climate, skills, geography, and natural resources. A nation produces and exports what it can produce most efficiently relative to other nations, even if another nation is a more efficient producer of those goods in an absolute sense. Implicitly, the classical view is a static theory of trade. It

2. Steven Globerman, *Foreign Ownership in Telecommunications: A Policy Perspective*, 19 TELECOM. POL'Y 21, 25 (1995) [hereinafter *Foreign Ownership*].

3. Market Entry and Regulation of Foreign-Affiliated Entities, Notice of Proposed Rulemaking, IB Dkt. No. 95-22, 10 F.C.C. Rcd. 5256 (1995) [hereinafter *Market Entry and Regulation NPRM*].

4. *See* GOTTFRIED HABERLER, THEORY OF INTERNATIONAL TRADE 121–98 (William Hodge & Co. 1965) (German ed. 1933); ADAM SMITH, AN INQUIRY INTO THE NATURE AND CAUSES OF THE WEALTH OF NATIONS 447–96 (University of Chicago Press 1976) (1776); DAVID RICARDO, ON THE PRINCIPLES OF POLITICAL ECONOMY AND TAXATION 77–93 (Guernsey Press Co. 1992) (1817). For contemporary surveys and assessments of the arguments for free trade, *see* DOUGLAS A. IRWIN, AGAINST THE TIDE: AN INTELLECTUAL HISTORY OF FREE TRADE (Princeton University Press 1996); Jagdish Bhagwati, *Free Trade: Old and New Challenges*, 104 ECON. J. 231 (1994).

assumes, moreover, perfect competition—that many small firms produce a homogeneous product and act as price takers.[5]

The classical theory of trade provided reasonably good predictions of trade flows before World War II: Skill-intensive, advanced economies exported manufactured goods, and land-abundant economies exported raw materials and agricultural products. Since the war, however, a large share of trade cannot be attributed to the differences in endowments between countries. Among developed countries, trade in manufactured goods has increased between countries with similar endowments. Economists believe that, in many technology-intensive industries, competitive advantage today arises from a firm's research and development and its accumulated learning or experience.[6] By 1986 those trends had prompted the noted trade economist Paul Krugman to observe that "trade seems to reflect arbitrary or temporary advantages resulting from economies of scale or shifting leads in close technological races."[7] A decade later, one could be even more confident of that assessment.[8]

STRATEGIC TRADE THEORY AND ITS CRITICS

Strategic trade theory presents an alternative to classical trade theory and its prescription for free trade. Employing tools from industrial organization and game theory, strategic trade theorists emphasize the effect on domestic economic welfare of oligopolistic markets' earning supracompetitive profits, increasing returns to scale in production, and positive spillover effects in high-technology industries. Strategic trade theory is relevant to section 310(b) because it is likely to appeal to, and thus be cited as intellectual support by, those who argue that foreign direct investment in U.S. telecommunications firms should be

5. For a more detailed exposition of classical trade theory, see KRUGMAN & OBSTFELD, *supra* note 1, at 11–37.

6. *See* MICHAEL E. PORTER, THE COMPETITIVE ADVANTAGE OF NATIONS 33–68 (Free Press 1990); SHARON M. OSTER, MODERN COMPETITIVE ANALYSIS 289–308 (Oxford University Press, 2d ed. 1994).

7. Paul R. Krugman, *Introduction: New Thinking About Trade Policy, in* STRATEGIC TRADE POLICY AND THE NEW INTERNATIONAL ECONOMICS 9 (Paul R. Krugman ed., MIT Press 1986).

8. *See* PAUL R. KRUGMAN, PEDDLING PROSPERITY (MIT Press 1994); PORTER, *supra* note 6, at 12.

made conditional on the openness of the foreigner's telecommunications market to direct investment by U.S. firms.

Oligopoly Rents

Strategic trade theorists assert that many markets are not perfectly competitive. Oligopolistic firms can earn supracompetitive returns for an extended period of time, and nations compete to capture a larger share of those economic rents. That line of analysis builds on the work of James Brander and Barbara Spencer, who developed a game-theoretic model in which a government's strategic management of trade can benefit domestic firms.[9]

Brander and Spencer consider an industry having two firms, each in one country, that export their output to a third country. To simplify, they assume no domestic demand for that good in any country. The choice variable is the quantity produced by each firm; the firms determine prices once the total quantity produced in the industry has been determined. In other words, each firm takes the quantity produced by the other firm as given and then determines its own profit-maximizing output level. Such behavior is called a Cournot oligopoly.[10] In equilibrium (called a Nash equilibrium), each firm maximizes its profits, given the quantity that the other firm produces.[11] If in the first round the domestic firm were to increase its output from the Nash equilibrium, then the foreign firm would decrease its output to prevent the price of its good from falling. Given the Cournot assumption, the domestic firm would decrease output in the second round to maximize its profit, given the reduced output of the foreign firm in the first round. That process would lead the foreign firm to change its output, and successive rounds of output adjustments would occur until the firms returned to a Nash equilibrium.

---

9. James A. Brander & Barbara J. Spencer, *Export Subsidies and International Market Share Rivalry*, 16 J. INT'L ECON. 83 (1985).

10. For a discussion of Cournot and other oligopoly models, see DENNIS W. CARLTON & JEFFREY M. PERLOFF, MODERN INDUSTRIAL ORGANIZATION 229-80 (HarperCollins, 2d ed. 1994); HAL R. VARIAN, MICROECONOMIC ANALYSIS 285-88 (W. W. Norton & Co., 3d ed. 1992); JEAN TIROLE, THE THEORY OF INDUSTRIAL ORGANIZATION 209-44 (MIT Press 1988).

11. *See, e.g.*, DAVID R. KREPS, A COURSE IN MICROECONOMIC THEORY 427-43 (Princeton University Press 1990).

If, after increasing its output in the first round, the domestic firm did *not* change its output in the second round, then neither would the foreign firm; there would result, relative to the Nash equilibrium, higher output and profits for the domestic firm at the expense of the foreign firm. If the government gave its domestic firm a subsidy to increase its output in the first round, the domestic firm would have the incentive not to change its output in the second round. The government's strategic trade policy would thereby make the domestic firm better off than it would be under free trade.

In the Brander-Spencer model it is therefore possible, at least under one set of assumptions regarding noncooperative oligopolistic behavior, for a government to alter the rules of the game and shift excess returns in an oligopolistic industry from foreign to domestic firms. That intervention would increase the national income of the country imposing the subsidy and decrease the national income of the foreign firm's country.

Although one may cite the Brander-Spencer model to support strategic intervention by the government, slight variations on the model eliminate or even reverse its salutary conclusions. The predictions of strategic trade theory depend critically on the assumption of Cournot competition. If one instead assumes Bertrand competition to exist—firms choose price instead of quantity as the choice variable[12]—then, it can be shown analytically that the government's optimal strategy is to impose an export *tax*, not an export subsidy.[13] Of course, that analysis is not an argument in favor of export taxes. Rather, it supports the conclusion that free trade is the superior policy because strategic trade policy can be shown to improve economic welfare only under very rigid conditions that one does not normally observe in real markets.

Furthermore, the Brander-Spencer model ignores the effect of a subsidy to the exporting firm on other domestic firms.[14] A subsidy will lead the targeted firm to increase its output, thereby shifting resources away from other firms in the economy. Domestic firms will experience an increase in their marginal costs and will become less

---

12. On Bertrand competition, see VARIAN, *supra* note 10, at 291–94.

13. Jonathan Eaton & Gene M. Grossman, *Optimal Trade and Industrial Policy Under Oligopoly*, 101 Q.J. ECON. 383 (1986).

14. Avinash K. Dixit & Gene M. Grossman, *Targeted Export Promotion with Several Oligopolistic Industries*, 21 J. INT'L ECON. 233 (1986).

competitive. To determine whether the subsidy would increase domestic economic welfare, one must have detailed knowledge about the targeted firm, its foreign competitor, and the other domestic firms that compete with the targeted firm for resources. Timely information of that sort is unlikely to be available to any government.

Finally, the entry of new firms into the subsidized industry might dissipate the benefits of strategic trade policy by creating excess capacity.[15] Inefficient entry causes the average costs of production to rise, thereby leading to a welfare loss. That effect works in exactly the opposite direction of the no-entry case, which can hardly be guaranteed.[16]

The empirical studies that test strategic trade theory have had mixed results. A survey of the empirical research in strategic trade theory as of 1992 found that the available evidence did not undermine the case for free trade; the benefits of regulating trade were uncertain or even negative.[17] Ordinarily, one would use econometric models to test the various claims that rely on imperfect competition. In practice, however, it is nearly impossible to identify (and quantify) the relevant firm behavior, and often the crucial data are unreliable, unavailable, or unobservable. Consequently, some studies have used a "calibration process" to test the predictions of strategic trade theory.[18] With that technique, one specifies a theoretical model with a reasonable number of parameters. Estimates of some of the parameters may be available from previous econometric estimates (if they exist) or engineering studies. By altering the parameter values, one can test whether the results (and hence the policy recommendations) of the strategic trade model are sensitive to any particular assumption made in the model. Richard Baldwin and Paul Krugman used the calibration technique to investigate whether strategic trade theory could explain international competition for jet aircraft. They concluded that models like that of

---

15. Ignatius J. Horstmann & James Markusen, *Up Your Average Cost Curves: Inefficient Entry and New Protectionism*, 20 J. INT'L ECON. 225 (1986).

16. *Id.* at 226.

17. Anette Gehrig & Klaus F. Zimmermann, *Recent Developments in Strategic Trade Policy and Empirical Evidence, in* EXPORT ACTIVITY AND STRATEGIC TRADE POLICY 9 (Horst Kräger & Klaus Zimmermann eds., Springer-Verlag-Heidelberg 1992).

18. *See* Avinash K. Dixit, *Optimal Trade and Industrial Policies for the U.S. Automobile Industry, in* EMPIRICAL METHODS FOR INTERNATIONAL TRADE 141, 144–48 (Robert C. Feenstra ed., MIT Press 1987).

Brander and Spencer "neglect consumer effects of strategic trade policy" and that "while useful for expositional purposes," those models were "likely to be misleading when applied to real situations in which a sizeable fraction of production goes to satisfy domestic demand."[19]

In another study, Robert Baldwin and Richard Green found in five U.S. industries that strategic trade protection did not promote domestic demand.[20] Finally, Robert Baldwin and Paul Krugman studied competition between the United States and Japan in random access memories and concluded that privileged access to a domestic market gave Japanese firms an advantage, although other parts of Japan's economy incurred losses.[21]

Increasing Returns to Scale

Perfect competition assumes constant average cost, which does not change with the scale of output. Increasing returns to scale are said to exist when average cost decreases with the quantity produced.[22] In the presence of increasing returns, a country in principle can help its domestic firms achieve a cost advantage by protecting the domestic market from foreign competition. That policy enables the domestic firm to attain a higher level of output and thus produce at a lower average cost, which benefits the firm when it competes with foreign firms in international markets.

As in the case of using strategic trade policy to capture oligopoly rents for a domestic firm, the use of government intervention to enable the domestic firm to achieve economies of scale is fraught with difficulties. Again, one would expect the domestic welfare effects to depend critically on whether the oligopoly conforms to Bertrand or

19. Richard Baldwin & Paul R. Krugman, *Industrial Policy and International Competition in Wide-Bodied Jet Aircraft, in* TRADE POLICY ISSUES AND EMPIRICAL ANALYSIS 45, 71 (Robert E. Baldwin ed., University of Chicago Press 1988).

20. Robert E. Baldwin & Richard K. Green, *The Effects of Protection on Domestic Output, in* TRADE POLICY ISSUES AND EMPIRICAL ANALYSIS, *supra* note 19, at 205, 223–24.

21. Robert E. Baldwin & Paul R. Krugman, *Market Access and International Competition: A Simulation Study of 16K Random Access Memories, in* EMPIRICAL METHODS FOR INTERNATIONAL TRADE 171, 194–95 (Robert C. Feenstra ed., MIT Press 1987).

22. *See, e.g.,* CARLTON & PERLOFF, *supra* note 10, at 58.

Cournot behavior. And, again, the use of a subsidy for the exporting industry is likely to divert resources from other domestic industries for the reasons discussed earlier.

Positive Externalities
in High-Technology Industries

Beneficial spillover effects in high-technology industries are another justification given for departing from free trade. A positive externality is a benefit arising from the production of a good that its producer cannot capture for himself.[23] When positive externalities are present, the social benefits of production exceed the private benefits. Producers maximize their returns by expanding output until the marginal cost of production equals the private benefit. The presence of positive externalities leads to market failure in the sense that firms produce less output than would be socially optimal.

Positive externalities may arise in certain high-technology industries because firms cannot appropriate all the benefits of their investment in knowledge. Some of those benefits accrue to other sectors of the economy. The social benefits of innovation therefore exceed the private benefits. Through subsidies, tariffs, or other trade protections, a government may be able to coax a higher level of output from the high-technology industry producing large spillovers and thus could increase that nation's aggregate welfare. Conversely, a foreign country's targeting of high-technology sectors through subsidies or protection of home markets could cause the shrinkage of U.S. industries that could have yielded valuable spillovers for the rest of the U.S. economy. By targeting important sectors, the strategic trade theorists conclude, other countries could drive the United States from key industries.[24]

That reasoning presents three practical problems. The first is the government's ability to target the right industries. Proponents of strategic trade theory argue that trade policy should encourage strategic industries—which, presumably, are industries that have high value added, linkages with the rest of the economy, and strong

---

23. *See, e.g.*, DANIEL F. SPULBER, REGULATION AND MARKETS 46–48 (MIT Press 1989).

24. *See* KRUGMAN & OBSTFELD, *supra* note 1, at 281–82.

growth potential. But if such industries exist, those distinguishing characteristics should be self-evident to the capital market, labor market, and goods market; if no market failure is present, resources will flow into those industries without government intervention. If externalities often arise in high-growth industries, speculation could compensate for the fact that the private investor might not capture the entire return. And even if the market failure can be shown to exist, the question then becomes whether, given the scarcity of information about the causes and consequences of externalities, the government is more able to correct the externality than the market. A solution could be for government to subsidize *all* high-technology industries, but that kind of general subsidy would be too blunt an instrument to be efficacious and, in any event, would be implausible in an era of tighter constraints on government spending. Alternatively, the government could subsidize industries requiring substantial expenditures on research and development, but the burden of administering such a policy would be substantial. Any definition of R&D entitled to government subsidy would induce firms to characterize as many costs as possible as R&D expenditures.

The second practical problem of attempting to use strategic trade policies to capture technological spillovers is that no one knows the magnitude of the relevant externalities. Without knowing the size of the external economies, policymakers cannot know the point at which subsidies cease to produce net benefits to society and instead begin to waste public resources.

Third, the positive externalities of high-technology industries are likely to spill across national boundaries to firms in the other countries. International diffusion of new technological knowledge should be especially likely to occur in telecommunications, where network externalities are commonly hypothesized to arise from higher levels of access to, and usage of, the network.[25] In the age of the Internet and multinational corporations, new technologies developed in Silicon Valley spill over to Bangalore almost as quickly as they do to Route 128 and the Research Triangle. The more that technological spillovers cross national boundaries, the less would any one country have the incentive to subsidize those industries in the misplaced belief that it could keep all the resulting benefits within its borders.

---

25. Lester D. Taylor, Telecommunications Demand in Theory and Practice 9, 212–40, 348–69 (Kluwer Academic Publishers, rev. ed. 1994).

Is Free Trade Passé?

The preceding discussion suggests that the existence of positive externalities, increasing returns, and oligopoly represents factors that require qualification of the assumption of perfect competition that underlies the classical theory of free trade. Should we therefore reject the policy prescriptions of free trade? The answer is an unequivocal no. Although theoretical models can identify potential gains from strategic behavior under certain sets of assumptions, in practice not enough information is available for the government to choose the correct model and implement its policy prescriptions with any reasonable degree of confidence that the outcome will be superior to that which would obtain under free trade.

Summarizing the literature on strategic trade policy, Paul Krugman has observed that, although modest tariffs and subsidies, imposed unilaterally, can benefit the country imposing them, the same models producing those results also suggest that there would be large costs to a trade war and large gains from mutually removing trade barriers.[26] At a minimum, more empirical work on strategic trade theory needs to be done, and the debate remains unsettled in the academic literature. "One can always do better than free trade," note Elhanan Helpman and Paul Krugman, "but the optimal tariffs and subsidies seem to be small, the potential gains tiny, and there is plenty of room for policy errors that may lead to eventual losses rather than gains."[27] The strategic traders emphasize only the first part of that statement.

In short, free trade represents the best policy prescription given the uncertainty about the welfare implications of alternative trade policies, the difficulty of managing political intervention, and the need to avoid trade wars. "The case for free trade, as brought up to date from 1817," observed the eminent trade theorist Jagdish Bhagwati in 1987, "is . . . alive and well."[28] The same assessment seems justified today

---

26. Paul R. Krugman, *Introduction, in* EMPIRICAL STUDIES OF STRATEGIC TRADE POLICY 5 (Paul R. Krugman ed., National Bureau of Economic Research 1994).

27. ELHANAN HELPMAN & PAUL R. KRUGMAN, TRADE POLICY AND MARKET STRUCTURE 186 (MIT Press 1989).

28. Jagdish Bhagwati, *Is Free Trade Passé After All?, in* POLITICAL ECONOMY AND INTERNATIONAL ECONOMICS (Douglas A. Irwin ed., MIT Press 1991).

and, significantly, holds as well under a newer theory of trade propounded by William J. Baumol and Ralph E. Gomory, which economists may eventually regard as surpassing strategic trade theory in its ingenuity and analytical rigor.[29]

THE ALLURE AND DANGER
OF RECIPROCITY

The proposals to engraft a reciprocity standard onto section 310(b) are emblematic of the general fondness among Congress and regulators for trade policies that condition access to the U.S. market on the extent to which America's trading partners open their markets to American firms.[30] Reciprocity may appeal to one's sense of fair play. But it is a costly indulgence to base policy on that attitude, for American consumers benefit from import competition *whether or not* American firms can export to the markets from which those imports originate. Moreover, even if the apparent fairness of reciprocity is politically satisfying, that condition does not explain why the appeal of reciprocity rises and falls over time, as it has in U.S. trade policy. What, then, explains the rising popularity of reciprocity, as reflected, for example, in efforts in Congress and at the FCC in 1995 to condition foreign investment in U.S. telecommunications firms on market access abroad?

The growing appeal of reciprocity in U.S. trade policy may stem from the increasing U.S. trade deficit. In 1979 and 1980, the domestic economy witnessed double-digit inflation due to the second oil price shock and growing fiscal deficit. The government responded with a tight monetary policy that increased interest rates. The combination of rising interest rates and a growing fiscal deficit led to a massive inflow of capital and an overvaluation of the exchange rate

---

29. WILLIAM J. BAUMOL & RALPH E. GOMORY, ON EFFICIENCY AND COMPARA-TIVE ADVANTAGE IN TRADE EQUILIBRIA UNDER SCALE ECONOMIES (C. V. Starr Center for Applied Economics, New York University, Working Paper RR#94-13, Apr. 1994); RALPH E. GOMORY & WILLIAM J. BAUMOL, SHARE OF WORLD OUTPUT, ECONOMIES OF SCALE, AND REGIONS FILLED WITH EQUILIBRIA (C. V. Starr Center for Applied Economics, New York University, Working Paper RR#94-29, Oct. 1994).

30. *See generally* THE FOREIGN INVESTMENT DEBATE: OPENING MARKETS ABROAD OR CLOSING MARKETS AT HOME? (Cynthia A. Beltz ed., AEI Press 1995).

for the dollar. An overvalued dollar and a sluggish European economy after 1982 expanded the U.S. trade deficit. The overvalued exchange rate was ignored until the Plaza Agreement in 1985, when the United States finally devalued the dollar.[31]

The devaluation was too late. When it failed to reduce the U.S. trade deficit, some argued that the trade deficit had resulted from U.S. markets' being substantially more open than markets in Japan and Europe. Hence the idea gained currency that, to reduce the U.S. current account deficit, the U.S. government had to use both multilateral and bilateral negotiations to open foreign markets—if necessary, using reciprocity backed by the threat of retaliation.

After 1985, the U.S. government used the provisions of section 301 of the Trade Act of 1974[32] and "Super 301"—section 301 of the Omnibus Trade and Competitiveness Act of 1988[33]—to open product markets in other countries, especially Japan.[34] Another instrument increasingly used for opening foreign product markets is the voluntary import expansion, which mandates that a country import a specific quantity of a foreign good in a specific industry (usually by setting a minimum import market share) and often is backed by the threat of retaliation.[35] Jagdish Bhagwati has called that development in U.S. trade policy "aggressive unilateralism" because the United States unilaterally determined whether a country had violated U.S. trade laws and then announced its judgment, followed by punishment or the threat of punishment.[36]

Critics of reciprocity have argued that a policy of unilateral

---

31. *See* THOMAS O. BAYARD & KIMBERLY ANN ELLIOTT, RECIPROCITY AND RETALIATION IN U.S. TRADE POLICY 16–18 (Institute for International Economics 1994); YOICHI FUNABASHI, MANAGING THE DOLLAR: FROM THE PLAZA TO THE LOUVRE (Institute for International Economics, 2d ed. 1989).

32. 19 U.S.C. § 2411.

33. Pub. L. No. 100-418, 102 Stat. 1107, 1176–79, tit. I, § 1302(a) (codified at 19 U.S.C. § 2420). President Clinton renewed Super 301 by executive order in 1994. Exec. Order No. 12,901, 59 FED. REG. 10,727 (1994).

34. *See* ANNE O. KRUEGER, AMERICAN TRADE POLICY: A TRAGEDY IN THE MAKING 64–67, 78 (AEI Press 1995).

35. Jagdish Bhagwati, *Aggressive Unilateralism: An Overview, in* AGGRESSIVE UNILATERALISM: AMERICA'S 301 TRADE POLICY AND THE WORLD TRADING SYSTEM 1, 32–36 (Jagdish Bhagwati & Hugh Patrick eds., University of Michigan Press 1990); *see also* DOUGLAS A. IRWIN, MANAGED TRADE: THE CASE AGAINST IMPORT TARGETS 1 (AEI Press 1994).

36. Bhagwati, *Aggressive Unilateralism: An Overview, supra* note 35, at 2.

aggression would more likely close markets than open them and could ignite trade wars. William Cline, for example, argued in the early 1980s: "Such action, which may be called 'aggressive reciprocity' (as opposed to 'passive' reciprocity whereby new concessions are not granted in the absence of reciprocal liberalization), would run a serious risk of counterretaliation, with increased protection and reduced welfare on all sides."[37] Even if reciprocity did not provoke trade wars, critics argued, it would produce trade diversion rather than the trade liberalization that U.S. trade negotiators hoped to achieve: If U.S. firms increased their share of Japan's market through the U.S. government's threats of retaliation, Japan could respond by increasing its share of imports to the United States at the expense of other trading partners, such as the Europe Community. In response, Europe would demand an increased quota for its exports to Japan. World trade would depend on the relative bargaining strength of nations rather than on their comparative advantages.

Whatever market opening occurred because of bilateral negotiations would come at the expense of the international trading system. Bhagwati has warned that a bilateral trade policy predicated on threats of sanctions under section 301 could undermine the achievements of the Uruguay Round: "Whether . . . targeting . . . countries would make them more, rather than less, recalcitrant by compounding their sense of unfair U.S. play in trade negotiations remains to be seen."[38] Stanford professor Anne Krueger similarly has warned that "result-oriented aggressive bilateralism has scope for big disruptions of the international trading system and little potential for enhancing the efficient flow of goods and services in the international economy."[39]

Even if foreign countries did not retaliate, is trade policy an efficacious means to correct an adverse U.S. trade balance? After the Plaza Agreement, U.S. government spending persisted, and no significant reduction in private spending occurred. Under those circumstances, Bhagwati has observed, expenditure-switching policies such as devaluation fail to produce any appreciable effect on the trade

---

37. WILLIAM R. CLINE, "RECIPROCITY": A NEW APPROACH TO WORLD TRADE POLICY? 35–36 (Institute for International Economics 1982).

38. Bhagwati, *Aggressive Unilateralism: An Overview, supra* note 35, at 33.

39. Anne O. Krueger, *Free Trade Is the Best Policy, in* AN AMERICAN TRADE STRATEGY: OPTIONS FOR THE 1990s, at 68, 91 (Robert Z. Lawrence & Charles L. Schultze eds., Brookings Institution 1990).

deficit.[40] That assessment comports with the view of other respected economists that the trade balance is fundamentally a macroeconomic phenomenon that is not significantly affected by trade policy.[41] If they are correct, then resorting to trade reciprocity would not reverse an increasing trade deficit.

THE ECONOMICS OF RESTRICTING FOREIGN DIRECT INVESTMENT IN TELECOMMUNICATIONS

Does it benefit the United States to open its telecommunications markets to foreign investment even when American firms do not receive the same treatment from foreign countries? No, say proponents of reciprocity. They argue that the United States should use the offer to reduce its restrictions on foreign investment as a bargaining chip to induce other nations to lift their restrictions on American direct investment in their telecommunications industries.[42] Yet proponents of reciprocity for section 310(b) noticeably fail to cite economic evidence that reciprocal treatment has successfully acted in the past or will successfully act in the future to lower barriers to U.S. direct investment in overseas telecommunications markets.

Market Access Abroad

As noted earlier in the case of trade in goods, it is equally likely that a reciprocity policy will cause one's trading partner to harden its position rather than to reduce barriers. Similarly, the outright elimination of section 310(b) would deny foreign governments a convenient excuse for limiting the extent of U.S. direct investment in their

---

40. Jagdish Bhagwati, *U.S. Trade Policy at the Crossroads, in* POLITICAL ECONOMY AND INTERNATIONAL ECONOMICS, *supra* note 28, at 35, 39.

41. *See* KRUEGER, *supra* note 34, at 66–67; IRWIN *supra* note 35, at 18–24; C. FRED BERGSTEN & MARCUS NOLAND, RECONCILABLE DIFFERENCES? UNITED STATES–JAPAN ECONOMIC CONFLICT 52–54 (Institute for International Economics 1993).

42. *E.g.*, ERIK OLBETER & LAWRENCE CHIMERINE, CROSSED WIRES: HOW FOREIGN POLICIES AND U.S. REGULATORS ARE HOLDING BACK THE U.S. TELE-COMMUNICATIONS SERVICES INDUSTRY (Economic Strategy Institute 1994); *Hearing on Telecommunications Policy Reform: Hearing on S. 253 Before the Subcomm. on Telecommunications of the Sen. Comm. on Commerce, Science, and Transportation*, 104th Cong., 1st Sess. 236 (1995) (testimony of Eli M. Noam).

telecommunications firms. For example, Sam Ginn, the chief executive officer of AirTouch Communications, one of the largest cellular telephony service providers in the United States, observed in 1994:

> [W]hen AirTouch goes into a foreign cellular consortium, we bring the technology, the skills, the operating systems, the billing systems, and the training, and my local partner is there to run political interference for the license. What I hear time after time is, "You can only own up to 25% because that's the limitation that your government places on our companies."[43]

The ability of foreign governments to raise that objection disproportionately harms U.S. firms, Ginn argued, "because the most effective players in the international expansion of wireless today are U.S. companies."[44] The 25 percent de facto limitation on ownership had caused AirTouch to resort to a decidedly inferior structure for the ownership and control of the foreign consortium in which the U.S. company was investing:

> Because I cannot own a share of a company in another country greater than 25%, what I have to do is negotiate contractually for board representation. I can then fill certain slots in the management team and veto the business plan. Often, one cannot obtain these conditions through negotiation.[45]

---

43. Sam Ginn, *Restructuring the Wireless Industry and the Information Skyway*, 4 J. ECON. & MGMT. STRATEGY 139, 144–45 (1995) [hereinafter *Information Skyway*].

44. *Id.* at 145. "There are almost no other players. The active companies include BellSouth, AirTouch, U S West, and Southwestern Bell." *Id.*

45. *Id.* On another occasion, Ginn testified:

> Foreign ownership restrictions limit our opportunities. U.S. restrictions inhibit our ability to invest and expand abroad. Foreign governments tend to mirror U.S. government treatment of their firms doing business in the U.S. Therefore, U.S. restrictions on foreign investment for wireless telecommunications create difficulties for U.S. firms trying to invest abroad.

Trade Implication of Foreign Ownership Restrictions on Telecommunications Companies: Hearings Before the Subcomm. on Commerce, Trade and Hazardous Materials of the House Commerce Comm., 104th Cong., 1st Sess. 35 (1995) (testimony of Sam Ginn, Chairman and Chief Executive Officer, AirTouch Communications) [hereinafter *Ginn Testimony*].

What Ginn described was plainly an unintended consequence overseas of the foreign ownership restrictions in U.S. telecommunications law. It has two economic implications that are detrimental to U.S. interests.

First, the situation that Ginn described increases financial risk for U.S. firms investing in foreign markets. A firm like AirTouch cannot monitor its investments so effectively as it could in the absence of ownership restrictions' being placed on them. Although chapter 4 explains that the problems of ownership and control also arise under section 310(b) in the structuring of foreign investments in U.S. telecommunications firms, the problem is more severe when the investor is an American firm and the recipient of the investment is in another country. Even with the new telecommunications legislation of 1996, the U.S. market and regulatory environment are surely more settled than those of virtually any other nation. The reason is that the United States has not undergone a privatization and deregulation of a government-owned telecommunications monopoly. Many countries that are the targets of U.S. direct investment are in the midst of just such monumental regulatory transitions. Changes of such magnitude present enormous hazards and opportunities—which is to say, that the quality of managerial decisions will influence shareholder value more than would be the case in a firm with a mature product market and regulatory environment.[46]

The second economic implication of the situation that Ginn described is that it implicitly transfers rents from U.S. investors to domestic investors in the foreign market. The value of AirTouch's investment in the Portuguese wireless market, for example, is not limited to the U.S. firm's supply of financial capital. As Ginn made clear, U.S. wireless companies bring technological and managerial expertise, and those resources have in relative terms far fewer alternative sources of supply than the necessary financial capital—which, after all, is entirely fungible and can be raised through competing investment and commercial banks around the world. Conceivably, AirTouch could separately contract with the foreign

---

46. An analogous situation in the United States involved the growing returns to managerial decision making following the deregulation of railroads. *See* Ann F. Friedlaender, Ernst R. Berndt & Gerard McCullough, *Governance Structure, Managerial Characteristics, and Firm Performance in the Deregulated Rail Industry*, 1992 BROOKINGS PAPERS ON ECON. ACTIVITY: MICROECONOMICS 95.

consortium to supply such intellectual and managerial capital; but that alternative would be cumbersome, have high transactions costs, and be potentially regarded by foreign partners as overbearing. Consequently, AirTouch is limited to capturing at most 25 percent of the value created by the consortium's use in Portugal of AirTouch's unique intellectual and managerial capital. Stated differently, AirTouch must supply 25 percent of the financial capital and essentially 100 percent of the intellectual and managerial capital so that it can have the opportunity to earn 25 percent of the residual net cash flows of the foreign consortium.

Ginn consequently argued that, "from the standpoint of U.S. companies trying to penetrate these world markets," it is important to reform section 310(b).[47] The potential gains are substantial:

> The Japanese government has been willing to open its wireless telecommunications markets to our investment. . . . In my opinion, the Japanese government would be open to permitting greater levels of foreign investment in wireless telecommunications if U.S. policymakers liberalized our limits on Japanese investment. This would allow companies like ours to increase our equity in a very attractive market.[48]

Although reciprocity is one approach, it seems more plausible, in light of the difficulty of predicting the strategic responses of other governments, that reducing U.S. investment barriers would actually set in motion a political dynamic in foreign countries that would cause more liberal investment policies there. As Ginn noted, the United Kingdom's experience with unilateral reduction in barriers to foreign direct investment in its telecommunications market provided actual evidence that such an approach can benefit domestic interests:

> The United Kingdom provides an example of an alternate approach. Its government has declared unilateral openness. There are no foreign ownership restrictions. As a result, there has been an influx of new competitors and the creation of a very competitive market. Their economy has ample capital investment, their wireless market new service offerings and broad partnerships.[49]

---

47. Ginn, *Information Skyway, supra* note 43, at 145.
48. *Ginn Testimony, supra* note 45.
49. *Id.*

In short, the prescriptions for reciprocity in foreign investment policies rest on questionable assumptions concerning the strategic response of other nations. A policy of reciprocity could easily backfire and produce results that are inferior to what has *actually been observed* to result in other industrialized nations from unilateral liberalization of restrictions on foreign direct investment in telecommunications.

The potential for a U.S. policy of reciprocity to provoke other countries was evident in a January 1995 letter from Horst G. Krenzler, the director-general I of the European Commission, to an American business group, the EC Committee of the American Chamber of Commerce in Belgium, that had expressed concern over suggestions in late 1994 that the EC might impose foreign ownership limits in telecommunications.[50] The director-general's letter was blunt:

> I was particularly concerned with the statement made by the EC Committee in its position paper that there is no need to overstate the reach of foreign ownership limits in US legislation applying to the telecommunications sector. The EC Committee argues that these limits, and particularly those foreseen under [section 310], do not affect the whole range of telecommunications services in the US, in particular those that do not rely on radio facilities, those provided over the infrastructure of another operator (i.e., resold) or private services.
>
> I could not share such an interpretation on the reach of the restrictions imposed under Section 310 of the US Communications Act. . . .
>
> In practice, since most carriers rely on the use of radio transmission stations, satellite earth stations and in some cases, microwave towers, the majority of foreign-owned carriers are unable to compete in the long distance market (the local market being subject to the monopoly of the Bell Operating Company in most States), and only through a minority share holding in the mobile market (AMSC, a consortium comprising most of the US

---

50. Letter from Horst G. Krenzler, Director-General I, European Commission, to Mr. Oliver, EC Committee of the American Chamber of Commerce in Belgium, Jan. 18, 1995.

companies involved in the sector, enjoys exclusive monopoly rights for the provision of US mobile satellite services, thus excluding any foreign competition).[51]

The director-general also disputed that the FCC had exercised or would exercise its discretion under section 310(b) in a manner that would lessen the practical impediment that the statute presented:

> The EC Committee seems also to consider that there is sufficient flexibility in US legislation not to prevent the required foreign investment in the sector and that, in the specific case of indirect ownership, the FCC would be in a position to waive or extend the limits to foreign participation if it finds that this would be in the public interest. *The FCC, however, rarely uses this possibility. I would argue, on the contrary, that there is a lack of principles on which to base FCC decisions, and therefore insufficient predictability to justify foreign investment.*[52]

Chapter 4 shows that, as Director-General Krenzler correctly noted, the principles the FCC used to waive application of section 310(b)(4) are utterly amorphous and thus amenable to any result that the FCC desires in a given case. The agency's 1995 rulemaking on foreign entry and investment is not likely to improve matters. The director-general also expressed concern that American proposals for reciprocity "could . . . endanger[ ] progress in current multilateral negotiations."[53] That concern is well founded. As we shall see shortly, given that the FCC's waiver decisions under section 310(b)(4) have lacked consistency and logic, one can only expect that experience to be repeated in a more byzantine manner as the FCC undertakes to enforce its 1995 foreign ownership policy predicated on bilateral reciprocity.

Capturing Economic Rents and Technological Spillovers in the United States

Steven Globerman analyzed how restrictions on foreign ownership of telecommunications firms may serve other economic objectives

---

51. *Id.* at 1–2.
52. *Id.* (emphasis added).
53. *Id.* at 2.

besides the securing of reciprocal treatment of direct investment in the foreign firm's home market.[54] Along the lines of the oligopoly rent arguments made in strategic trade theory, one can theorize that economic rent may accrue to "first mover" advantages or the establishment of early incumbency in an industry. The incumbent firm may use its status to establish a dominant position that may last for years. Entry barriers like section 310(b) may facilitate that establishment of incumbent advantage and reserve for domestic owners of firms (rather than foreigners) the privilege of capturing economic rents.[55]

The argument does not withstand scrutiny, however, because restrictions on foreign investment in telecommunications are, Globerman argued, the least desirable means to capture economic rent in that industry because "they encourage production by less efficient domestic suppliers at the expense of more efficient foreign suppliers."[56] Only extraordinary circumstances, such as largely undeveloped markets or markets with high anticipated growth (such as personal communications services), could possibly justify investment barriers.[57] In the case of the U.S. domestic telecommunications market, the first condition plainly is not satisfied. Few rural markets in the United States are undeveloped. With respect to the second condition, anticipation of high growth is better reserved for the emerging markets in Asia, Latin America, and eastern and central Europe.

Globerman suggested other means, less damaging to the general welfare than restrictions on foreign direct investment, by which the host government could capture economic rent. If the foreign firm is known to operate more efficiently, then the host government could auction to the foreign firm the rights to supply telecommunications services.[58] Alternatively, the host government could tax away the economic rent earned by suppliers, although such a policy would impose a deadweight loss on the host economy.[59]

---

54. Globerman, *Foreign Ownership, supra* note 2, at 23–25.

55. *Id.* at 23.

56. *Id.* at 24.

57. *Id.* at 23.

58. *Id.* at 23 n.8 (citing Oliver E. Williamson, *Franchise Bidding for Natural Monopolies—In General and with Respect to CATV,* 7 BELL J. ECON. 73 (1976)).

59. *Id.* at 24.

Another possible economic rationale that Globerman identified for foreign ownership restrictions in telecommunications is the existence of positive externalities along the lines hypothesized by the strategic trade theorists.[60] U.S. carriers may be more likely than foreign carriers to carry out their research and development in the United States; if so, then more spillover benefits arguably would accrue in the United States if investment by foreign carriers were restricted.

That rationale, Globerman noted, is subject to two large caveats. First, technological externalities result from the diffusion of technology and not merely from the location of R&D activities.[61] Globerman argued that

> the introduction and spread of already existing technology will generate external benefits for consumers and related suppliers. The potential entry of foreign suppliers should be a spur to existing domestic suppliers to adopt best-practice technology in a timely fashion, while the actual entry of more technologically advanced foreign suppliers will directly promote the diffusion of technology in the domestic industry.[62]

Globerman's second caveat is that restricting foreign ownership in any one country will "almost certainly retard technological changes in that country to the extent that they beget similar restrictions in other countries."[63] Should such retaliation arise, it would be especially costly in the specific context of telecommunications services because of the network externalities that one would expect to observe on an international basis.

Globerman concluded that "constraining foreign ownership in any segment of the telecommunications industry does not seem to be good public policy. Either such constraints will not contribute to the

---

60. *Id.*

61. *Id.*

62. *Id.* Similarly, Sam Ginn of AirTouch testified before Congress: "Foreign investment in U.S. companies is also directly beneficial to the U.S. national interest. Foreign investment provides additional capital available for R&D and the build-out of networks, jobs for American workers, new technology, and exposure to skills and best practices we can use to improve our methods of doing business." *Ginn Testimony, supra* note 45, at 4.

63. Globerman, *Foreign Ownership, supra* note 2, at 24.

realization of their putative objectives, or there are equally effective instruments that will do less damage to the competitive process."[64] We shall now examine proposals to amend or interpret section 310(b) to incorporate the principle of bilateral reciprocity and consider whether those proposals are likely to realize their objectives of opening foreign markets to investment by U.S. firms.

CONGRESS'S UNSUCCESSFUL ATTEMPT TO MAKE SECTION 310(b) A RECIPROCITY TEST

In 1995 both the Senate and House telecommunications deregulation bills, S. 652 and H.R. 1555, contained provisions to amend section 310(b) to embody the concept of bilateral reciprocity.[65] Also in 1995, the FCC proposed and adopted a rule that would engraft a reciprocity standard onto the public interest determination that the agency would make when enforcing the then-existing section 310(b).[66] Those approaches to alter the foreign ownership restrictions were inferior to outright repeal of section 310(b), which Representatives Michael Oxley and Rick Boucher proposed unsuccessfully in 1995.[67] But the reciprocity proposals nonetheless were politically attractive for the reasons that reciprocity proposals generally find favor in Washington. Although those legislative proposals did not become part of the Telecommunications Act of 1996, as chapter 3 explains, it is nonetheless useful to analyze them in some detail. Future proposals to amend section 310(b) will no doubt begin with elements of the reciprocity proposals deleted from the 1996 legislation.

THE SENATE BILL

In 1995 the Senate passed legislation that would subject foreign investment in U.S. radio licensees to a test of bilateral reciprocity if

---

64. *Id.* at 28.

65. S. 652, § 105, 104th Cong., 1st Sess. (1995); H.R. 1555, § 302, 104th Cong., 1st Sess. (1995).

66. *Market Entry and Regulation NPRM,* 10 F.C.C. Rcd. at 5257 ¶ 4; Market Entry and Regulation of Foreign-Affiliated Entities, Report and Order, IB Dkt. No. 95-22, 11 F.C.C. Rcd. 3873, 3890 ¶ 43 (1996) [hereinafter *Market Entry and Regulation Report and Order*].

67. H.R. 514, 104th Cong., 1st Sess. (1995); *see* 141 CONG. REC. H241 (daily ed. Jan. 13, 1995).

the investment were to exceed the benchmarks contained in section 310(b). The provision was poorly conceived from the start and, even after considerable revision, remained highly problematic in the form that passed the Senate.

The Discussion Draft of the Senate's
1995 Telecommunications Bill

On January 31, 1995, Senator Larry Pressler, chairman of the Telecommunications Subcommittee of the Senate Commerce Committee, circulated for public comment his "Telecommunications Competition and Deregulation Act of 1995," a draft bill to rewrite the Communications Act of 1934.[68] As midnight approached on January 30, the Senate staff still did not have language concerning reform of section 310(b). As a "place holder," one staffer who had worked on the telecommunications section of the 1988 Trade Act—which addressed trade in telecommunications equipment but not foreign direct investment—suggested that the provision in the 1988 legislation be borrowed.[69] The suggestion was accepted and the draft bill included a provision that would amend section 310 by adding a new subsection 310(f). As we shall see, however, the statutory language would have created anomalies and perverse economic incentives when applied to flows of financial and managerial capital rather than to flows of goods. Unfortunately, those oddities, to the extent that they would reduce the likelihood of foreign direct investment in U.S. telecommunications carriers, appear in retrospect to have made the borrowed language unexpectedly appealing to protectionists in the Senate.

The new subsection would have made section 310(b) inapplicable in cases where reciprocal policies among two countries permitted more liberal foreign direct investment in telecommunications than the 25 percent level mentioned in section 310(b). Proposed subsection 310(f)(1) stated that section 310(b)

shall not apply to any license held, or for which application is

---

68. Telecommunications Competition and Deregulation Act of 1995 (discussion draft, Jan. 31, 1995) [hereinafter *Senate Draft*].

69. Interview with Keith E. Bernard, Vice President of International and Regulatory Affairs, Cable & Wireless, Inc. (Aug. 30, 1995).

made, after the date of enactment of the Telecommunications Act of 1995 with respect to any alien (or representative thereof), corporation, or foreign government (or representative thereof) if the United States Trade Representative has determined that the foreign country of which such alien is a citizen, in which such corporation is organized, or in which such foreign government is in control provides mutually advantageous market opportunities for broadcast, common carriers, or aeronautical enroute or fixed radio station licenses to citizens of the United States (or their representatives), corporations organized in the United States, and the United States Government.[70]

That provision was the carrot: If, for example, the United Kingdom allowed 100 percent foreign ownership of wireless common carriers (such as providers of cellular telephony), then a U.K. investor could buy 100 percent of a U.S. cellular carrier.

The January 31 draft also contained a stick, known as the "snapback for reciprocity failure."[71] Proposed section 310(f)(2) provided:

If the United States Trade Representative determines that any foreign country with respect to which it has made a determination under paragraph (1) ceases to meet the requirements for that determination, then—

(A) subsection [310](b) shall apply with respect to such aliens, corporations, and government (or their representatives) on the date on which the Trade Representative publishes notice of its determination under this paragraph, and

(B) any license held, or application filed, which could not be held or granted under subsection [310](b) shall be withdrawn, or denied, as the case may be, by the Commission under the provisions of subsection [310](b).[72]

Despite the plethora of jargon and acronyms used in the daily parlance of telecommunications law, the term "snapback" was entirely foreign to lawyers specializing in the regulation of telecommunications services and facilities. Their unfamiliarity underscored the fact

---

70. *Senate Draft*, § 105, at 39, ll. 4–19 (proposed 47 U.S.C. § 310(f)(1)).

71. *Id.* at 39, l. 20 (proposed 47 U.S.C. § 319(f)(2)).

72. *Id.* at 39, l. 21, to 40, l. 10 (proposed 47 U.S.C. § 310(f)(2)).

that the Senate language was a square peg in a round hole—a provision drafted to regulate trade in telecommunications equipment now being used to regulate trade in financial and managerial capital. "Snapback" meant something to international trade lawyers, but not to telecommunications lawyers.

Given the forced nature of the Senate language, it is not surprising that the snapback provision was deeply problematic for several reasons. First, key words did not match the existing terminology of telecommunications law. The FCC does not "withdraw" a license or an application. The agency can revoke a license or decline to renew it, or it can refuse to grant an application for a license. But the withdrawal of an application is something that an *applicant* would have to do, and there is no concept at all corresponding to a licensee's "withdrawing" a license already granted to him.

Second, the snapback provision in proposed subsection 310(f)(2)(B) would have been draconian and would have run counter to common sense. Suppose that Canada allows 33 percent foreign ownership and that Bell Canada is therefore allowed to purchase 33 percent of a U.S. cellular telephone company. Suppose that the Canadian government subsequently reduces its foreign ownership limit to 25 percent. The proposed subsection 310(f)(2)(B) would have caused Bell Canada to have the license for the cellular carrier "withdrawn . . . or denied." The remedy, in other words, would not have been that Bell Canada would have had to reduce its ownership of the cellular company to 25 percent; rather, Bell Canada would have had to forfeit its license entirely.

There is an economic rationale that one could impute to the harshness of the snapback provision. (Whether the statutory language was ever intended by its sponsors to effect that purpose is another matter.) It is widely recognized in economic theory that commitments made in bargaining situations influence the behavior of other actors only to the extent that the person making such commitments is credibly bound (by himself or others) to honoring them.[73] In effect, the forfeiture feature of proposed subsection 310(f)(2)(B) would have made Bell Canada the guarantor of the foreign investment policies of

---

73. *See, e.g.,* PAUL R. MILGROM & JOHN ROBERTS, ECONOMICS, ORGANIZATION AND MANAGEMENT 131 (Prentice-Hall 1992); OLIVER E. WILLIAMSON, THE ECONOMIC INSTITUTIONS OF CAPITALISM 167 (Free Press 1985); THOMAS C. SCHELLING, THE STRATEGY OF CONFLICT (Oxford University Press 1960).

the Canadian government. In the jargon of economics and game theory, the Canadian government would have made a "credible commitment," and Bell Canada's investment in the U.S. cellular licensee would have become a "hostage." The Canadian government would know that the consequences of lowering its foreign ownership limit, having once raised it, would be that Bell Canada's investments (indeed *all* Canadian investments) in U.S. radio common carriers that exceeded 25 percent ownership of the carrier would be lost. Consequently, the Canadian government would have a strong incentive not to lower the foreign ownership limit.

Four insurmountable problems with that argument cause one to suspect that the forfeiture provision resulted from sloppy draftsmanship rather than from a subtle attempt to create binding commitments to reduce barriers to foreign investments. First, if the United States can draft such a provision, so can other nations. Any firm, U.S. or foreign, would be exceedingly reluctant to invest in a country that permitted the expropriation of one's investment without cause. Thus, the prospect of retaliatory enactment of snapback laws in other nations would likely *reduce* the amount of foreign direct investment both in the United States and in the nations where U.S. telecommunications firms would like to invest. Second, the prospect that Canada could never lower its foreign ownership limit without sacrificing Canadian investments in wireless in the United States would create a powerful incentive for Canada never to raise its foreign ownership limits in the first place. Third, knowing that dynamic, the Canadian government could evade the bite of the snapback provision by secretly notifying Canadian companies with wireless investments in the United States that the Canadian government intended to lower the foreign ownership ceiling and by advising those companies to disinvest to the 25 percent level in the United States before the Canadian government announced its new policy. Fourth, a company cannot surrender part of a license. How could a company partly owned by Canadian investors have its license "withdrawn" or "denied" unless the other shareholders—the company's *American* shareholders—also were made to forfeit their license at the same time? But if that happened, U.S. companies would have a strong incentive *never* to accept any foreign investment in excess of 25 percent—for if they did, all shareholders of U.S. citizenship would be turned into guarantors of the foreign ownership policies of the Canadian government. In short, the snapback provision in S. 652 was poorly conceived and drafted; it would have been disastrous in the counterproductive incentives that it would have created for foreign direct investment.

On February 14, 1995, Senator Ernest Hollings of South Carolina, the ranking minority member of the Senate Commerce Committee, circulated a Democratic draft bill in response to Chairman Pressler's.[74] It contained no provisions addressing foreign ownership.

The Foreign Ownership Provisions in S. 652

On March 29, 1995, Senator Pressler introduced S. 652, the revised Telecommunications Competition and Deregulation Act of 1995, which reflected certain provisions sought by Senator Hollings. In its preamble the bill stated: "The efficient development of competitive United States communications markets will be furthered by policies which aim at ensuring reciprocal opening of international investment opportunities."[75] The foreign ownership provision was substantially the same as that in the January 31 draft bill.[76] There were five differences between the two bills.

First, whereas the January 31 draft bill encompassed "any license," the March 29 bill narrowed proposed section 310(f) to "any common carrier license." The practical import of that change was to exclude radio and television broadcast licenses from the amendments to section 310(b).

Second, the March 29 bill changed, from the U.S. Trade Representative to the FCC, the government body that would determine access to foreign markets.

Third, the bill redefined market-access determination. In the January 31 discussion draft it had been whether the foreign government provided "mutually advantageous market opportunities for broadcast, common carriers, or aeronautical enroute or fixed radio station licenses." In the March 29 bill the test became whether the foreign government provided "equivalent market opportunities for common carriers." Thus, the March 29 bill excluded broadcast

74. Universal Service Telecommunications Act of 1995 (staff working draft, Feb. 14, 1995).

75. S. 652, § 5(14), 104th Cong., 1st Sess. (1995).

76. *Id.* § 105 (creating 47 U.S.C. § 310(f)).

licenses from the reciprocity test, along with aeronautical enroute or fixed radio station licenses.

Fourth, the March 29 bill added the following sentence to the end of proposed subsection 310(f)(1): "The determination of whether market opportunities are equivalent shall be made on a market segment specific basis." The committee report gave a curious interpretation of that sentence:

> The FCC must enforce the provision on a market segment by market segment basis. For instance, if a foreign company wishes to acquire a common carrier license, the openness of the foreign market to U.S. communications equipment manufacturers is not the relevant market to examine. If a foreign company wishes to acquire a common carrier license, the FCC should examine the openness of the foreign country's common carrier market to U.S. investment.[77]

It is difficult to take that report language at face value, for it imparts so lax an interpretation to the statute that the concept of a "market segment" becomes larger than that of simply a "market." The report's interpretation could conceivably allow comparisons of cellular telephony in one country with long-haul fiber-optic transport in another. A more literal reading of the statutory text suggests that the Senate Commerce Committee intended to create something closer to a "mirror reciprocity" standard for section 310(b)(4).[78]

Fifth, the March 29 bill replaced USTR with the FCC as the government body that would determine whether the snapback provision would be triggered by a failure of reciprocity.

The full Senate passed S. 652 by a vote of 81 to 18 on June 15,

---

77. S. REP. NO. 23, 104th Cong., 1st Sess. 34 (1995).

78. The committee report also misstated the existing law relating to section 310(b)(4): "Existing section 310(b) of the 1934 Act provides in relevant part . . . that an alien *may not own more than 25%* of any corporation that directly or indirectly owns or controls any corporation to which a common carrier license is granted." *Id.* at 33 (emphasis added). As chapter 3 documented, that misinterpretation of section 310(b)(4) correctly summarizes the FCC's reading of the statute; but the agency's version of section 310(b)(4) is not what the plain language of the statute says. Ironically, the House committee report accompanying H.R. 1555 chastised the FCC for that misinterpretation of the statute. H.R. REP. NO. 204, 104th Cong., 1st Sess. 120–21 (1995).

1995.[79] The version of proposed section 310(f) had changed significantly since the March 29 bill and in its entirety now read:

(f) TERMINATION OF FOREIGN OWNERSHIP RESTRICTIONS.

(1) RESTRICTION NOT TO APPLY WHERE RECIPROCITY FOUND. Subsection (b) shall not apply to any common carrier license held, or for which application is made, after the date of enactment of the Telecommunications Act of 1995 with respect to any alien (or representative thereof), corporation, or foreign government (or representative thereof) if the Commission determines that the foreign country of which such alien is a citizen, in which such corporation is organized, or in which such foreign government is in control provides equivalent market opportunities for common carriers to citizens of the United States (or their representatives), corporations organized in the United States, and the United States Government (or its representative): *Provided*, That the President does not object within 15 days of such determination. If the President objects to a determination, the President shall, immediately upon such objection, submit to Congress a written report (in unclassified form, but with a classified annex if necessary) that sets forth a detailed explanation of the findings made and factors considered in objecting to the determination. The determination of whether market opportunities are equivalent shall be made on a market segment specific basis within 180 days after the application is filed. While determining whether such opportunities are equivalent on that basis, the Commission shall also conduct an evaluation of opportunities for access to all segments of the telecommunications market of the applicant.

(2) SNAPBACK FOR RECIPROCITY FAILURE. If the Commission determines that any foreign country with respect to which it has made a determination under paragraph (1) ceases to meet the requirements for that determination, then—

(A) subsection (b) shall apply with respect to such aliens, corporations, and government (or their representatives) on the date on which the Commission publishes notice of its determination under this paragraph, and

(B) any license held, or application filed, which could

---

79. 141 CONG. REC. S737 (daily ed. June 15, 1995). Democrats voted for the bill 30 to 16, and Republicans voted for it 51 to 2.

> not be held or granted under subsection (b) shall be
> withdrawn, or denied, as the case may be, by the Commis-
> sion under the provisions of subsection (b).[80]

The new version modified proposed section 310(f)(1) in three respects. First, it introduced the president's power to "object" to a
market-opportunities determination by the FCC, although the new text
only implied silently that the president's objection would trump the
FCC's determination. Second, it specified that the FCC's market-
opportunities determination must be made within 180 days after an
application had been filed. Third, it added the provision that "the
Commission *shall* also conduct an evaluation of opportunities for
access to *all* segments of the telecommunications market of the
applicant."[81] That last change would substantially increase the factual
and economic complexity of an FCC proceeding to determine whether
to allow foreign ownership exceeding 25 percent. In addition to
modifying proposed section 310(f), the bill that passed the Senate also
clarified that the amendment of section 310(b) would not implicitly
amend or repeal the applicability of the Exon-Florio Amendment to
a telecommunications transaction.[82]

THE HOUSE BILL

On May 3, 1995, Representative Thomas J. Bliley, Jr., chairman of
the House Commerce Committee, introduced H.R. 1555, the
"Communications Act of 1995."[83] His cosponsors included Representatives John Dingell, the ranking Democrat on the committee, and
Jack M. Fields, Jr., the chairman of the Subcommittee on Telecommunications and Finance. As introduced, the bill did not address
foreign ownership restrictions.

On May 20, 1995, the subcommittee reported the text of H.R.
1555. By that time, the subcommittee had incorporated into H.R.
1555[84] a separate bill previously introduced by two of its members,

---

80. S. 652, § 105(a), 104th Cong., 1st Sess. (1995) (proposed 47 U.S.C. § 310(f)).

81. *Id.* (emphasis added).

82. "Nothing in this section (47 U.S.C. 310) shall limit in any way the application
of the Exon-Florio law (50 U.S.C. App. 2170) to any transaction." *Id.* § 105(c).

83. H.R. 1555, 104th Cong., 1st Sess. (1995).

84. *Id.* § 302(a).

Representatives Michael G. Oxley and Rick Boucher, that would simply repeal section 310(b).[85] In addition, H.R. 1555 amended section 310(a) to allow the FCC to license foreign governments to operate mobile earth stations for satellite news gathering.[86]

The Oxley Amendment

In markup on May 25, 1995, the foreign ownership provisions in H.R. 1555 changed entirely. Recognizing after discussions with the White House that his proposal to repeal section 310(b) would fail, Representative Oxley instead offered an amendment, which passed. The Oxley Amendment consisted of two parts. The first part was the minor provision, already contained in H.R. 1555, exempting foreign governments from section 310(a) for purposes of news gathering by satellite.[87]

The second, and more important, part of the Oxley Amendment was the House's alternative to the Senate's proposal to amend section 310(b)(4) to impose a bilateral reciprocity test. Like S. 652, the Oxley Amendment would create a new section 310(f). The mechanics of Representative Oxley's amendment would be entirely different from its counterpart in S. 652, however. In its entirety, the Oxley Amendment's version of section 310(f) would read:

(f) TERMINATION OF FOREIGN OWNERSHIP RESTRICTIONS.
(1) RESTRICTIONS NOT TO APPLY. Subsection (b) shall not apply to any common carrier license granted, or for which

---

85. H.R. 514, 104th Cong., 1st Sess. (1995).
86. H.R. 1555, § 302(b), 104th Cong., 1st Sess. (1995).
87. *Id.* § 302(a). The provision would amend section 310(a) to read:

(a) GRANT TO OR HOLDING BY FOREIGN GOVERNMENT OR REPRESEN-TATIVE. No station license required under title III of this Act shall be granted to or held by any foreign government or any representative thereof. This subsection shall not apply to licenses issued under such terms and conditions as the Commission may prescribe to mobile earth stations engaged in occasional or short-term transmissions via satellite of audio or television program material and auxilliary [*sic*] signals if such transmissions are not intended for direct reception by the general public in the United States.

Id.

application is made, after the date of enactment of this subsection with respect to any alien (or representative thereof),
corporation, or foreign government (or representative thereof)
if—

(A) the President determines that the foreign country
of which such alien is a citizen, in which such corporation
is organized, or in which the foreign government is in
control is party to an international agreement which
requires the United States to provide national or most-
favored-nation treatment in the grant of common carrier
licenses; or

(B) the Commission determines that not applying
subsection (b) would serve the public interest.

(2) COMMISSION CONSIDERATIONS. In making its determination, under paragraph (1)(B), the Commission may consider,
among other public interest factors, whether effective competitive opportunities are available to United States nationals
or corporations in the applicant's home market. In evaluating
the public interest, the Commission shall exercise great
deference to the President with respect to United States
national security, law enforcement requirements, foreign
policy, the interpretation of international agreements, and trade
policy (as well as direct investment as it relates to international
trade policy). Upon receipt of an application that requires a
finding under this paragraph, the Commission shall cause
notice thereof to be given to the President or any agencies
designated by the President to receive such notification.

(3) FURTHER COMMISSION REVIEW. Except as otherwise
provided in this paragraph, the Commission may determine
that any foreign country with respect to which it has made a
determination under paragraph (1) has ceased to meet the
requirements for that determination. In making this determination, the Commission shall exercise great deference to the
President with respect to United States national security, law
enforcement requirements, foreign policy, the interpretation of
international agreements, and trade policy (as well as direct
investment as it relates to international trade policy). If a
determination under this paragraph is made then—

(A) subsection (b) shall apply with respect to such
aliens, corporation, and government (or their representatives) on the date that the Commission publishes notice of
its determination under this paragraph; and

(B) any license held, or application filed, which could

not be held or granted under subsection (b) shall be reviewed by the Commission under the provisions of paragraphs (1)(B) and (2).

(4) OBSERVANCE OF INTERNATIONAL OBLIGATIONS. Paragraph (3) shall not apply to the extent the President determines that it is inconsistent with any international agreement to which the United States is a party.

(5) NOTIFICATION TO CONGRESS. The President and the Commission shall notify the appropriate committees of Congress of any determinations made under paragraph (1), (2), or (3).

Compared with the proposed version of section 310(f) in S. 652, the Oxley Amendment was vastly superior in terms of the incentives that it would have created for mutual reduction in barriers to foreign direct investment. In that respect, the Oxley Amendment, which was subsequently improved during floor debate of H.R. 1555, represented a good second-best alternative to repeal of section 310(b).

Several points are immediately apparent from examination of section 310(f)(1) in the Oxley Amendment. First, the lifting of foreign ownership restrictions applied only to common carrier licenses; as in S. 652, television and radio broadcast licenses and aeronautical enroute or fixed radio station licenses would not have been affected. Second, either the president or the FCC could have made the determination that section 310(b) should not apply. Third, the lifting of the foreign ownership restrictions by the president or the FCC encompassed not only section 310(b)(4), but also sections 310(b)(1), 310(b)(2), and 310(b)(3); in other words, the president and the FCC would have been empowered to waive restrictions not only on stock ownership, but also on citizenship requirements for officers and directors. Fourth, the lifting of section 310(b) would have occurred automatically, without need for any additional public interest determination, if the president determined that the investor's country was a party to an agreement (presumably a future multilateral agreement flowing from talks conducted pursuant to the General Agreement on Trade in Services, or GATS) requiring the United States to grant common carrier licenses on a national or most-favored-nation basis. Fifth, the FCC would have had the discretion—which it lacks under sections 310(b)(1), 310(b)(2), and 310(b)(3)—*not* to limit foreign investment if it determined that so doing would serve the public interest.

The Oxley Amendment also differed significantly from S. 652 in the nature of the market-access determination that the FCC would make. Section 310(f)(2) in the Oxley Amendment instructed the FCC to examine "effective competitive opportunities" for U.S. nationals or corporations in the foreign investor's "home market." S. 652, in contrast, spoke of "equivalent market opportunities" and did not specifically limit the consideration of such opportunities to those in the foreign investor's home market. The committee report elaborated on the meaning of "effective competitive opportunities":

> It is the Committee's intent that by applying a "reciprocity" approach, U.S. markets will be open to foreign investment from another country, to the same extent that country's market is open to U.S. investment. Thus, in making such determinations, it is the Committee's intent that the Commission focus principally on the effective competitive opportunities. In other words, absent the unusual circumstance of a serious national security or law enforcement consideration, if an applicant is otherwise well-qualified, a finding of adequate reciprocity in the relevant country should result in a grant of a license.[88]

The relevance of the Oxley Amendment's specific reference to "home market" will become clearer when we examine the FCC's proposed rule and final rule for interpreting section 310(b)(4) shortly.

The Oxley Amendment was more specific than S. 652 in stating that the FCC shall "exercise great deference" to the president on "United States national security, law enforcement requirements, foreign policy, the interpretation of international agreements, and trade policy (as well as direct investment as it relates to international trade policy)." As examination of the FCC's proposed rule will show, that language can be seen as a rebuke of the FCC's attempt to expand its role in matters concerning international trade and foreign affairs.

The most significant difference between the Oxley Amendment and S. 652 was that the former would have avoided the ill-conceived "snapback for reciprocity failure." Section 310(f)(3) of the Oxley Amendment would have empowered the FCC to make a determination that a country had "ceased to meet the requirements" for the lifting of the foreign investment restrictions in section 310(b). According to the committee report, the FCC would rarely find justification for resorting to that power:

---

88. H.R. REP. NO. 204, *supra* note 78, at 121.

> The Committee anticipates that this provision would be utilized only where the policies and practices of a foreign country are egregious and would result in significant harm to U.S. companies, *e.g.*, where national security and law enforcement concerns would require such action.[89]

Rather than provide that any license held by the foreign entity "shall be withdrawn . . . or denied" by the FCC, as S. 652 provided in its section 310(f)(2)(B), the Oxley Amendment provided only that such license "shall be reviewed by the Commission." In other words, loss of the license would not have been automatic; the Oxley Amendment would thus have permitted the FCC to pursue the more sensible remedy of ordering the foreign investor, in the case of section 310(b)(4), to reduce its holdings to 25 percent. Section 310(f)(4) of the Oxley Amendment, however, provided that the FCC shall not have the power to make such a determination "to the extent the President determines that it is inconsistent with any international agreement to which the United States is a party."

Floor Debate and House Passage

The full House passed H.R. 1555 by a vote of 305 to 117 on August 4, 1995.[90] On the House floor, members approved several changes to the Oxley Amendment that would have made the provision even more hospitable to foreign direct investment. Representative Bliley, chairman of the House Commerce Committee, offered those changes in the floor manager's amendment and was joined by Representative Dingell and Representative Henry Hyde, chairman of the House Judiciary Committee.[91]

The first change that the floor manager's amendment made to the Oxley Amendment broadened the scope of proposed section 310(f)(1) so that its exemption of foreign investment from section 310(b)(4) would have applied not only to any common carrier license "granted,

89. *Id.* at 122.
90. 141 CONG. REC. H8506-7 (daily ed. Aug. 4, 1995).
91. *Id.* at H8444, H8451. The floor manager's amendment passed by a vote of 256 to 149. *Id.* at H8459.

or for which application is made, after the date of enactment of this subsection," but also to any common carrier license "held" after that date.[92] That change would have ensured that any existing foreign investor in a U.S. radio common carrier could use section 310(f)(1) to request the president or the FCC to permit him to raise his level of investment in the U.S. carrier without regard to the benchmarks in section 310(b) or the public interest showing that the FCC previously had required of a foreign investor and U.S. radio licensee seeking approval of investment exceeding the 25 percent benchmark in section 310(b)(4).

The floor manager's second change rewrote subsection 310(f)(1)(A) to clarify that the president's grant of authority for higher levels of foreign investment would not compromise national security or law enforcement. The change divided section 310(f)(1)(A) into two conjunctive elements and required, in new section 310(f)(1)(A)(ii), that the president determine "that not applying subsection [310](b) would be consistent with national security and effective law enforcement."[93]

The third change clarified that the FCC would determine, under section 310(f)(1)(B), whether *not* applying section 310(b) to a foreign investment would serve the public interest:

> (2) COMMISSION CONSIDERATIONS. In making its determination under paragraph (1), the Commission shall abide by any decision of the President whether application of section (b) is in the public interest due to national security, law enforcement, foreign policy or trade (including direct investment as it relates to international trade policy) concerns, or due to the interpretation of international agreements. In the absence of a decision by the President, the Commission may consider, among other public interest factors, whether effective competitive opportunities are available to United States nationals or corporations in the applicant's home market. Upon receipt of an application that requires a determination under this paragraph, the Commission shall cause notice of the application to be given to the President or any agencies designated by the President to receive such notification. The Commission shall not make a determination under paragraph (1)(B) earlier than 30 days

---

92. *Id.* at H8449.
93. *Id.*

after the end of the pleading cycle or later than 180 days after the end of the pleading cycle.[94]

The effect of that substitute language would have been to increase significantly, and to make more precise, the role of the president or his designee (most likely, the U.S. trade representative) in scrutinizing foreign direct investment that would exceed the benchmarks in section 310(b). Gone was the language directing the FCC to "exercise great deference to the President" on matters of national security, foreign policy, and the like. The new language commanded the FCC not to defer, but to obey: "the Commission *shall abide by* any decision of the President" on such matters.[95]

In the absence of a presidential decision on the proposed investment, the FCC would have been allowed to consider "among other public interest factors, whether effective competitive opportunities are available to United States nationals or corporations in the applicant's home market."[96] But what "other public interest factors" would then have been permissible for the FCC to consider? The clear implication of the floor manager's language was that the FCC would be *precluded from considering* those factors described in the original language of the Oxley Amendment that the floor amendment had stricken. In other words, the FCC was not to consider "United States national security, law enforcement requirements, foreign policy, the interpretation of international agreements, and trade policy (as well as direct investment as it relates to international trade policy)." Instead, the House directed the FCC to defer *entirely* to the president on such matters and not to undertake its own independent evaluation of them.

As in the original Oxley Amendment, the FCC would be required to notify the president of any application that it received requesting an exemption from section 310(b). But unlike the Oxley Amendment, the floor manager's substitute language would have ensured that the president would have the opportunity to preempt any FCC decision on such an application. The FCC would have been forbidden to make a determination until thirty days after the end of the pleading cycle—during which time, obviously, the president

94. *Id.*
95. *Id.* (emphasis added).
96. *Id.*

would have been free to issue his own determination and thus moot the FCC's consideration of the application.[97] In addition, the floor manager's substitute added that the FCC would have to make its determination within six months of the filing of the last round of pleadings.[98]

The floor manager's fourth change rewrote the process by which the FCC would have been permitted to review whether the foreign investment in question should subsequently be subjected to section 310(b) because of changed circumstances:

> (3) FURTHER COMMISSION REVIEW. The Commission may determine that, due to changed circumstances relating to United States national security or law enforcement, a prior determination under paragraph (1) ought to be reversed or altered. In making this determination, the Commission shall accord great deference to any recommendation of the President with respect to United States national security or law enforcement. If a determination under this paragraph is made then—
> (A) subsection (b) shall apply with respect to such aliens, corporation, government (or their representatives) on the date that the Commission publishes notice of its determination under this paragraph; and
> (B) any license held, or application filed, which could not be held or granted under subsection (b) shall be reviewed by the Commission under the provisions of paragraphs (1)(B) and (2).[99]

Compared with the original Oxley Amendment, that language would have greatly reduced the discretion of the FCC subsequently to rescind the authorization given a foreign investor to exceed the benchmarks in section 310(b). The FCC would have been permitted to base its determination only on "changed circumstances relating to United States national security or law enforcement." That language differed completely from the Oxley Amendment, which would have based the FCC's review on a "foreign country . . . [having] ceased to meet the requirements for [the earlier] determination" that one of its citizens should be allowed to invest in a U.S. radio licensee in

---

97. *Id.*
98. *Id.*
99. *Id.* at H8449–50.

excess of the benchmarks in section 310(b). In other words, the FCC's review would have been confined to national security and law enforcement considerations, not international trade policy concerning foreign direct investment. That conclusion finds further support in the fact that the floor manager's substitute language deleted the Oxley Amendment's clarification that, in making a determination under that review process, the FCC "shall exercise great deference to the President with respect to . . . foreign policy, the interpretation of international agreements, and trade policy (as well as direct investment as it relates to international trade policy)." Under the floor manager's substitute, there would have been no need to instruct the FCC to defer to the president on such matters for the simple reason that the FCC would not have been authorized to consider them in the first place when making a review determination.[100]

The floor manager's fifth change was to delete, because it was no longer relevant, the requirement in the Oxley Amendment that the provision authorizing the FCC to undertake a subsequent review of a particular foreign investment "shall not apply to the extent the President determines that it is inconsistent with any international agreement to which the United States is a party." Because the floor manager's substitute authorized the FCC subsequently to review a foreign investment only on national security or law enforcement grounds, the agency would never be in the position of making potentially embarrassing pronouncements about international trade policy of the sort that the deleted provision in the Oxley Amendment was obviously intended to prevent.

Sixth, the floor manager's substitute created a new section 310(f)(5), which provided: "Any Presidential decisions made under the provisions of this subsection shall not be subject to judicial review."[101] Any final decision by the FCC under proposed section 310(f) would be appealable to the U.S. Court of Appeals for the D.C. Circuit under section 402(b) of the Communications Act.[102] In any FCC matter, the prospect of appeal raises the likelihood of remand to

---

100. No congressional debate or committee report language explained the insertion of "law enforcement" as a factor justifying FCC consideration. Perhaps Congress was concerned about wealthy foreign drug lords' acquiring control over sophisticated means of radio communications in the United States.

101. 141 CONG. REC. H8450 (daily ed. Aug. 4, 1995).

102. 47 U.S.C. § 402(b).

the agency and, consequently, a protracted administrative process. The inability of losing parties (such as competitors of the U.S. radio licensee receiving the foreign direct investment) to appeal a presidential determination would have greatly expedited the process by which a major foreign investor could begin operations in the United States. For that reason, a foreign investor would vastly prefer for the president, rather the FCC, to determine that the proposed investment should be allowed to exceed the benchmarks of section 310(b).

The seventh and final change that the floor manager made to the Oxley Amendment was to specify: "The amendments made by this section shall not apply to any proceeding commenced before the date of enactment of this Act."[103] In practical effect, that provision would have excluded the investment in Sprint by France Télécom and Deutsche Telekom from the liberalized standards that section 310(f) would have established.

During floor debate on H.R. 1555 on August 4, 1995, Representative Oxley expressed "firm support" for the floor manager's amendment, noting that it made "some important refinements regarding foreign ownership."[104] Oxley's principal point was "to clarify the committee report language . . . concerning how the [Federal Communications] Commission should determine the home market of an applicant."[105] He elaborated:

> It is the committee's intention that in determining the home market of any applicant, the Commission should use the citizenship of the applicant—if the applicant is an individual or partnership—or the country under whose laws a corporate applicant is organized. Furthermore, it is our intent that in order to prevent abuse, if a corporation is controlled by entities—including individuals, other corporations or governments—in another country, the Commission may look beyond where it is organized to such other country.[106]

Representative Oxley concluded his floor remarks by emphasizing that

---

103. 141 CONG. REC. H8450 (daily ed. Aug. 4, 1995).
104. *Id.* at H8458.
105. *Id.*
106. *Id.* Those remarks reiterated a similar statement made in the committee report. H.R. REP. NO. 204, *supra* note 78, at 122.

the amendments to section 310(b), including the floor manager's modifications of the Oxley Amendment, "have the support of the administration and the ranking members of the Committee on Commerce."[107] The foreign ownership provisions of H.R. 1555 elicited no further floor debate from any member.

THE FCC'S PROPOSED RECIPROCITY RULE

As debate on amending section 310(b) was beginning in the 104th Congress in 1995, the FCC proposed to exercise its existing authority under the Communications Act to impose a reciprocity test for section 310(b).[108] The agency proposed to incorporate an "effective market access standard" into the public interest determination that it conducts under section 310(b)(4) in situations where foreign ownership would exceed 25 percent.[109]

Under the proposed rule, the FCC would "consider whether the foreign entity's primary markets pass the effective market access test" "when an applicant in whom foreign ownership in the parent holding company exceeds the 25 percent benchmark seeks a common carrier radio license, or when a U.S. licensee seeks to increase the level of foreign ownership in its parent holding company beyond the 25 percent benchmark or previously authorized levels of foreign ownership."[110] The likely operation of the rule was suggested by the following question posed by the FCC:

> Thus, for example, if a foreign entity seeks to invest in the parent holding company of an applicant for authority to provide Personal Communication Services ("PCS"), should we consider whether U.S. companies can provide PCS, or its functional equivalent, in the foreign entity's primary market?[111]

The question suggested that the FCC expected to compare the regulatory treatment of foreign investment on a service-by-service basis.

---

107. 141 CONG. REC. H8458 (daily ed. Aug. 4, 1995).
108. *Market Entry and Regulation NPRM*, 10 F.C.C. Rcd. at 5257 ¶ 4.
109. *Id.* at 5293 ¶ 92.
110. *Id.* at 5295 ¶ 95.
111. *Id.* ¶ 96.

The FCC's notice of proposed rulemaking was reminiscent of the manner in which the agency has granted waivers under section 310(b)(4). There was always some additional fact that the agency felt free to consider as pertinent to its public interest determination. That added fact became the *deus ex machina* that provided a ready escape hatch for any desired outcome:

> We also seek comment on whether . . . we should find that our effective market access finding under Section 310(b)(4) is not dispositive of our decision to license a particular entity. For instance, once we have reviewed the effective market access element of our public interest analysis, should we also assess other public interest factors which might weigh in favor of, or against, allowing entry into the U.S. market? Such factors in this context could include the state of liberalization in the foreign country's other radio-based service markets, national security, or the competitiveness of the applicant's target market in the United States. Finally, we seek comment on whether, if we do consider effective market access, this would be a more tailored and predictable application of Section 310(b)(4) that will assist us in encouraging and recognizing foreign countries' efforts to liberalize their communications market.[112]

---

112. *Id.* at 5295–96 ¶ 96. The FCC's notice of proposed rulemaking also addressed section 214 certification of international carriers. The factors that the agency considered potentially relevant to its effective market access standard were numerous, and they suggested ways in which the FCC could be expected to broaden its assessment of effective market access for purposes of section 310(b)(4):

> We propose to define effective market access as the ability for U.S. carriers, either currently or in the near future, to provide basic, international telecommunications facilities-based services in the primary markets served by the foreign carrier seeking entry. . . . We would consider the following factors, none of which would be dispositive, to determine whether effective market access exists: (1) whether U.S. carriers can offer in the foreign country international facilities-based services substantially similar to those the foreign carrier seeks to offer in the United States; (2) whether competitive safeguards exist in the foreign country to protect against anticompetitive and discriminatory practices, including cost allocation rules to prevent cross-subsidization; (3) the availability of published, nondiscriminatory charges, terms and conditions for interconnection to foreign domestic carriers' facilities for termination and origination of international services; (4) timely and nondiscriminatory disclosure of technical information needed to use or interconnect with

The FCC also suggested that it would be more inclined, under its bilateral reciprocity model, to allow more than 25 percent foreign ownership in television and radio broadcasters.[113] Again, consistent with agency practice, the FCC appeared willing to allow higher levels of foreign investment in broadcasters so long as the agency could continue to justify any desired result in a given case by selectively imputing significance to facts *extraneous to the level of foreign ownership*: "[E]ven if we incorporate the effective market access standard in our evaluation of broadcast applications, the nature of the case-by-case review conducted under Section 310(b)(4) is such that we retain the discretion to deny particular applications if warranted by the facts of a specific case."[114]

The FCC's proposed rule was problematic for all the reasons that reciprocity rules are. But there was an additional feature of the FCC rule that would have made it highly anticompetitive. The agency's examination of market access would not be confined to the home market of the foreign carrier, but rather would encompass its "primary markets." A primary market was defined to be a market "where a carrier has a significant facilities-based presence."[115] Cable & Wireless, for example, is a British company whose home market, the United Kingdom, is completely open to foreign investment and indeed is characterized by extensive direct investment by U.S. telephone and cable television companies. But Cable & Wireless also has a major interest in Hongkong Telecom, the incumbent provider of telephone service in Hong Kong, which until 1995 enjoyed a monopoly franchise. The FCC's proposed rule would have treated Hong Kong

---

carriers' facilities; (5) the protection of carrier and customer proprietary information; and (6) whether an independent regulatory body with fair and transparent procedures is established to enforce competitive safeguards.

Id. at 5271 ¶ 40. Again, in a manner that would maximize its discretion, the agency envisioned considering those factors on a selective basis: "In considering these indicators to determine whether effective market access exists, we will not necessarily require that each factor be present in order to make a favorable finding, particularly if there is evidence that the market is fully competitive. Rather, we will look to the arguments of the applicant and commenting parties as to the appropriate weight of each factor in a particular market." *Id.*

113. *Id.* at 5296–98 ¶¶ 99–103.
114. *Id.* at 5298 ¶ 102.
115. *Id.* at 5271 ¶ 40.

as one of Cable & Wireless's "primary markets" and thus could have (and presumably would have) considered the company to be in violation of section 310(b)(4) if its U.S. holding company held more than 25 percent of an American radio licensee subject to that statute.

The incentives that such a rule would create are doubly perverse. First, the FCC's proposed rule would insulate existing U.S. carriers from competition taking the form of entry by established international carriers. It would reduce the likelihood or delay the arrival of a fourth international full-service network that could compete against the three global networks already being formed by AT&T, BT and MCI, and Sprint in conjunction with France Télécom and Deutsche Telekom.

Second, the "primary markets" test would discourage major foreign carriers—such as Cable & Wireless, Telefónica de España, Canada's BCE, and Japan's NTT—from investing in significant foreign markets that currently are closed to competition. In virtually every privatization, however, the government's sale of ownership in the state-owned telephone company includes a statutory monopoly for a limited term of years, during which time the private owner must significantly upgrade infrastructure. The FCC's proposed rule would have put the following choice to the world's largest overseas telecommunications firms: Even if your home market is open to U.S. foreign direct investment, if you want to invest more than 25 percent in a U.S. radio common carrier, you must forgo the opportunity to participate in significant privatizations around the world on terms that would include the enjoyment of a temporary statutory monopoly. If foreign carriers chose to invest in high-growth markets in Asia or Latin America, for example instead of in the United States, then American consumers would suffer. On the other hand, if foreign carriers invested in the United States and consequently declined to invest (or disinvested) in emerging markets elsewhere in the world, those carriers would retard the development of a modern global telecommunications infrastructure. Even under the second scenario, the United States would indirectly suffer harm in the sense that American corporations have substantial needs for telecommunications services in foreign countries. Of course, one cynical explanation for the proposal of "primary markets" test is that, by putting large foreign carriers to that choice, the FCC would increase the likelihood that U.S. telecommunications firms would win the opportunity to invest in foreign privatizations that would confer a temporary monopoly privilege. In other words, it would be perfectly fine for

U.S. telecommunications firms to secure the same temporary monopolies in emerging markets that would provide the FCC's rationale for disqualifying foreign carriers from competing in the U.S. market. It is preposterous that a rule having those perverse consequences could be "in the public interest."

Although it was fortunate for American economic welfare that the FCC ultimately retreated from its proposal of the "primary markets" test, Reed E. Hundt, chairman of the FCC, nonetheless strongly supported the proposed rule and the general principle of bilateral reciprocity in testimony before Congress in May 1995:

> Section 310 is a most powerful lever in opening restricted overseas markets to U.S. investment. *But, it would be a mistake simply to repeal Section 310(b).* Any change should be flexible enough to be market opening, not market closing. The Commission has instituted a proceeding proposing that the public interest standard it uses in determining whether to apply Section 310 take into account the reciprocal openness of the market in the nation from which a potential foreign owner comes. *Any revision of Section 310 should embody this reciprocity principle.*[116]

Chairman Hundt reached that assessment despite his concession in congressional testimony in March 1995 that "foreign governments view Section 310 as closing the U.S. market to their companies," that the statute "has become a metaphor for a closed U.S. market," and that he "seldom attend[s] an international gathering or bilateral negotiation without hearing the United States criticized for Section 310."[117] Evidently alluding to Director-General Krenzler's January 1995 letter, Chairman Hundt noted:

> The European Union . . . has recently argued that, since most

---

116. *Communications Law Reform: Hearings on H.R. 1555 Before the Subcomm. on Telecommunications and Finance of the House Commerce Comm.*, 104th Cong., 1st Sess. 266, 269 (1995) (statement of Reed E. Hundt, Chairman, Federal Communications Commission) (emphasis added).

117. *Trade Implication of Foreign Ownership Restrictions on Telecommunications Companies: Hearings on Section 310 of the Communications Act of 1934 Before the Subcomm. on Commerce, Trade, and Hazardous Materials of the House Commerce Comm.*, 104th Cong., 1st Sess., 15, 18 (1995) (statement of Reed E. Hundt, Chairman, Federal Communications Commission).

> U.S. carriers use some form of radio facility to supplement their
> wireline telecommunications facilities, any foreign equity invest-
> ment will be subject to the restrictions of Section 310. The Euro-
> pean Union, therefore, views the U.S. communications market as
> essentially closed. This dramatically overstates the truth. But it
> does not dramatically overstate the problem we face.[118]

The problem, in the chairman's view, was a "negative foreign perception."[119] The controversy, it would seem, stemmed not from the
reality of U.S. policy, but from the failure of Director-General
Krenzler to perceive it correctly and comprehend its magnanimity.
Chairman Hundt's comment was ironic, for he evidently did not
perceive the arrogance with which his statement could be taken by
Europeans, even though he acknowledged in the same breath that
"certain foreign governments have incorporated, or are proposing to
incorporate, parallel investment limitations in their own regulatory
frameworks."[120]

To the extent that economic analysis has informed the FCC
proposal for bilateral reciprocity at all—of which there was absolutely
no evidence in its notice of proposed rulemaking—Chairman Hundt's
reasoning appeared to be the following: Unilateral repeal of section
310(b) would constitute a concession to foreign countries that, by
sacrificing a bargaining chip, would be less likely than a reciprocity
rule to elicit a corresponding reduction in barriers to foreign direct
investment abroad. In that respect, unilateral repeal must be dismissed
as strategically naïve—not to mention politically unmarketable.

But that reasoning is itself flawed. If the Europeans in 1995 had
an incorrect "foreign perception" of FCC enforcement of section
310(b), why would there have been any reason to expect that their
perceptions would be more astute under a far more complex policy of
service-by-service bilateral reciprocity in which the FCC scrutinized
"effective market access" in "primary markets"? Even if the FCC
preferred its proposed policy because it appeared to be more strategically sophisticated in some game-theoretic sense than either the
existing statute or its outright repeal, it was still necessary for the
agency to proceed with realistic assumptions about how the strategic

118. *Id.*
119. *Id.*
120. *Id.*

responses of other governments could be based on misperception of U.S. law, nationalism, or domestic political pressures. For Chairman Hundt and the FCC to have neglected to do so was itself strategically naïve.

THE FCC'S 1995 REPORT AND ORDER
IMPOSING BILATERAL RECIPROCITY

Despite the flaws in the FCC's proposed rule on foreign investment, by November 1995 the agency had embraced a bilateral reciprocity test for both sections 310(b)(4) and 214 in its report and order establishing standards for regulating the entry of foreign carriers into the U.S. market for international telecommunications services and foreign investment in U.S. radio licensees.[121] The underlying policy goal of the report and order was to promote effective competition in the U.S. telecommunications services market, particularly the market for global, seamless network services.[122] The FCC believed that its new entry standards would prevent anticompetitive conduct in the provision of international services and facilities and encourage foreign governments to open their communications markets.[123] The report and order established entry criteria that the FCC believed were "necessary to promote effective competition in the U.S. market for international telecommunications services."[124] In seeking to ensure the public interest benefits of effective competition, the report and order added a new criterion—the "effective competitive opportunities (ECO) test"—to the public interest analysis conducted under both sections 214 and 310(b)(4) of the Communications Act.[125] The report and order also established a new notification requirement.

Notification Requirement

The FCC's report and order requires any U.S. carrier subject to section 214 or 310(b)(4) to notify the FCC sixty days before a foreign carrier's current or planned investment, direct or indirect, in that

---

121. *Market Entry and Regulation Report and Order*, 11 F.C.C. Rcd. at 3873.
122. *Id.* at 3877 ¶ 6.
123. *Id.*
124. *Id.* at 3875 ¶ 1.
125. 47 U.S.C. §§ 214, 310(b)(4).

carrier equaling or exceeding 10 percent or resulting in effective control at any level.[126] According to the FCC, the purpose of the notification requirement is to determine the regulatory status of the U.S. carrier and the applicability of the ECO test.[127] The FCC will presume that such an investment is in the public interest unless it raises a "substantial and material question of fact as to whether the investment serves the public interest, convenience and necessity."[128] For example, a foreign investor who owns a 10 percent interest would necessarily be benign unless he used his interest in concert with other foreigners.[129]

Like any prenotification requirement, that new obligation will signal to competitors a firm's strategic move before it can be executed. To the extent that speed and surprise are admirable qualities of rivalry in competitive markets, the FCC's notice requirement harms competition. Because the FCC has given itself the discretion to block foreign investment at levels as low as 10 percent, a foreign investor is unlikely to commit capital without first receiving a declaratory ruling from the FCC finding that his investment would be consistent with the public interest. The burden involved in proving that such foreign investment would serve the public interest merely increases the cost of capital to U.S. telecommunications firms. The FCC in effect has rewritten section 310(b)(4) to lower its benchmark from 25 to 10 percent. In the guise of making markets more competitive, the FCC has provided incumbent American telecommunications firms greater insulation from competitive entry by foreign carriers.

The Effective Competitive Opportunities Test

The most significant aspect of the FCC's report and order is the addition of the ECO test to the public interest analysis conducted under sections 214 and 310(b)(4).

Public Interest Analysis under Section 214. Under section 214 all carriers seeking to provide U.S. international telecommunications

---

126. *Market Entry and Regulation Report and Order*, 11 F.C.C. Rcd. at 3910 ¶ 97.
127. *Id.* ¶ 98.
128. *Id.* ¶ 97.
129. *Id.* ¶ 98.

services must obtain FCC authorization through a process of prior application and approval.[130] A foreign carrier seeking to provide U.S. international telecommunications services, either directly or through "affiliation" with a U.S. carrier, must obtain FCC authorization.[131] The FCC defines affiliation "as an ownership interest of greater than 25 percent, or a controlling interest at any level, in a U.S. carrier by a foreign investor."[132] Nonequity arrangements do not constitute affiliation for the purposes of the market-entry standard.[133] To determine affiliation, the FCC will aggregate multiple foreign carrier interests if the foreign carriers are likely to act in concert—where a contractual relation, such as a joint venture or marketing alliance, between two or more foreign carrier investors may affect the U.S. market for basic telecommunications services.[134] The FCC will also review a foreign carrier investment below the 25 percent threshold if it may have a significant impact on competition in the U.S. market for international telecommunications services.[135]

Under the rules promulgated in the order, the FCC, as part of its section 214 public interest analysis, must examine whether effective competitive opportunities currently exist, or with reasonable certainty will exist in the near future, for U.S. carriers in the "destination markets" of those foreign carriers seeking to enter the U.S. international services market, either directly or through an affiliation with either a U.S. facilities-based or resale carrier.[136] The FCC defines destination market to be not only the foreign carrier's home market, but also any market where the foreign carrier has the ability to leverage market power.[137] The agency applies the ECO test, on a route-by-route basis,[138] only to foreign carriers that have market power over the services and facilities in the destination markets that potentially can be leveraged on international routes to the detriment of unaffiliated U.S. carriers.[139] The FCC defines market power, in

---

130. *Id.* at 3881 ¶ 18.
131. 47 U.S.C. § 214.
132. *Market Entry and Regulation Report and Order*, 11 F.C.C. Rcd. at 3896 ¶ 4.
133. *Id.* at 3909 ¶ 95.
134. *Id.* at 3907–08 ¶ 92.
135. *Id.* at 3906 ¶ 89.
136. *Id.* at 3891 ¶ 46.
137. *Id.* at 3917 ¶ 116.
138. *Id.* at 3881 ¶ 21.
139. *Id.* at 3912 ¶ 102.

turn, not in the conventional antitrust sense,[140] but rather "as the ability of the carrier to act anticompetitively against unaffiliated U.S. carriers through the control of bottleneck services or facilities on the foreign end" of a specific route.[141]

The factors of the ECO test apply to a foreign carrier's, or its U.S. affiliate's, application to provide international facilities-based or resale services.[142] In an application for facilities-based services, the FCC, under the order, will first examine the legal, or de jure, ability of U.S. carriers to enter the foreign market, and whether U.S. carriers are legally permitted to offer international message telephone service in the destination foreign country.[143] If no explicit legal restrictions on entry exist, then the FCC will consider other factors of the ECO test to determine whether carriers have the practical, or de facto, ability to enter the foreign market.[144] The first such factor is "whether there exist reasonable and nondiscriminatory charges, terms and conditions for interconnection to a foreign carrier's domestic facilities for termination and origination of international services."[145] Moreover, the FCC requires adequate means to monitor compliance with the expressed standards for interconnection.[146]

The second de facto consideration is whether competitive safeguards exist to protect against anticompetitive or discriminatory practices against U.S. telecommunications firms already in the foreign market or attempting to enter the country's market.[147] Important safe-

---

140. *See* William E. Landes & Richard A. Posner, *Market Power in Antitrust Cases*, 94 HARV. L. REV. 937 (1981).

141. *Market Entry and Regulation Report and Order*, 11 F.C.C. Rcd. at 3197 ¶ 116. Bottleneck services or facilities are defined as "those that are necessary for the provision of international services, including inter-city or local access facilities on the foreign end." *Id.*

142. *Id.* at 3890 ¶ 42. A U.S. carrier would be facilities-based if it "holds an ownership, indefeasible-right-of-user, or leasehold interest in an international facility, regardless of whether the underlying facility is a common or non-common carrier submarine cable, or an INTELSAT or separate satellite system." *Id.* at 3923 ¶ 130. A carrier that leases capacity would only be considered facilities-based if it operates the U.S. half-circuit under its own operating agreement with the carrier that provides the corresponding underlying capacity on the foreign end. *Id.* at 3920–21 ¶ 125.

143. *Id.* at 3891 ¶ 47.

144. *Id.* at 3890 ¶ 44.

145. *Id.* at 3892 ¶ 49.

146. *Id.*

147. *Id.* at 3893 ¶ 51.

guards noted by the FCC include: (1) the existence of cost-allocation rules to prevent cross-subsidization, (2) the timely and nondiscriminatory disclosure of technical information needed to use or interconnect with carriers' facilities, and (3) the protection of carrier and customer proprietary information.[148] The FCC's final de facto consideration is whether the destination country has an effective regulatory framework to develop, implement, and enforce legal requirements, interconnection arrangements, and other safeguards.[149] That factor focuses on the independence of the regulatory authority—the separation between the foreign regulator and the foreign carrier, and the fairness and transparency of the regulatory procedures.[150]

The FCC will apply a similar ECO analysis to foreign carriers seeking to enter the U.S. international market through resale. The FCC will apply the ECO test to a foreign carrier's section 214 application to provide service through switched resale and resale of noninterconnected private lines.[151] First, the agency will consider the legal ability to resell switched services in the destination country where the applicant possesses market power. Then the FCC will consider any practical barriers to entry, including "the existence of reasonable and nondiscriminatory charges, terms and conditions for the provision of such resale services, competitive safeguards to protect against anticompetitive and discriminatory practices affecting resale, fair and transparent regulatory procedures, and separation between the regulator and operator of international facilities-based services."[152]

If, on the other hand, the resale is of interconnected private lines for the provision of switched services, under the order the FCC will conform its equivalency requirement with the ECO analysis.[153] But the two standards will still differ on two grounds. First, whereas the equivalency requirement applies to all routes, the ECO test is only applicable to routes where a foreign carrier has market power.[154]

---

148. *Id.*

149. *Id.* at 3894 ¶ 54.

150. *Id.*

151. *Id.* at 3923 ¶ 132, 3926 ¶ 139.

152. *Id.* at 3929 ¶ 146.

153. *Id.* at 3923 ¶ 132. The equivalency requirement was established in FCC's Regulation of International Accounting Rates, First Report and Order, CC Dkt. No. 90–337, 7 F.C.C. Rcd. 559 (1991).

154. *Market Entry and Regulation Report and Order*, 11 F.C.C. Rcd. at 3926

Second, under the ECO test the de jure and de facto considerations must be satisfied in the "near future," while the equivalency test requires that they be satisfied at the time of the FCC's equivalency finding. The FCC will not demand that foreign carrier affiliates adopt cost-based accounting rates as a condition for authorizing private-line resale.[155] The FCC also modified its private-line resale policy to allow U.S. carriers to provide switched services over their authorized facilities-based U.S. private-line half-circuits.[156] The FCC will allow such service without additional section 214 authorization or equivalency demonstration where the private lines are connected to the public switched network at one end only and where the U.S. carrier does not correspond with the dominant foreign carrier that owns the foreign facilities.[157] Where circuits are interconnected on both ends, or where the U.S. carrier corresponds with a dominant foreign carrier that owns the foreign facilities, the FCC will require prior section 214 authorization and a demonstration of equivalency.[158] Finally, the FCC will allow inbound and outbound "hubbing" of switched services over resold private lines through equivalent countries to and from "points beyond," provided that the carrier takes at published rates and resells the international message toll service of a carrier in the equivalent country.[159]

The FCC will not impose the ECO analysis on other forms of market entry such as domestic interexchange or enhanced services, separate satellites, or other noncommon carrier facilities.[160] Although some commenting parties advocated adopting a standard for U.S.-licensed carriers to use foreign satellite systems, the FCC found the record inadequate to adopt such a standard and deferred consideration of the issue to the proceeding on international and domestic satellite policies.[161]

Under its section 214 public interest analysis, the FCC will

¶ 138.

155. *Id.* at 3930 ¶ 150.
156. *Id.* at 3934–35 ¶ 161.
157. *Id.* at 3931 ¶ 153.
158. *Id.* at 3934–35 ¶ 161.
159. *Id.* at 3938 ¶ 169.
160. *Id.* at 3939 ¶ 172.
161. *Id.* at 3940 ¶ 176 (citing Amendment to the Commission's Regulatory Policies Governing Domestic Fixed Satellites and Separate International Satellite Systems, Notice of Proposed Rulemaking, IB Dkt. No. 95-41, 10 F.C.C. Rcd. 7789 (1995)).

consider, in addition to the ECO test, other public interest factors that might bear upon the entry decision. Those factors include: the general significance of the proposed entry to promoting competition in the U.S. communications market; national security, law enforcement, foreign policy, or trade concerns raised by the executive branch; and the presence of cost-based accounting rates.[162]

Public Interest Analysis under Section 310(b)(4). In considering foreign investment in U.S. carriers in excess of the 25 percent benchmark in section 310(b)(4), the FCC now examines whether foreign markets offer effective competitive opportunities to U.S. entities.[163] The FCC uses the same analysis under both sections 214 and 310(b)(4) to determine whether foreign ownership or affiliation, under specific circumstances, is consistent with the public interest.[164] The order includes the ECO test in the public interest analysis under section 310(b)(4) for common carrier licenses, but not for broadcast or aeronautical licenses.[165] The FCC ruled that if U.S. companies can acquire a controlling interest in the home market of a foreign investor, then effective competitive opportunities exist that justify allowing foreign ownership at any level above the 25 percent benchmark, absent other public interest factors. In identifying the appropriate market segment for comparison, the FCC uses a service-by-service approach.[166] It considers restrictions on U.S. participation in the home market for the particular service in which the foreign applicant seeks to participate in the U.S. market.[167] For example, the degree to which a foreign government permits U.S. cellular providers to offer cellular services in its home market determines the extent to which that country's cellular providers may enter the U.S. cellular market.

---

162. *Id.* at 3897 ¶ 62.

163. *Id.* at 3875 ¶ 2.

164. *Id.*

165. The FCC decided not to apply the ECO test to broadcast or aeronautical licenses. *Id.* at 3943 ¶ 182. The agency excluded aeronautical licenses because it "lack[ed] any historical guidance with respect to foreign ownership of [such] licenses." *Id.* The FCC distinguished broadcast licenses from other radio licenses on the ground that "foreign control of a broadcast license confers control over the content of widely available broadcast transmissions." *Id.* at 3946 ¶ 192.

166. *Id.* at 3948 ¶¶ 211–12.

167. *Id.*

As with section 214, the ECO analysis under section 310(b)(4) consists of a four-part analysis. The first considers legal, or de jure, barriers to entry. The second evaluates the price terms and conditions of interconnection. The third examines the presence or lack of competitive safeguards. The fourth analyzes the regulatory framework of the home market.[168] The FCC stated that it would examine additional public interest factors under section 310(b)(4), including (1) the general significance of the proposed entry to promoting competition in the U.S. telecommunications market, (2) any national security, law enforcement, foreign policy, and trade concerns raised by the executive branch, and (3) the extent of foreign participation in the applicant's parent corporation, especially the presence of foreign officers or directors in excess of the statutory benchmarks that existed at the time of the order.[169] It would be unlawful, of course, for the FCC to rely upon this third factor in light of the repeal in 1996 of the restrictions on foreign officers and directors formerly contained in section 310(b)(4). The ECO test, as applied under both sections 214 and 310(b)(4) public interest inquiries, weighs in favor of approval if the test reveals that competitive opportunities exist or if it is reasonably certain that they will exist in the near future.[170] Although the FCC declined to make any one factor of the ECO test dispositive, the agency did decide, in the interest of improving predictability for potential entrants, to place more emphasis on the first factor of the test—the legal ability of U.S. firms to provide similar services in the relevant foreign market.[171] Although the absence of any one factor of the ECO test will weigh against the transaction, the test is still only one element of the overall public interest analysis.[172] Other countervailing public interest factors may weigh in favor of approval. Even if the FCC determines that the public interest hurdle is met and approves the foreign investment, the agency reserves the right to

---

168. *Id.* at 3954 ¶ 213.

169. *Id.* at 3955 ¶ 216.

170. In the order, the FCC recognized "that requiring all the factors of the ECO test to be present at the time of entry would be counterproductive." Nonetheless, "the foreign carrier must demonstrate that necessary measures will be adopted and implemented in the near future for [the FCC] to reach a favorable determination about the destination country." *Id.* at 3891 ¶ 46.

171. *Id.* at 3890 ¶ 43.

172. *Id.* ¶ 44.

impose conditions and safeguards on the foreign-affiliated U.S. carrier or the foreign carrier.[173]

Regulatory Status of Carriers

Dominant Carrier. The FCC may regulate a U.S. carrier affiliated with a foreign carrier as a dominant carrier in a particular international route as a means of conditioning the agency's approving the foreign carrier's entry. After determining that entry of a foreign carrier is in the public interest, the FCC determines the carrier's regulatory status.[174] A U.S. carrier is regulated either as a dominant or nondominant carrier on a particular international route. A U.S. carrier is regulated as a dominant carrier if it is "affiliated" with a foreign carrier that has market power in the particular international route.[175] The definition of "affiliation" for the purpose of postentry regulation is the same as for entry. Therefore, for purposes of postentry regulation, a foreign-affiliated U.S. carrier is one where a foreign carrier holds more than 25 percent or where its holding, even if less than 25 percent, is a controlling interest.[176] But rules are meant to be broken, it would seem, and the FCC might regulate a U.S. carrier as dominant even though its foreign investment is *below* the 25 percent threshold if the agency believes that the investment "presents a significant potential impact on competition."[177] Moreover, the FCC intends to apply dominant carrier regulation to nonequity business relationships.[178] The agency regulates a U.S. carrier as a dominant carrier on routes where it has a "non-exclusive co-marketing or other arrangement with a dominant foreign carrier [that] presents a substantial risk of anticompetitive effects in the U.S. international services market."[179]

A dominant carrier is subjected to additional burdens under the order. The requirement that a dominant, foreign-affiliated U.S. carrier obtain section 214 authorization before adding or deleting

---

173. *Id.* at 3897 ¶ 63.
174. *Id.* at 3966 ¶ 245.
175. *Id.*
176. *Id.* at 3967–68 ¶ 249.
177. *Id.* at 3968 ¶ 250.
178. *Id.* at 3969 ¶ 253.
179. *Id.*

international circuits on routes for which it is regulated as dominant remains unchanged under the order. The dominant U.S. carrier must also file quarterly traffic and revenue reports for such routes.[180] The FCC also requires that a dominant, foreign-affiliated U.S. carrier "maintain complete records of the provisioning and maintenance of network facilities and services it procures from its foreign carrier affiliate, including, but not limited to, those it procures on behalf of customers of joint ventures for the provision of U.S. basic or enhanced services."[181] The FCC declined to require the filing of a complete list of accounting rates that a foreign affiliate of a dominant, facilities-based U.S. carrier maintains with other countries.[182] Instead, the FCC in its public interest analysis under section 214 will consider the disclosure of the foreign affiliate's accounting rates to be a favorable factor.[183]

Exclusive Arrangements. The FCC may prohibit any exclusive arrangement between a U.S. and a foreign carrier that, either on its face or in practice, grants exclusive rights to the U.S. carrier for providing basic telecommunications services originating or terminating in the United States.[184] Similarly, all U.S. carriers, regardless of their regulatory status or affiliation, are prohibited from agreeing to accept special concessions from any foreign carrier.[185] The FCC waives that provision if the U.S. carrier can demonstrate that the foreign carrier granting the concession lacks the ability to discriminate against U.S. international carriers in providing facilities or services.[186]

Conclusions

The FCC contended that the new ECO test would be "predictable yet flexible."[187] A more pragmatic assessment is that the ECO test is malleable to the point of being utterly unpredictable. Like the

---

180. *Id.* at 3974 ¶¶ 263–64.
181. *Id.* at 3975 ¶ 266.
182. *Id.* at 3976 ¶ 267, 3977 ¶ 271.
183. *Id.*
184. *Id.* at 3971–72 ¶¶ 256–58.
185. *Id.*
186. *Id.* at 3971–72 ¶ 257.
187. *Id.* at 3881 ¶ 20.

hodgepodge of decisions on section 310(b) that preceded it, the ECO test is so byzantine that it can produce any result that the FCC desires. The FCC predicted that the new ECO test would increase competition owing to the specificity of the critical entry factors. Just the opposite will occur in the United States in light of the prenotification warning that incumbents will receive and the complex factors that will provide them grist for strategic litigation to retard entry. The benefits from competition that American consumers will be forced to forgo will dwarf the benefits that the FCC believes will redound from preventing foreign carriers from leveraging their market power against U.S. affiliates or competitors.[188]

THE INEFFICACY OF IMPOSING BILATERAL RECIPROCITY ON SECTION 310(b)

Suppose that, despite all the reasons given in this chapter for eschewing reciprocity, a future Congress seeking to reform section 310(b) preferred a service-by-service bilateral policy of reciprocal market access, as the FCC opted for in its *Foreign Market Entry Report and Order*, over a unilateral policy of unrestricted foreign investment in American radio licensees. The actual implementation of such a reciprocity rule would still encounter several practical complications that would make the approach intractable and therefore inefficacious.

The asymmetry of regulatory institutions across countries would frustrate the comparisons that would be necessary to determine whether a foreign country had offered U.S. investors reciprocal market access. In the United States, until February 1996 entry into telecommunications markets was regulated by the FCC, by the state public utilities commissions, by municipalities in the case of cable television franchising, and by the U.S. District Court in Washington, D.C., by virtue of its jurisdiction over the Modification of Final Judgment. Those multiple layers of regulation have left their mark on the organization of the U.S. telecommunications industry; yet they are unlikely to have counterparts in a foreign country whose citizens seek to invest in American radio licensees. Furthermore, it would be regrettable if the United States pressured other nations to imitate the

---

188. *Id.* at 3884–85 ¶¶ 28–29.

way in which it has categorized telecommunications services by their regulatory treatment, for the principal effect of that regulatory pigeon-holing has been to allocate markets and suppress competition to the detriment of consumers.[189] In short, the unfortunate oddities of U.S. telecommunications law that have accreted since 1927 imply that a reciprocity analysis for section 310(b) would result in a comparison of apples with oranges.

A related, and more serious, problem is that the domestic experience in the United States gives no basis for optimism that bilateralism would produce mutually advantageous reductions in the barriers to foreign investment in telecommunications. For years, the regional Bell operating companies and the interexchange carriers (principally AT&T, MCI, and Sprint) were unable to agree on legislation that would simultaneously lift the interLATA restriction in the MFJ while eliminating barriers to entry into local exchange and intraLATA toll markets. Similarly, local exchange carriers resisted the entry of cable television companies into telephony, and cable has resisted the entry of telephone companies into video. The enactment of major telecommunications legislation in 1996 was truly extraordinary in light of the repeated inability of previous Congresses to pass such a bill. Yet that legislation did not end the battle over market entry, as the litigation over the FCC's rules on local competition attests.[190] Those experiences testify to the failure of bilateral schemes to reduce entry barriers in domestic telecommunications markets. When bilateralism is taken to the international scale, the additional considerations of culture, language, and nationalism would decrease further the likelihood that reciprocity would successfully reduce restrictions on American firms seeking to invest in foreign telecommunications markets.

Finally, even a relatively simple reciprocity test for section 310(b) would lend itself to incumbent firms' strategic use of the regulatory process. Foreign direct investment by BT makes MCI a

---

189. *See, e.g.,* Robert W. Crandall & J. Gregory Sidak, *Competition and Regulatory Policies for Interactive Broadband Networks*, 68 S. CAL. L. REV. 1203 (1995); J. Gregory Sidak, *Telecommunications in Jericho*, 81 CAL. L. REV. 1209, 1227–34 (1993).

190. Implementation of the Local Competition Provisions in the Telecommunications Act of 1996, First Report and Order, CC Dkt. No. 96-98, 11 F.C.C. Rcd. 13042 (1996), *stayed pending appeal sub nom.* Mississippi Pub. Serv. Comm'n *v.* FCC, No. 96–3608 (8th Cir. Oct. 15, 1996).

stronger competitor against other carriers, which already are able to oppose that form of market entry by urging the FCC not to waive section 310(b) or to issue a declaratory ruling that the investment complies with that statute. When that regulatory approval process must also consider whether reciprocal market access exists on a service-by-service basis in the foreign carrier's home market (let alone in each of its "primary markets"), foreign direct investment by a prominent overseas carrier becomes more costly and risky. Lawyers for the domestic carriers opposing the investment could easily elevate the factual complexity of the proceeding by disputing the characterization of the regulatory environment for telecommunications in the investor's home markets. Economists would be necessary to opine on what is the "market" to which access is or is not reciprocally offered. Each expert's direct testimony would necessitate another expert's rebuttal testimony. Much like the securities or antitrust litigation precipitated by a hostile tender offer,[191] the reciprocity litigation before the FCC would be high-stakes posturing in which the substance of the legal and economic arguments made would be incidental to the question of greatest relevance to the public interest: Will a market serving U.S. consumers be subjected to greater competition and innovation through the entry of a major foreign carrier making a direct investment?

THE SECOND-BEST SOLUTION

We have considered the uncompromising case against bilateral reciprocity. But, within the set of politically feasible alternatives, which statutory provision concerning foreign ownership would pose the least risk to consumer welfare? Stated differently, how should a Congress bent on regulating foreign direct investment fashion a foreign ownership statute that represents the second-best solution?

The starting place is the 1995 House version of proposed section 310(f). That provision, however, should incorporate language from the Senate bill to give the FCC the authority to approve foreign investment on a bilateral basis while awaiting the outcome of multilateral trade negotiations. Whether or not the House language as

---

191. *See, e.g.,* J. Gregory Sidak, *Antitrust Preliminary Injunctions in Hostile Tender Offers,* 30 KAN. L. REV. 491 (1982).

passed in 1995 would increase foreign investment in the United States would depend critically on the policies of the sitting president and his U.S. trade representative. Depending on the president, the FCC may be more or less open to foreign direct investment than are the president and the U.S. trade representative. It would be naïve to presume that the president will always be more inclined to free trade than the FCC, and that the 1995 House version of section 310(f) therefore need not grant the FCC approval powers apart from those which the president could effectively supersede by intervening on day 179 of a 180-day pleading cycle.

In addition, the legislation would benefit from adding to the language of the 1995 House bill the following provision contained in section 310(f)(1) of the 1995 Senate bill: "While determining whether such opportunities are equivalent on that basis, the Commission shall also conduct an evaluation of opportunities for access to all segments of the telecommunications market of the applicant." Again, one must read that provision without imputing to it one's own predilection for or against unrestricted foreign direct investment. A free trader, for example, might jump to the conclusion that domestic incumbents would use that passage from the 1995 Senate bill as a vehicle to raise the cost and complexity of FCC proceedings for the purpose of slowing or deterring foreign carriers' entry. Such strategic use of the regulatory process, however, might be more likely to occur in the absence of an overall evaluation of the openness of the foreign investor's home market. For example, the United Kingdom in general has a very open market to foreign investment, and the FCC should not ignore that fact when considering whether some specific segment of that market is as open to foreign investment in the United Kingdom as it is in the United States. In short, a future Congress should write the law to grant as much latitude as possible to the president and the FCC to allow foreign investment that is likely to enhance economic welfare in the United States.

THE WORLD TRADE ORGANIZATION NEGOTIATIONS ON TELECOMMUNICATIONS SERVICES

The FCC anticipated future multilateral agreements in its *Market Entry and Regulation Report and Order* and said that the agency "will defer to the Executive Branch on any matter involving the interpre-

tation of international agreements."[192] Moreover, the agency emphasized that the policies and rules set forth in its order would not interfere with efforts by the executive branch to forge a multilateral agreement to open telecommunications markets to global competition.
Nonetheless, the bilateral reciprocity policy outlined in the FCC's
order had a substantial impact on negotiations on basic telecommunications conducted under the auspices of the World Trade
Organization (WTO).

Many negotiating parties, particularly those from the European
Union, had felt that the application of the FCC's ECO test was not
consistent with either the WTO's principle of "national" treatment or
the extension of equal treatment to all foreign parties.[193] Yet, the
centerpiece of the U.S. position in the talks was its determination to
use the ECO test to encourage other parties to make advance commitments on liberalization, market access, and foreign investment. Scott
Blake Harris, chief of the FCC's International Bureau, told Congress
that "the Commission's efforts in its rulemaking and the trade talks
have been integral in getting as many offers from foreign countries as
we did prior to the April 1996 deadline" and that "as we demonstrate
through application of our new entry rules a commitment to liberalization in a fair environment we will see stronger offers being made
before . . . February" of 1997, the deadline for the conclusion of the
basic telecommunications negotiations.[194] The chief U.S. negotiator
at the talks said that the United States would continue to insist that
"foreign market opening[s] come[ ] on a fair [and] reciprocal basis,
especially among . . . peer countries in the industrial world."[195]

---

192. *Market Entry and Regulation Report and Order*, 11 F.C.C. at 3945 ¶ 188.

193. The FCC conceded: "If the U.S. schedules commitments for market access
and national treatment at the conclusion of the work of the Negotiating Group on Basic
Telecommunications, the [FCC] may be obliged to revisit these rules at that time." *Id.*
at 3966 ¶ 244.

194. *Hearing on Foreign Ownership Restrictions Before the Subcomm. on Com-
merce, Trade, and Hazardous Materials of the House Comm. on Commerce*, 104th
Cong., 2d Sess. 12 (1996) (statement of Scott Blake Harris, Chief of International
Bureau, FCC).

195. *Hearings on International Telecommunications Trade Before the Subcomm.
on Commerce, Trade, and Hazardous Materials of the House Comm. on Commerce*,
104th Cong., 2d Sess. 9 (1996) (statement of Jeffrey M. Lang, Deputy U.S. Trade
Representative) [hereinafter *Lang Testimony*]. *See also* Cynthia Beltz, *Talk Is Cheap*,
REASON, Aug.–Sept. 1996, at 45, 46.

The General Agreement on Trade in Services

The Uruguay Round[196] of multilateral trade negotiations not only established the World Trade Organization,[197] but also gave birth to the first set of multilaterally agreed-upon and legally enforceable rules ever negotiated to cover international trade in services. The General Agreement on Trade in Services (GATS) adapted and applied to all trade in services the principles of market access, national treatment, and nondiscrimination that had been the basis of the GATT trading system governing the trade in goods for the preceding forty-five years.[198] The agreement consisted of two parts. The first part was a framework of obligations and disciplines that applied to all services, such as obligations of nondiscrimination among trading partners and transparency in national rules and regulations.[199] The second part was the national schedules of commitments.[200] The 114 national schedules of the GATS specified the services for which governments had guaranteed market access and national treatment, and at what level. The commitments, like the rules of the framework, were legally binding and enforceable under the WTO dispute settlement mechanism.[201]

The GATS covered all existing and future telecommunications

---

196. Eight rounds of multilateral trade negotiations were held under the auspices of the General Agreement on Tariffs and Trade (GATT), each with the aim of liberalizing trade between contracting parties by reducing trade barriers and other measures impeding free trade. The most ambitious of those rounds was the Uruguay Round (1986–94). In addition to establishing the WTO, the Uruguay Round extended the multilateral trading system to trade in services. *See* Cathryn J. Prince, *Umpire for World Trade*, CHRISTIAN SCI. MONITOR, Jan. 17, 1996, at 1.

197. The WTO was established on January 1, 1995, and acts as a single "umbrella" over the GATT and the multilateral agreements that resulted from the Uruguay Round. It provides both a code of rules and a forum in which countries address their trade problems and negotiate to enlarge world trading opportunities. The WTO embodies certain fundamental principles such as trade without discrimination, protection through tariffs, the binding of tariffs at levels negotiated among members, national treatment, and consultations on the basis of equality.

198. *See The Final Act and Agreement Establishing the World Trade Organization*, Apr. 15, 1994, General Agreement on Trade in Services, Annex 1B, 33 I.L.M. 1168 (1994) [hereinafter *GATS*]. In particular, see arts. II–V.

199. *Id.*

200. *GATS, supra* note 198, arts. XIX–XXI.

201. *Id.*

services.[202] During the Uruguay Round some fifty-eight countries made market access commitments on sophisticated value-added services, but no agreement was reached on basic telecommunications services because of the complex and politically sensitive issues of privatization of government telecommunications monopolies arising in many countries.[203] Instead, those governments agreed to continue negotiations on basic telecommunications services.[204] Consequently, the negotiations undertaken by some fifty-three nations after the Uruguay Round focused on basic services rather than value-added services.[205]

Basic Telecommunications Services

In April 1994, pursuant to a declaration issued at the time of the signing of the Uruguay Round agreement, the Negotiating Group on Basic Telecommunications (NGBT) was formed under the auspices of the WTO with the mandate to conclude negotiations on basic telecommunications services leading to new national commitments and a set of basic rules for liberalization. The scope of the negotiations was comprehensive, with no basic telecommunications service excluded. The negotiators also chose not to develop a definitive listing of what constituted basic telecommunications services but instead

---

202. *GATS, supra* note 198, *Annex on Telecommunications* § 2.

203. *Lang Testimony, supra* note 195, at 2–3. Value-added services include such services as voice-mail and electronic mail and account for a rapidly growing but still small share of the telecommunications market. The lion's share of the market belongs to basic telecommunications services, such as local voice telephony. Value-added services are products of and are based on digital technology. The Uruguay Round Agreement included a "Telecommunications Annex" that guaranteed access to infrastructure for value-added service suppliers. But it did not contain provisions addressing monopoly pricing of the leased infrastructure needed to supply value-added services. *Id.*

204. Basic telecommunications services include the regulated voice telephony and other services that can be supplied using analog (as well as digital) technologies. In many countries, basic services are still supplied only by a single telephone company holding a legal monopoly covering all telecommunications services and infrastructure.

205. Even though the 53 countries taking part in the negotiations made up less than half the total WTO membership of 120 countries, they accounted for 93 percent of the world's telecommunications revenues, 82 percent of the world's telephone lines, and 84 percent of its traffic. *See U.S. Pressing for Telecom Concessions as Deadline Nears,* Agence France-Presse, Apr. 26, 1996.

agreed that the negotiations would cover any and all telecommunications services with the exception of value-added ones. Such services included international and domestic voice telephony, data transmission, telex, telegraph, facsimile, private leased circuits, satellite services, mobile services, and video transport services.

Those negotiations aimed to secure market-opening commitments for as many countries as possible.[206] Those commitments involved the introduction of competition into the sector by removing restrictions that prevented or impeded the supply of services by foreign companies.[207] Every mode of supply was to have been covered—cross-border trade and foreign direct investment, resale, and ownership and operation of networks and infrastructure. In addition, the negotiators discussed commitments that concerned the telecommunications regulatory environment. Those commitments included safeguards on interconnection rights and ways to prevent the abuse of market power.[208] The NGBT's mandate expired on April 30, 1996. Nonetheless, the fifty-three telecommunications negotiating parties of the

---

206. Bhushan Bahree, *U.S. Takes a Hard Line on Telecom*, WALL ST. J., Apr. 30, 1996, at A3. Deputy U.S. Trade Representative Jeffrey Lang characterized the U.S. objectives in those negotiations as follows:

> [O]ur strategic goal has been to combine the classic principles of trade negotiations—market access and national treatment—with an emphasis on the need for pro-competitive regulatory principles. Over the past two years of negotiations the U.S. has sought (1) the adoption of fair and effective pro-competitive principles, including cost-based prices for interconnection, competition safeguards, regulatory transparency and independence of the regulator; and, (2) firm meaningful commitments by the participants in these negotiations to open their markets. The U.S. has included cost-based pricing of interconnection as one its key requirements, to overcome the problem of over-priced leased lines that has been prevalent in foreign monopoly markets.

Lang Testimony, supra note 195, at 4–5.

207. "In the basic telecommunications services, the United States maintains the world's most open and competitive market. Our objective in this negotiation is to obtain similar levels of openness to foreign investment and competition in the markets of other major trading partners, most of which are dominated by monopoly suppliers." *Id.* at 2.

208. The United States in particular feared that national telecommunications monopolies or dominant providers could use the comparatively high profits they derived from their market power in home markets to subsidize their efforts to compete overseas.

NGBT agreed that day to extend the deadline to February 15, 1997.[209] The NGBT parties also agreed that each country's then-current offer would be "frozen" until January 15, 1997, after which date any party could modify or withdraw its offer. In addition, the parties agreed not to enact legislation or implement new rules that would be inconsistent with the frozen offers until the renewal of negotiations. Despite the need to extend their talks, the parties still intended to implement a telecommunications agreement on January 1, 1998.

Offers Tabled. At the close of negotiations on April 30, 1996, forty-seven of the fifty-three participating countries had tabled offers.[210] Thirty-three countries adopted as binding commitments the competition rules that the negotiators had developed. Twenty-one offers provided open market access for international services and facilities. Another twelve offers would phase in the opening of international services markets. Five more offers made more limited international services commitments. Market access for use of satellite facilities to provide international services was open and unqualified in fifteen offers and phased-in for eight offers. It was limited for another twelve offers. Twenty-five offers allowed 100 percent foreign investment in all basic telecommunications services and facilities. Another eighteen had either limitations with respect to the basic telecommunications sectors in which foreign investment was allowed or limitations restricting the foreign investor to a minority holding. Five offers had made no commitments regarding foreign investment.

Japan maintained restrictions on foreign ownership of its dominant telephone companies, NTT and KDD. Seven of the fifteen members of the European Union made offers with limitations that made them relatively less open than the other eight member states. Canada's offer maintained a 46 percent limit on foreign investment. Every Asian offer would limit foreign investment to minority levels for all telecommunications services firms.

The American Position. Remarking on the tabled offers, acting U.S.

209. Paul Lewis, *Telecommunications Talks Postponed as U.S. Balks*, N.Y. TIMES, May 1, 1996, at D4.

210. *Status of WTO Telecom Offers* (Office of the United States Trade Representative, Press Release No. 96-40, Apr. 30, 1996). All of the facts cited in this discussion are drawn from the USTR press release.

Trade Representative Charlene Barshefsky said:

> [U]nfortunately, we have not yet reached a critical mass of quality offers from our trading partners. Rather than accept a bad deal—or walk away from the good offers tabled by many countries—the United States won support for the extension of the telecom talks to February 17, 1997.[211]

She noted that, under the offers tabled by April 30, 1996, 40 percent of world telecommunications revenues and 34 percent of world telecommunications traffic would not have been covered by "acceptable" offers, an amount she said was well short of the "critical mass" of quality offers that the Clinton administration was seeking.[212] The American negotiators in Geneva complained that they had not received enough offers for market access to make it worthwhile for the United States to open its own telecommunications sector, although by the terms of its tabled offer the United States had pledged to do so. The original U.S. offer covered local, long-distance, and international telecommunications services and would have effectively removed restrictions on foreign indirect ownership of common carrier radio licenses if the United States had obtained sufficient reciprocal commitments from its major trading partners.[213] Any final U.S. commitment, however, was qualified by the rights of the United States and other WTO members, under the GATS, to take action in the interests of national security or maintenance of public order.

In a surprising move, the weekend before the April 30, 1996, deadline date, the United States abruptly withdrew its offer of complete and free access to the American market in international services and satellite communications, reasoning that other countries had failed to offer sufficient reciprocity in those areas.[214] The removal

---

211. *Basic Telecom Negotiations, Statement of Ambassador Charlene Barshefsky* (Office of the United States Trade Representative Press Release No. 96-40, Apr. 30, 1996).

212. *Id.*

213. Ambassador Lang said of the U.S. offer: "In February 1995, Vice President Gore announced that the United States was prepared to open its $215 billion telecommunications market if other nations would do the same. In July 1996, the United States followed through on that pledge by making the most comprehensive market opening offer of any nation." *Lang Testimony, supra* note 195, at 5.

214. Lewis, *supra* note 209.

of international services drew immediate and harsh responses from the European and Japanese negotiators. The EU representative found the exclusion of international services unacceptable.[215] And a senior Japanese trade negotiator was quoted as saying that the U.S. position was "irresponsible, counterproductive, even threatening."[216] One reason for the U.S. removal of international services from the table was "[t]he problem of above-cost settlement payments."[217]

The U.S. negotiators and telecommunications firms both believed that the problem "could be exacerbated if the United States made an MFN-based commitment to open the U.S. market indiscriminately to foreign international service suppliers."[218] The American representatives were especially concerned that U.S. facilities-based entry by foreign carriers with market power in countries not allowing full access by U.S. carriers would permit those carriers to use their high accounting rates to subsidize their operations in the U.S. market. A foreign carrier with market power, they argued, could capture significant U.S. outbound traffic by selling its services at incremental cost, while U.S. carriers could not survive without some margin of contribution for common or fixed network costs. That development, in turn, would distort the global telecommunications market. Consequently, U.S. negotiators insisted on retaining domestic licensing discretion to inhibit anticompetitive conduct in the U.S. market.[219] Otherwise, they thought, some foreign carriers "would have a variety of options for entering the U.S. market and could choose to manipu-

---

215. *Id.*

216. *Id.*

217. *Lang Testimony, supra* note 195, at 4. A U.S. carrier makes "settlement payments" to foreign carriers for their provision of terminating access for international services. Telephone companies agree on prices for joint supply of international services, or "accounting rates." U.S. carriers have generally run a significant deficit in settlement payments because most foreign telecommunications operators are monopolists or otherwise enjoy market power in their home markets and charge high prices for terminating access. In 1995, for example, the imbalance in settlement payments cost U.S. carriers over $4.4 billion in net payments to foreign carriers. *Hearing on International Telecommunications Trade Before Subcomm. on Commerce, Trade, and Hazardous Materials of the House Comm. on Commerce*, 104th Cong., 2d Sess. (1996) (statement of Reed E. Hundt, Chairman, FCC) [hereinafter *Hundt Testimony*]. The foreign carrier's market power allows it to maintain above-cost accounting rates with U.S. and other carriers with respect to the market that it dominates.

218. *Lang Testimony, supra* note 195, at 5.

219. *Id.; see also* Bahree, *supra* note 206.

late the existing system."[220] Those negotiators saw the FCC's ECO test, outlined in the FCC's *Market Entry and Regulation Report and Order* of November 1995, as preventing such anticompetitive conduct. A senior U.S. negotiator said that preserving "the FCC's ability to preclude such manipulation in [the U.S.] market and to promote greater competition in international services markets" was one of the key objectives of the U.S. delegation.[221]

A week after the NGBT's deadline had passed, FCC Chairman Hundt testified before Congress that "if the talks had ended in an agreement, these countries would have been eligible for full market access rights in the United States under the Most Favored Nation clause of the WTO."[222] That outcome, he believed, would "have meant that foreign carriers whose markets [were] not open to U.S. competitors could have engaged in anti-competitive behavior in the U.S. market for international services."[223] He asserted that "tremendous opportunities for anti-competitive behavior in international markets" existed because "accounting rates [were] substantially above costs and entry [was] limited."[224] Chairman Hundt further testified that during the NGBT talks the United States had "argued for the right to retain use of competitive safeguards as a tool for addressing these problems" and that at one point in April 1996 other industrialized nations had generally agreed that safeguards were necessary and legal under the WTO.[225] Nevertheless, he stated that the United States disagreed with those countries over the range of safeguards and the degree of discretion that the FCC would have.

CONCLUSION

The FCC's 1995 proposal for bilateral reciprocity based on a "primary markets" test would not have increased foreign market access for U.S. firms relative to the status quo. If anything, the proposal would have encouraged other governments to retaliate with equally overbearing attempts to regulate the foreign direct investments of

---

220. *Id.*
221. *Id.*
222. *Hundt Testimony, supra* note 217, at 15.
223. *Id.*
224. *Id.* at 14.
225. *Id.* at 15.

U.S. telecommunications firms. Neither the domestic nor the foreign consequences of such a rule would serve the public interest of American consumers and producers. It remains to be seen whether the "home market" test that FCC instead adopted in 1995 will eliminate or merely lessen those detrimental effects.

Only slightly better results would obtain if Congress were to impose a bilateral reciprocity standard on section 310(b)(4), as the Senate envisioned in 1995 in its passage of S. 652. If outright repeal of section 310(b) is politically infeasible—and experience during consideration of H.R. 1555 in 1995 and 1996 suggests that it is—then Congress could achieve the greatest success in opening foreign telecommunications markets to direct investment by U.S. firms by adopting the Oxley Amendment in its amended form, which the House passed on August 4, 1995, as part of H.R. 1555—subject, however, to the inclusion of the key language from the 1995 Senate bill discussed earlier.

Thereafter, those whose enforce that new foreign ownership provision should heed Jagdish Bhagwati's renewed warning in 1995 concerning what he called the "unhealthy obsession with reciprocity" in U.S. trade policy: "Especially in industries such as finance and telecommunications, only openness and deregulation can create enduring competitiveness."[226]

---

226. Jagdish Bhagwati, *An Unhealthy Obsession with Reciprocity*, FIN'L TIMES, Aug. 24, 1995.

7

Free Speech

THE RESTRICTIONS on foreign ownership in the Communications Act of 1934 limit the ability of aliens to speak to Americans and the ability of American citizens to hear aliens. Do those restrictions violate the freedom of speech guaranteed by the First Amendment? Of course not, the FCC has said.

In 1987 Seven Hills Television, a Spanish-language broadcaster in the Southwest that had a wealthy Mexican investor, argued that section 310(b)(3) "not only limits the class of persons who are entitled to express their views in this country," but also "abridges the public's right to receive information, commentary, and ideas."[1] The statute "restricts rather than promotes freedom of speech" and, "unless correctly construed," violates the First Amendment.[2] The FCC's Review Board responded that it could not declare an act of Congress unconstitutional.[3] True, but that observation was irrelevant

---

1. Seven Hills Television Co., 2 F.C.C. Rcd. 6867, 6876 ¶ 33 (Rev. Bd. 1987), *recons. dismissed*, 3 F.C.C. Rcd. 826 (Rev. Bd. 1988), *recons. denied*, 3 F.C.C. Rcd. 879 (Rev. Bd. 1988), *rev. dismissed*, 4 F.C.C. Rcd. 4062 (1989).

2. *Id*. at 6876 ¶ 33.

3. *Id*. at 6876 ¶ 34 (citing Meredith Corp. *v*. FCC, 809 F.2d 863, 872 (D.C. Cir. 1987); Plano *v*. Baker, 504 F.2d 595 (2d Cir. 1974); Dowen *v*. Warner, 481 F.2d 642 (9th Cir. 1973)).

to Seven Hills' assertion that the FCC's *interpretation* of section 310(b)(3) violated the First Amendment. Obviously, the agency did possess the power to correct its own interpretation of the statute. The FCC then concluded by quoting the Supreme Court's decision in *Red Lion Broadcasting Co.* v. *FCC* for the proposition that Seven Hills had raised no First Amendment issue whatsoever, for "no one has a First Amendment right to a license."[4]

It is fitting that the FCC's logic in *Seven Hills* rested on *Red Lion*. As a matter of First Amendment law, Seven Hills actually had a strong argument in 1987. A decade later, the argument had become compelling. In this chapter we shall see why.

Courts have historically assigned different levels of protection to speech disseminated by disparate technologies. The convergence of radio and television broadcasting, cable, telephony, and interactive information services has rendered obsolete the rationales for protecting speech differentially depending on its mode of transmission.[5] We shall see that those technological advances also have rendered obsolete the rationales for the disparate protection of foreign speech by electronic means.

As chapters 2 and 3 show, the rationales for the foreign ownership restrictions in section 310(b) of the Communications Act explicitly rest on restricting foreign speech. It was, fundamentally, a concern over the *content* of the messages that foreigners might disseminate by electronic means that animated congressional restrictions on foreign ownership or control of American wireless companies. The principal national security goal of the restrictions was to prevent, during wartime, point-to-point wireless communication with the enemy; a second goal was to protect the United States from foreign propaganda during wartime. Those goals, though understandable in times of military conflict, are not necessary or even desirable during times of peace and international cooperation. In such times, the foreign ownership restrictions infringe the First Amendment.

---

4. *Id.* at 6876 ¶ 35 (quoting Red Lion Broadcasting Co. *v.* FCC, 395 U.S. 367, 388 (1969)).

5. *See* Thomas G. Krattenmaker & Lucas A. Powe, Jr., *Converging First Amendment Principles for Converging Communications Media*, 104 YALE L.J. 1719 (1995); J. Gregory Sidak, *Telecommunications in Jericho*, 81 CAL. L. REV. 1209 (1993); Note, *The Message in the Medium: The First Amendment on the Information Superhighway*, 107 HARV. L. REV. 1062 (1994).

The foreign ownership restrictions can be understood as an anachronism dating from times less hospitable to free expression and foreigners. Following World War I, the Supreme Court upheld numerous restrictions on speech that were designed to protect national security.[6] Such measures had a significant impact on the recent immigrants residing in the United States and may have been rooted at least partly in fear of them. In many instances the laws of the United States have been discovered to pose constitutional problems that were not recognized years before, when such laws were enacted and the Court's case law might have been less developed on a particular subject. Telecommunications is surely one such subject, and it is therefore appropriate now to scrutinize more closely how the foreign ownership restrictions inhibit freedom of speech. We shall see that, because they threaten fundamental First Amendment rights under established Supreme Court precedent, the foreign ownership restrictions are probably unconstitutional in at least some familiar circumstances.

DIMENSIONS OF
TELECOMMUNICATIONS REGULATION

Courts and the FCC have made the five following legal distinctions when justifying the regulatory treatment of telecommunications firms: (1) the speaker's right to speak versus the listener's right to hear; (2) wireline transmission versus radio transmission; (3) point-to-point communications versus broadcasting; (4) content-based regulation versus content-neutral regulation; and (5) common carriage versus private carriage. Those five categorizations reveal some of the facets of the foreign ownership restrictions and suggest the ways in which those restrictions on speech are overinclusive or underinclusive as a matter of First Amendment jurisprudence.

---

6. *See, e.g.*, Whitney *v.* California, 274 U.S. 357 (1927); Gitlow *v.* New York, 268 U.S. 625 (1925); Abrams *v.* United States, 250 U.S. 616 (1919); Debs *v.* United States, 249 U.S. 211 (1919); Schenck *v.* United States, 249 U.S. 47 (1919).

The Speaker's Rights
versus Listener's Rights

Few would dispute that protecting free speech is sacrosanct to American society. America's political and legal traditions rest on the belief that free speech produces better ideas with which to live and to govern society than does regulated speech. By participating in a "marketplace of ideas," citizens will use their critical faculties to select the "best" or "truest" ideas from among diverse competitors. That competition in turn enriches the stock of beneficial ideas available to all of society. As Judge Learned Hand wrote during World War II, the First Amendment "presupposes that right conclusions are more likely to be gathered out of a multitude of tongues, than through any kind of authoritative selection. To many this is, and always will be, folly; but we have staked upon it our all."[7]

There is a powerful economic corollary to Hand's thesis that emerged at the same point in world history. The great Austrian economist Friedrich Hayek argued that the most profound function of a market economy is not the efficient allocation of scarce resources but the creation and exploitation of knowledge that could not be replicated through the conscious efforts of a central planner for the economy.[8] The protection of speech facilitates the private flows of information that enable markets to produce the spontaneous and decentralized order that Hayek described. As the evidence from corporate finance and other fields of economics attests, as markets become more efficient processors of information, assets become more liquid and more accurately valued. Uncertainty diminishes, and wealth is created. Thus, free speech can have ripple effects that benefit the participants in markets far beyond those in which the speech immediately occurs.

Further, free speech is central to America's democratic political system. Free debate informs voters about their own preferences and the proposals of candidates for office, and informs the right to participate in the electoral process. Free speech also has a political

---

7. United States *v.* Associated Press, 52 F. Supp. 362, 372 (S.D.N.Y. 1943).

8. Friedrich A. Hayek, *The Use of Knowledge in Society*, 35 AM. ECON. REV. 519 (1945). For a succinct summation of Hayek's theories of information and markets written near the end of his long and productive life, see FRIEDRICH A. HAYEK, THE FATAL CONCEIT: THE ERRORS OF SOCIALISM (W. W. Bartley III ed., Routledge 1988).

"checking" value.[9] To the extent that the press can investigate and publish information concerning misconduct or negligence by government officials, such behavior is more likely to be deterred. On the individual level, free speech helps guarantee a person autonomy and aids him in developing his mental and moral capabilities. Speaking, listening, reading, and writing are central elements of thinking and making moral judgments. Without those freedoms, a person cannot fully develop as a sentient being. The First Amendment protects that capacity for intellectual maturation.

The First Amendment rights tread upon by the foreign ownership restrictions are of two sorts. The first is the right of the foreigner to speak. Though not citizens, foreigners still benefit from the First Amendment's guarantee of freedom of expression.[10] To the extent that the foreign ownership restrictions curtail or prohibit foreigners from speaking, by denying them control and limiting their ownership of electronic media through which ideas may be expressed, the restrictions impinge upon foreigners' First Amendment rights. Throughout the analysis in this chapter it will be useful to repeat the question whether analogous restrictions on the ownership and control of American newspapers could withstand scrutiny under the First Amendment.

The First Amendment also protects the concomitant right to listen to foreigners' speech. Obviously, the right to speak would be useless if the government could ban others from listening. Yet that outcome has been precisely the effect of the foreign ownership restrictions. By closing channels of "foreign" speech, the restrictions artificially limit the marketplace from which listeners may select competing ideas.

Wireline Transmission
versus Radio Transmission

The method of effecting telecommunications is critical to understanding the rationales and implications of regulation, even if, as a technical matter, the choice of transmission medium does not dictate whether the sender will be transmitting voice, video, or data. Cable

---

9. *See* Vincent Blasi, *The Checking Value in First Amendment Theory*, 1977 AM. BAR FOUND. RES. J. 521 (1977).

10. *See* Bridges *v.* Wixon, 326 U.S. 135 (1945).

television operators use wires to send video programming; telephone companies use wires to transmit voice and data. Cable television and telephone wires are strung aerially from pole to pole or are buried in underground conduits. The placement of those wires raises regulatory questions of whether those companies use public property—either by digging up streets to lay cable or by using the roads to string it across poles—in a way that justifies regulation to compensate the municipality for the use of public property. If regulation is justified, of course, the question whether the regulation infringes freedom of speech still remains. We shall return to those questions later in this chapter.

In contrast to telephony and cable television, radio and television stations transmit information by emitting energy at specified frequencies over the electromagnetic spectrum, which the federal government allocates and licenses for use. The supposedly finite nature of the spectrum has given rise to regulation rooted in notions of scarcity and subsidy. As we shall examine at length later in this chapter, scarcity arguments posit that, because the spectrum is a finite public good that government allocates, the government must regulate its use in the public interest. The scarcity argument is untenable, however, because the spectrum is not scarce, nor are the consequences of a free market for its allocation apocalyptic, as the proponents of regulation assert. The subsidy argument is that, because the government privileges some citizens by permitting them to use valuable spectrum, government has the concomitant right to regulate that use. Both scarcity and subsidy arguments fail to resolve First Amendment issues satisfactorily.

Finally, there is now a substantial overlap between the once disparate wireline and radio technologies. Telephone companies, for example, are now capable of sending cable television programs to their customers. That development raises the question whether bans on such video programming violate the First Amendment. Further, telephone companies have for years used the spectrum for cellular telephony and long-distance microwave. At least one telephone company has invested in a microwave technology at 28 GHz that may be so spectrally capacious as to offer the consumer the full range of voice, data, and interactive broadband video services.[11] Those

---

11. *See, e.g.,* Daniel Pearl, *FCC Resolves Radio-Spectrum Dispute, Giving Big Boost to Wireless Cable TV,* WALL ST. J., July 14, 1995, at B3.

wireless services of telephone companies may someday raise First Amendment issues as well.

In short, the differentiated technologies of telecommunication have converged in a way that makes irrelevant the pigeonholes for wireless and wireline services that characterized the first seven decades of federal telecommunications regulation. Congress recognized as much when enacting the Telecommunications Act of 1996. To the extent that the federal government retains a regulatory apparatus for telecommunications, it must be one that will be adaptable to inevitable scientific breakthroughs. The experience leading to the 1996 legislation was that regulation had been slow to respond to progress.

Point-to-Point Communications
versus Broadcasting

Whether communication is point-to-point or point-to-multipoint (that is, broadcasting) is probably the dimension of telecommunications regulation that implicates individual liberties to the greatest extent. Point-to-point telecommunication is the sort that occurs through cellular telephone calls, paging, dispatch services, or fixed-link microwave transmissions. As the name suggests, point-to-point telecommunication occurs between a transmitter of voice, video, or data and a specific intended recipient. Broadcasting, on the other hand, is telecommunication that radiates from a central source to potentially millions of consumers within a given area or "service contour." Thus, in many countries (including Canada and the United Kingdom) the term "broadcasting" includes radio, television, and cable television transmissions. In the United States, however, cable television is not generally called a form of broadcasting because its mass distribution of programming does not rely on an omnipoint radio transmission.

The distinction between point-to-point and broadcasting telecommunications raises obvious First Amendment questions. Should point-to-point communications be free from regulation on the rationale that such speech is simple conversation between individuals? Does the anonymity of the recipient of broadcasting alone justify onerous regulation "in the public interest"? Should broadcasting be regulated differently from the print media, which also disseminate information among the general public? In this chapter we shall not delve deeply into the regulation of point-to-point telecommunications; we shall, however, consider whether the public-good nature of broadcasting fails to justify much of the regulation that currently purports to protect the public interest, particularly to the extent that parallel regulation of the print media would violate the First Amendment.

Point-to-point delivery networks, broadcasting, and the print media may eventually converge into a general electronic information network. Even now, books and newspapers are digitized on CD-ROMs and are accessible over the Internet and other on-line services. Eventually, the federal government and the courts also will recognize that convergence and adapt current regulations to accommodate it. Otherwise, the law will be overtaken by technology and become irrelevant at best and injurious to consumer welfare and free speech at worst.

Content-Based Regulation versus Content-Neutral Regulation

Whether a restriction on speech is content-based or content-neutral determines the level of constitutional scrutiny a court will apply in reviewing the restriction.[12] Content-based laws are ones that discriminate between types of speech on the basis of content or are motivated by the desire to discriminate on that basis. Courts generally apply "strict scrutiny" to content-based laws, and often use that standard of review to strike down such laws. Strict scrutiny requires that a law be justified by a "compelling" state interest and that the law be narrowly tailored to achieve that compelling interest.[13] To survive strict scrutiny, a regulation must be extremely important and effective.

Content-neutral laws may incidentally affect speech but must not discriminate between types of speech nor be rooted in a motivation to discriminate on that basis.[14] Courts apply "intermediate scrutiny" to content-neutral laws. That standard of review is more relaxed than strict scrutiny. To pass intermediate scrutiny, a law must be supported by an "important" or "substantial" state interest and must not

---

12. *See, e.g.*, Turner Broadcasting Sys., Inc. *v.* FCC, 114 S. Ct. 2445 (1994).
13. *See* Widmar *v.* Vincent, 454 U.S. 263 (1981).
14. *Turner Broadcasting*, 114 S. Ct. at 2458–59.

"burden substantially more speech than is necessary to further the government's legitimate interests."[15]

The Supreme Court has found content neutrality in some odd places. In *Meese* v. *Keene*,[16] the Court upheld, against a First Amendment challenge, the Foreign Agents Registration Act, which empowered the U.S. Attorney General to classify certain foreign-made films as propaganda and required the producers of those films to label them as such before distribution. Astonishingly, the Court found that *propaganda* was neutral and not a pejorative term at all and characterized the statute as an innocuous disclosure requirement.[17] That decision illustrates the contortions to which courts subject the distinction between content specificity and content neutrality.

The restrictions contained in section 310 of the Communications Act were an attempt to prevent foreigners from transmitting propaganda or other messages inimical to the national interest. A court, therefore, could hardly conclude that such a restriction is content-neutral. Nevertheless, the Supreme Court has done so before with respect to other statutes. That record makes the outcome of the analysis of section 310 as content-neutral or content-specific, which we shall take up again below, difficult to predict.

Common Carriage versus Private Carriage

The distinction between common carriage and private carriage seems simple enough. Common carriers hold themselves out for use by everyone on a nondiscriminatory basis at reasonable rates.[18] Common carriers exercise no editorial content or control and have limited liability for the messages they transmit. Private carriers are telecommunications firms that do not have those obligations and privileges. Roughly speaking, telephony firms are considered common carriers,

15. United States *v.* O'Brien, 391 U.S. 367 (1968).

16. Meese *v.* Keene, 107 S. Ct. 1862 (1987).

17. *See* Rodney A. Smolla & Stephen A. Smith, *Propaganda, Xenophobia, and the First Amendment*, 67 OR. L. REV. 253 (1988).

18. MICHAEL K. KELLOGG, JOHN THORNE & PETER W. HUBER, FEDERAL TELECOMMUNICATIONS LAW 112–13 (Little, Brown & Co. 1992); DANIEL L. BRENNER, LAW AND REGULATION OF COMMON CARRIERS IN THE COMMUNICATIONS INDUSTRY 35–36 (Westview Press 1992).

whereas cable television operators and radio and television broadcasters are considered private carriers.

The distinction between common carriage and private carriage, however, is not so simple as the preceding discussion suggests, and the limited case law on the subject fails to clarify the matter.[19] Moreover, technological advances are blurring whatever remains of the hoary distinction between common carriers and private carriers. The provision of cable television programs over telephone lines, for example, blurs the distinction because the telephone company will exercise editorial control over which programs to send. Thus, a common carrier will become, at least in part, a noncommon carrier—and a speaker, for First Amendment purposes. How that manifestation of convergence will affect telecommunications regulation cannot be known until the Supreme Court rules on the First Amendment status of bans on such technology.

The dichotomy between private and common carriage has important implications for our analysis of the application of section 310, which we shall take up again below. One important implication is that a common carrier that challenged the ban might find that the courts are not ready to recognize it as a speaker.

THE CONSTITUTIONAL TREATMENT
OF BROADCASTING AND PRINT

We turn now to an examination of how the Supreme Court has applied the First Amendment to various electronic media, starting with radio and television broadcasting. That examination will reveal the intellectual fragility of existing decisions concerning the First Amendment protections afforded electronic speech. As the Court's interpretation of the First Amendment inevitably encompasses the reality of electronic communications, the constitutional infirmity of the foreign ownership restrictions will become more evident.

---

19. National Ass'n of Regulatory Commissioners *v.* FCC, 525 F.2d 380 (D.C. Cir.), *cert. denied,* 425 U.S. 999 (1976); Amendment of the Commission's Rules to Establish Rules and Policies Pertaining to a Non-Voice, Non-Geostationary Mobile-Satellite Service, Report and Order, CC Dkt. No. 92-76, 8 F.C.C. Rcd. 8450 (1993).

The Greater Protection Afforded Print

Print media, particularly newspapers, receive sweeping First Amendment protection from the Supreme Court. In contrast, since the birth of broadcasting, Congress, the FCC, and the courts have refused to extend to that electronic medium the same constitutional protections accorded to the print media. Federal regulation of the broadcast industry has repeatedly entailed direct government control over program content. Congress and the FCC have consistently attempted to impose their notion of proper programming on television and radio broadcasters. In contrast, print media in the United States are generally unregulated. The courts have permitted, and even encouraged, that differential regulation of broadcast speech despite the uncategorical terms of the First Amendment. The dichotomy between print and broadcasting has its roots in the fallacious notion that broadcasting and print media differ in ways that justify their disparate constitutional status.

Prior Restraints. Print media are protected from prior restraints in a way that broadcasters are not. In *Near* v. *Minnesota*, the Supreme Court in 1931 invalidated a Minnesota law that permitted the state's courts to suppress the publication of any "malicious, scandalous or defamatory newspaper."[20] *Near* expanded the definition of "prior restraint" to include cases enjoining an individual from future speech on the basis of past speech.

Near stands in stark contrast to the *Brinkley*[21] and *Shuler*[22] cases of the same era. In *Brinkley* the Federal Radio Commission (the FCC's predecessor) denied renewal of the license of an individual who regularly broadcast radio programs discussing medical problems that listeners described to him by letter. The FRC found the program "inimical to the public health and safety" and thus "not in the public

---

20. 283 U.S. 697, 706, 722–23 (1931).

21. KFKB Broadcasting *v.* FRC, 47 F.2d 670 (D.C. Cir. 1931) (*Brinkley*).

22. Trinity Methodist Church *v.* FRC, 62 F.2d 850 (D.C. Cir. 1932), *cert. denied*, 288 U.S. 599 (1933) (*Shuler*). The definitive analysis of those cases, upon which the discussion here relies, is THOMAS G. KRATTENMAKER & LUCAS A. POWE, JR., REGULATING BROADCAST PROGRAMMING 24–28 (MIT Press & AEI Press 1994), and LUCAS A. POWE, JR., AMERICAN BROADCASTING AND THE FIRST AMENDMENT 13–27 (University of California Press 1987).

interest."²³ The D.C. Circuit upheld the FRC's action, asserting that
the agency "is necessarily called upon to consider the character and
quality of the service to be rendered" and that it thus had an un-
doubted right to look at past performance.²⁴ The difference, then,
between *Near* and the *Brinkley* case is that Minnesota had no such
right, while the FRC did.

The result in the *Brinkley* case comports with the *Shuler* case, in
which the FRC ordered a radio station off the air because its owner,
an evangelical preacher, had broadcast editorials attacking the
decadence of Los Angeles city government. The FRC's rationale was
that the Reverend Shuler's broadcasts were "sensational rather than
instructive."²⁵ The same could surely be said for the newspaper that
the Supreme Court the year before had protected from prior restraint
in *Near*, yet the state was forbidden to enjoin the publication of even
a single issue—no one even contemplated shutting the paper down.
The D.C. Circuit upheld the FRC's action in *Shuler*, granting the
agency nearly unbridled discretion to consider the character and
quality of programming. The court saw no denial of free speech, but
"merely the application of the regulatory power of Congress in a field
within the scope of its legislative power."²⁶ Thus, the difference
between *Near* and *Shuler* was that Congress had regulatory control
over broadcasting, while Minnesota had no such control over newspapers.

Taxes. Taxes on newspapers and magazines are more likely to be prohibited on First Amendment grounds than are taxes on broadcast
media. In *Grosjean* v. *American Press Co.*,²⁷ the Supreme Court in
1936 invalidated a 2 percent tax on gross receipts from advertising in
publications with a certain level of circulation. In *Minneapolis Star &
Tribune Co.* v. *Minnesota Commissioner of Revenue*,²⁸ the Court
invalidated a Minnesota tax on the paper and ink used by newspapers.
The Court reasoned that the tax singled out the press for special treatment and that "differential taxation" of the press is unconstitutional

23. *KFKB Broadcasting*, 47 F.2d at 671.
24. *Id.* at 672.
25. *Trinity Methodist Church*, 62 F.2d at 851.
26. *Id.*
27. 297 U.S. 233 (1936).
28. 460 U.S. 575 (1983).

because it threatened censorship.[29] Finally, in *Arkansas Writers' Project, Inc.* v. *Ragland*,[30] the Court struck down a sales tax that had an exemption for certain types of magazines. As a result of the exemption, only a few magazines in the state paid sales taxes. The Court invalidated the tax largely for the same reasons enunciated in *Minneapolis Star*—the fear of censorial abuse.

Those cases differ markedly from *Leathers* v. *Medlock*,[31] in which the Supreme Court upheld a 4 percent general sales tax that Arkansas had imposed on services including cable television. The tax did not target the media but did exempt newspapers and magazines from coverage. Despite the ample precedent of *Minneapolis Star*, *Grosjean*, and *Arkansas Writers' Project*, the Court found that the tax posed little threat of censorship because it singled out neither specific operators nor specific ideas.[32] Of course, neither had the taxes challenged in any of the three newspaper cases. Broadcasting was simply a poor relative to print media in the Court's First Amendment jurisprudence.

Right of Reply. Print media are not required to provide a right of reply to those who disagree with their editorial content. In *Miami Herald* v. *Tornillo*,[33] the Supreme Court struck down a Florida law that required newspapers to publish replies of political candidates to articles that attacked their character or performance. The Court reasoned that a right of reply might cause self-censorship among newspapers that wished to avoid the expense of printing replies to their editorials.[34] That incentive, in turn, would contravene the autonomy of newspaper editors, whose protection was paramount under the First Amendment.[35]

In *Red Lion Broadcasting Co.* v. *FCC*,[36] however, the Court upheld an FCC regulation that required radio stations to provide equal time to individuals who were editorially attacked by a program on the

29. *Id.* at 585.
30. 481 U.S. 221 (1987).
31. 499 U.S. 439 (1991).
32. *Id.* at 446–48.
33. 418 U.S. 241 (1974).
34. *Id.* at 257–58.
35. *Id.* at 258.
36. 395 U.S. 367 (1969).

air. *Red Lion* posed the same fundamental First Amendment issue that *Tornillo* did: Can the government force a speaker to say what he does not wish to say? The Court said no in *Tornillo*, but yes in *Red Lion*. The First Amendment, said the Court, does not preclude the FCC from requiring a broadcaster "to conduct himself as a proxy or fiduciary with obligations" to implement the "right of the public to receive suitable access to social, political, aesthetic, moral and other ideas and experiences."[37] The Court concluded that the public had a superior right to hear both sides of an issue. "It is the right of the viewers and listeners," the Court announced, "not the right of the broadcasters, which is paramount."[38]

In *Red Lion*, the Court elevated the notion that broadcast licensees had a duty to edify the public over the broadcasters' argument that their new duty would force them to self-censor and destroy their coverage of public issues.[39] The Court rejected that "chilling effect" argument, which had prevailed in *Tornillo*. There would be no such effect, the Court reasoned, because the government would prevent it: Should a timid broadcaster censor itself too severely, the FCC would simply decline to renew its license.[40]

Outright Bans on Content. Newspapers are not subject to outright bans on their content. Broadcasters are. For example, in *FCC* v. *Pacifica Foundation*,[41] the Supreme Court held that the FCC may ban speech from the airwaves that would be legal everywhere else. In *Pacifica*, the speech at issue was George Carlin's "filthy words" monologue. The Court explained that while the material would be permitted elsewhere, "of all forms of communication, it is broadcasting that has received the most limited First Amendment protection."[42] Therefore, censorship of the airwaves was permissible. Though *Pacifica* was eroded by the Court's subsequent decisions that destigmatized indecent material short of actual obscenity,[43] the

---

37. *Id.* at 389–90.
38. *Id.* at 390.
39. *Id.* at 393.
40. *Id.* at 394–95.
41. 438 U.S. 726 (1978)
42. *Id.* at 748.
43. *See* Cohen *v.* California, 403 U.S. 15 (1971); Hustler Magazine *v.* Falwell, 485 U.S. 46 (1988); Sable Communications of Cal., Inc. *v.* FCC, 492 U.S. 115

decision remains a fundamental example of the Court's differential treatment of print media and broadcasting.

Rationales for the Differential Treatment

Spectrum Scarcity. Spectrum scarcity has been an underlying premise of nearly all federal regulation of broadcasting. As chapter 2 explained, the ostensible purpose of the first significant radio regulation in the United States was to minimize interference between rival radio broadcasters in the early 1920s who lacked a system of enforceable property rights in the electromagnetic spectrum. Rather than permit private ownership of the spectrum, however, Congress enacted legislation in 1927 to nationalize the spectrum and license its use.[44] Economist Thomas Hazlett has shown that Congress fully understood in 1927 that a system of property rights in the broadcast spectrum was feasible and was already beginning to evolve through common law adjudication in state court in Illinois.[45] Congress chose, however, to allocate spectrum through a political process rather than through markets, and it restricted competition by limiting the supply of frequencies available for AM radio broadcasting below the level then technically feasible.[46]

The Federal Radio Commission, which became the FCC in 1934, erected an elaborate zoning system for the spectrum. By the early 1940s, though, the federal government's principal justification for regulating broadcasting had shifted away from preventing interference. The FCC and the Supreme Court, led by Justice Felix Frankfurter, maintained that the spectrum was finite and that the agency had to regulate the structure of the communications industry to prevent a monopoly in the marketplace of ideas.[47]

(1989).

44. Radio Act of 1927, ch. 169, 44 Stat. 1162 (1927), *repealed by* Communications Act of 1934, ch. 652, 602(a), 48 Stat. 1102 (current version at 47 U.S.C. §§ 151–613).

45. *See* Thomas W. Hazlett, *The Rationality of U.S. Regulation of the Broadcast Spectrum*, 33 J.L. & Econ. 133, 158–63 (1990).

46. *Id.* at 152–58.

47. The transformation began with Justice Frankfurter's opinion in FCC *v.* Pottsville Broadcasting Co., 309 U.S. 134, 137 (1940), and was complete with his opinion in National Broadcasting Co. *v.* United States, 319 U.S. 190, 215–17 (1943) *(NBC). See also* Associated Press *v.* United States, 326 U.S. 1, 20 (1945) (Black, J.);

"Diversity of Expression." Today, promoting efficient spectrum use and preventing interference are a very small part of the FCC's agenda. Instead, the FCC has become a forum for rent-seeking under the guise of promoting "diversity of expression." With respect to the creeping regulation of cable television during the 1960s, for example, the FCC construed its jurisdiction broadly to reach unregulated firms enabled by new technologies to compete with the agency's existing clientele.[48]

In 1990 the Supreme Court held in *Metro Broadcasting, Inc.* v. *FCC*[49] that the FCC did not violate the Equal Protection Clause of the Fourteenth Amendment by using racial preferences when awarding licenses to operate radio and television stations. Apart from its significance as an affirmative-action decision, which is minimal after the Court's 1995 decision in *Adarand Constructors, Inc.* v. *Pena*,[50] *Metro Broadcasting* was important for a reason that escaped notice: It indicated that the Court and the FCC as of 1990 were still willing to continue using specious scientific and economic arguments to justify denying the electronic media the full protection of the First Amendment. "Safeguarding the public's right to receive a diversity of views and information over the airwaves," because of the state of the Court's understanding of spectrum scarcity in 1990, was "an integral component of the FCC's mission."[51] Without any consideration of how the technology of telecommunications might have advanced since 1969 in such a way as to undercut the scarcity rationale, the Court quoted its opinion issued that year in *Red Lion*: "Because of the scarcity of electromagnetic frequencies, the Government is permitted to put restraints on licensees in favor of others whose views should be expressed on this unique medium."[52] Thus, the Court had no difficulty concluding, in the jargon of judicial

---

FCC *v.* Sanders Bros. Radio Station, 309 U.S. 470, 474 (1940) (Roberts, J.).

48. *See* KELLOGG, THORNE & HUBER, *supra* note 18, at 86, 695–96.

49. 497 U.S. 547, 579–601 (1990). *See* Neal Devins, Metro Broadcasting, Inc. *v.* FCC: *Requiem for a Heavyweight*, 69 TEX. L. REV. 125 (1990).

50. 115 S. Ct. 2097 (1995).

51. *Metro Broadcasting*, 497 U.S. at 567.

52. *Id.* at 566–67 (quoting *Red Lion*, 395 U.S. at 390). Justice White, author of *Red Lion*, was evidently the swing vote in *Metro Broadcasting*, a 5–4 decision. *See* Devins, *supra* note 49, at 125 n.6.

review, that "the interest in enhancing broadcast diversity is, at the very least, an important governmental objective."[53]

As an initial matter, "diversity of expression" is a remarkably vague objective for the U.S. government to pursue, considering that it directly touches freedom of speech. Sometimes the phrase connotes diverse ownership (but not too diverse, as section 310(b) suggests, lest the speech of foreigners fill our heads with foreign ideas). At other times, the phrase connotes a nannyish concern that listeners and viewers receive their recommended daily amount of various intellectual and cultural nutrients—the informational equivalent of the USDA listings found on the sides of cereal boxes. At still other times, "diversity of expression" is a shorthand for the underwhelming argument, seldom expressly articulated, that diverse content can result *only* from the diverse ownership of media companies (and hence the diverse control of FCC licenses).

All of that ignores a basic point: A government-approved menu of diverse programming is something less than freedom of speech. If "Congress shall make no law . . . abridging the freedom of speech, or of the press,"[54] how can it be the federal government's "important" function to judge whether electronic speech is sufficiently diverse? It is a formidable abridgment of speech when the government confers or withholds a person's opportunity to engage in electronic speech depending on whether his message or nationality or other lines of business comport with the government's preferred conception of "diversity." Only a Panglossian would suppose that an agency as politicized as the FCC would arrive at a definition of "diversity of expression" that was truly neutral with respect to content.

On engineering grounds, the spectrum-scarcity premise of *Metro Broadcasting*, *Red Lion*, and their predecessors is untenable. To the extent that it exists, the scarcity resulting from the finite supply of spectrum at any given moment is a problem that diminishes over time. The dynamic, as opposed to static, supply of usable spectrum depends on the state of communications technology, including the precision (and hence the cost) of transmitters and receivers. At any point in time, we could have more "diversity" if we were willing to pay the higher price to produce television receivers with more

---

53. *Metro Broadcasting*, 497 U.S. at 567.
54. U.S. CONST. amend. I.

demanding specifications, or if we were willing to degrade the quality of radio transmissions somewhat by assigning more broadcast licenses in a given region.

Spectrum becomes less scarce whenever new technologies permit transmissions to be packed more densely into a given bandwidth or to be transmitted by radio at higher frequencies that are generally considered to be less desirable. One spread-spectrum technology known as "frequency hopping multiple access" reportedly can achieve a 27-fold increase in the message-carrying capacity of a bloc of spectrum.[55] A more advanced technology known as "software radio" may one day offer virtually limitless spectrum capacity by enabling radio transmissions to shift continuously to unused frequencies across the entire spectrum.[56] Digital-compression technology already developed by companies such as General Instrument and Scientific-Atlanta permits a dozen motion pictures to be transmitted simultaneously in the bandwidth currently used by a single over-the-air television signal. It is startling to think that the scope of the First Amendment's protection of wireless electronic speech—so critical for the development of wireless telephony and wireless multichannel video, such as direct broadcast satellite service, "wireless cable," and local multipoint distribution service—could hang on a basic misconception of electrical engineering that could be corrected if the Justices were to peruse a random issue of *Broadcasting and Cable* magazine.

For a moment, however, assume counterfactually that not a single engineering breakthrough had been achieved in the spectral efficiency of radio transmission since 1934. The scarcity thesis still would be legally untenable because it relies on specious economic reasoning. All valuable goods are scarce. That is why the price of a product is almost always a positive number. Newsprint has a positive price because it too is scarce, but that characteristic in no way justifies regulating who may own a newspaper or what he may say in it, even if the newsprint is made from the pulp of trees harvested from federal forest land.

There is nothing new about that reasoning. Nobel laureate Ronald Coase had this insight in a famous article in 1959.[57] Judge

---

55. *See* GEOTEK, INC., 1993 SEC FORM 10-K, at 6–8 (1994); GEORGE CALHOUN, DIGITAL CELLULAR RADIO 344–51 (Artech House 1988).

56. Raymond J. Lackey & Donald W. Upmal, *Speakeasy: The Military Software Radio*, IEEE COMMUNICATIONS MAG., May 1995, at 56.

57. Ronald H. Coase, *The Federal Communications Commission*, 2 J.L. & ECON.

Robert Bork articulated it succinctly for the FCC's benefit in a decision for the D.C. Circuit in 1986.[58] And scholars in law, economics, and engineering before and since have explained the reasoning in exhausting detail.[59] Still, the Supreme Court and the FCC continue to eschew such logic—no doubt because it calls into question the constitutionality of virtually everything that the FCC does.[60]

Intrusion. The Supreme Court's second rationale for the second-class status of broadcasters under the First Amendment is the notion that broadcasting is an "intruder" that is "uniquely pervasive" and "uniquely accessible to children."[61] The intrusion rationale purports

---

1, 14 (1959).

58. Telecommunications Res. & Action Ctr. *v.* FCC, 801 F.2d 501, 508 (D.C. Cir.) (*TRAC*), *reh'g en banc denied*, 806 F.2d 1115 (D.C. Cir. 1986), *cert. denied*, 482 U.S. 919 (1987).

59. *See, e.g.*, KRATTENMAKER & POWE, *supra* note 22, at 204–18; ROBERT COOTER & THOMAS ULEN, LAW AND ECONOMICS 188–90 (Scott, Foresman & Co. 1988); DOUGLAS H. GINSBURG, REGULATION OF BROADCASTING 58–61 (West Publishing Co. 1979); HARVEY J. LEVIN, THE INVISIBLE RESOURCE: USE AND REGULATION OF THE RADIO SPECTRUM 111–12 (Johns Hopkins University Press 1971); BRUCE M. OWEN, ECONOMICS AND FREEDOM OF EXPRESSION: MEDIA STRUCTURE AND THE FIRST AMENDMENT (Ballinger Publishing Co. 1975); RICHARD A. POSNER, ECONOMIC ANALYSIS OF LAW 672–74 (Little, Brown & Co., 4th ed. 1992); POWE, *supra* note 22, at 199–209; MATTHEW L. SPITZER, SEVEN DIRTY WORDS AND SIX OTHER STORIES (Yale University Press 1986); Arthur S. De Vany, Ross D. Eckert, Charles J. Meyers, Donald J. O'Hara & Richard C. Scott, *A Property System for Market Allocation of the Electromagnetic Spectrum: A Legal-Economic-Engineering Study*, 21 STAN. L. REV. 1499 (1969); Jonathan W. Emord, *The First Amendment Invalidity of FCC Ownership Regulations*, 38 CATH. U. L. REV. 401 (1989); Hazlett, *supra* note 45, at 137; William T. Mayton, *The Illegitimacy of the Public Interest Standard at the FCC*, 38 EMORY L.J. 715, 718–19 (1989); Jora R. Minasian, *Property Rights in Radiation: An Alternative Approach to Radio Frequency Allocation*, 18 J.L. & ECON. 221 (1975); Daniel D. Polsby, *Candidate Access to the Air: The Uncertain Future of Broadcaster Discretion*, 1981 SUP. CT. REV. 223, 255–62; Matthew L. Spitzer, *The Constitutionality of Licensing Broadcasters*, 64 N.Y.U. L. REV. 990 (1989); Matthew L. Spitzer, *Controlling the Content of Print and Broadcast*, 58 S. CAL. L. REV. 1349, 1358–64 (1985); Leo Herzel, Comment, *"Public Interest" and the Market in Color Television Regulation*, 18 U. CHI. L. REV. 802 (1951); Abbott B. Lipsky, Jr., Note, *Reconciling* Red Lion *and* Tornillo: *A Consistent Theory of Media Regulation*, 28 STAN. L. REV. 563, 575–79 (1976).

60. The Supreme Court has discarded the notion of scarcity in the cable television arena, *Turner Broadcasting*, 114 S. Ct. at 2457, but has yet to do so elsewhere.

61. *Pacifica*, 438 U.S. at 748–49. For penetrating analysis of the intrusion rationale

to justify draconian regulations such as the ban on "filthy words" in *Pacifica*. It forces the conclusion that certain types of programming that some consumers desire may be barred from the air because some other consumers do not desire such programming. While the scarcity argument permits substitutions for what a broadcaster would prefer to air, the intrusion argument permits complete bans. Thus, intrusion supports direct censorship, while scarcity supports indirect censorship at worst.

Furthermore, is it even credible to portray the broadcast media as intruders? Radios and televisions, after all, are not forced on the public. We buy them willingly. Even if broadcast media could be equated with intruders, that association hardly would distinguish them from newspapers, magazines, or books. In each instance, the product is voluntarily brought into one's home yet may have scandalous contents. For example, many newspapers run photos and stories depicting risqué and bloody acts. Surely some newspaper readers find that content offensive. Yet such content obviously has not made those newspapers "intruders" subject to regulation.

Power. First Amendment scholars Thomas Krattenmaker and Lucas Powe suggest that *Pacifica* may imply a more fundamental justification for regulating broadcast speech: Broadcasting is too powerful a force to be left unregulated.[62] But what kind of power does broadcasting, and particularly television, actually possess? The basis of the power hypothesis today is the extraordinary amount of television that Americans watch, and its supposed credibility in the eyes of the public. The average person watches almost seven hours of television per day, almost two-thirds of the public uses it as their primary source of news, and almost half rank it as the most believable news source.[63] Yet such logic suggests that before the heyday of broadcast media, when most people read and trusted newspapers as their vital source of information, newspapers should have been entitled to lesser First Amendment status as well. The existence of *Near* and *Grosjean* in the 1930s refutes such reasoning.

---

upon which this discussion is based, see KRATTENMAKER & POWE, *supra* note 22, at 219–21; SPITZER, *supra* note 59, at 124–30.

62. KRATTENMAKER & POWE, *supra* note 22, at 221–24.

63. BROADCASTING YEARBOOK, at A–3 (Broadcasting Publications, Inc. 1991).

There are reasons to reject general regulation of broadcast media based on its supposed power. First, the notion of power does not distinguish between those outlets with power and those without it. Thus, the theory fails to explain why the smallest local television station is more powerful than, say, the *New York Times* or the *Washington Post*. Second, the power rationale seems rooted in the fear that those with power will abuse it, and it ignores that regulators can commit abuses in pursuit of illusory malefaction by broadcasters. Third, the theory is a sad commentary on our nation in that it suggests a populace of mindless automatons manipulated from afar by faceless executives who direct broadcast programming.[64] In short, the power hypothesis may explain why broadcast regulations exist, but it fails to justify them and to provide the constitutional rationale for the different protections of speech afforded print and broadcasting.

Public Property. The only rationale rooted in precedent and logic that might justify lesser First Amendment protection for broadcast media stems from government ownership of the broadcast spectrum. Since the Radio Act of 1927, the government has claimed ownership of the spectrum. From ownership follows control. As the Supreme Court mused, "it is hardly lack of due process for the government to regulate what it subsidizes."[65]

The government could be said to have subsidized broadcasters at one time. The initial licensee on any given frequency received a kind of government largesse. But the overwhelming majority of the radio and television station licenses held today were acquired in the secondary market at fair market value. And although broadcasters who purchase stations from previous holders enjoy FCC policies that limit competition by limiting spectrum usage, the value of that restraint on competition would have been incorporated into the purchase prices of stations.

Naturally, those subsidies, to the extent that they exist, may appear unjustified to those who do not enjoy them. But even if it is the case that only wealthy individuals, or large corporations, can own a station and communicate through broadcasting, that condition also

---

64. *See* Louis Jaffe, *The Editorial Responsibility of the Broadcaster: Reflections on Fairness and Access*, 85 HARV. L. REV. 768, 787 (1972).
65. Wickard *v.* Filburn, 317 U.S. 111, 131 (1942).

holds for newspaper ownership. In a capitalist economy, those with more resources have more choices. That fact should not affect whether one will receive constitutional protections.

That basic conclusion does not change even if one casts government ownership of the spectrum in constitutional terms. In particular, the spectrum may be viewed through traditional public forum analysis. "Traditional public fora" are places, such as parks and streets, that the public has long used for communication and expression.[66] Government regulations on the content of speech in a public forum are subject to strict scrutiny—that is, they must be narrowly drawn to advance a compelling governmental interest.[67] Content-neutral regulations (such as time, place, and manner restrictions) are subject to intermediate scrutiny. They are permissible if narrowly tailored to serve a significant governmental interest, and if they leave open sufficient alternative channels of communication.[68]

"Designated public fora" are created when the government makes property available for use in public expression.[69] Speech in such fora is protected identically with speech in traditional public fora. But the Supreme Court gives the government wide latitude in deciding what is a "designated" public forum.[70]

"Nonpublic fora" are governmental properties not intended for communicative purposes. Military bases are a prime example.[71] Speech regulations on such fora are subject to far less scrutiny. Time, place, and manner restrictions are permitted, and the government may also "preserve the forum for its intended purposes, communicative or otherwise, as long as the regulation of speech is reasonable and not an effort to suppress expression merely because public officials oppose the speaker's view."[72]

With that framework in mind, the broadcast spectrum can be viewed in public forum terms. Perhaps the spectrum is a designated

66. *See, e.g.*, Perry Educ. Ass'n v. Perry Local Educators Ass'n, 460 U.S. 37, 45 (1983); International Soc'y for Krishna Consciousness v. Lee, 505 U.S. 672 (1992).

67. *Perry*, 460 U.S. at 45.

68. Police Dep't of Chicago v. Mosley, 408 U.S. 92, 94–95 (1972).

69. *Perry*, 460 U.S. at 45; Widmar v. Vincent, 454 U.S. 263 (1981).

70. *See, e.g.*, Cornelius v. NAACP Legal Defense & Educ. Fund, 473 U.S. 788, 806 (1985); United States Postal Serv. v. Council of Greenburgh Civic Ass'ns, 453 U.S. 114, 126–31 (1981).

71. Greer v. Spock, 424 U.S. 828, 838 (1976).

72. *Perry*, 460 U.S. at 46.

public forum. Speech is permitted on the airwaves, if licensed, and the government may (at least by licensing) regulate the time, manner, and place of speech. Use of that doctrine blunts the argument that government ownership necessarily means government control. Thus, content-based regulations should be presumptively invalid.

But perhaps the spectrum is instead a nonpublic forum. In that case, the Court can sustain broadcast regulations on essentially any ground that Congress advances to justify them. The problem is that the Court has not attempted that forum-selection analysis, and precedent in the area is sufficiently murky to preclude us from predicting what type of forum the Court would designate the spectrum to be.

The biggest problem with the public ownership argument is that it proves too much. Government ownership of the airwaves stems purely from a legislative decision in 1927 to claim ownership, notwithstanding the prior use of the spectrum by "homesteaders." Some of those spectrum homesteaders challenged the nationalization of the spectrum as an uncompensated taking in violation of the Fifth Amendment; their legal theory was too far ahead of its time, however, and the homesteaders lost.[73] What if Congress were next to decide that it owns the *air* as well? Could the government then demand that all communication traveling through the air conform to rules? That prospect seems ludicrous, but it does not differ fundamentally from the status quo. By what right, after all, did Congress claim control of the broadcast spectrum? To answer that question, one must fall back to the scarcity argument, which has already been discredited.

Summary and Conclusions

We have come full circle. There are no relevant distinctions between broadcasting and print. None of the proffered rationales survives logical inspection. To be sure, the Court may be unlikely to undo nearly seventy years of constitutional error. "Although courts and commentators have criticized the scarcity rationale since its inception," the Court said in *Turner Broadcasting* in 1994, "we have

73. White *v.* Johnson, 282 U.S. 367 (1931); Trinity Methodist Church, South *v.* FRC, 62 F.2d 850 (D.C. Cir. 1932), *cert. denied*, 288 U.S. 599 (1933); City of New York *v.* FRC, 36 F.2d 115 (1929), *cert. denied*, 281 U.S. 729 (1930); United States *v.* Gregg, 5 F. Supp. 848 (S.D. Tex. 1934).

declined to question its continued validity for our broadcast jurisprudence."[74] Thus, the Court may simply continue to sustain broadcast
regulations on whatever grounds seem plausible at the time. That
possibility would require assessing the foreign ownership restrictions
on the assumption that the First Amendment is not a major barrier.
But it seems equally, if not more, likely that the Court will gravitate
away from its antiquated conception of wireless communications as
technological innovations in telecommunications produce a growing
number of examples familiar to the average consumer of how
spectrum is becoming abundant rather than scarce.

RADIO AND TELEVISION

Radio and television broadcasting have traditionally been subject to
regulation that aims at the lofty goals of promoting diversity of programming content and diversity of media ownership. It is useful to
review the zealotry with which the FCC has pursued its notion of
diversity, because such a review makes clear the agency's logical
inconsistency in limiting foreign ownership, which by its inherent
nature would increase the diversity of media ownership in the United
States.

Diversity of Programming

In 1960 the FCC issued its *Programming Statement*, which reflected
the agency's belief that radio and television licensees should offer the
public diverse programming.[75] The FCC listed the fourteen components of a balanced programming diet, ranging from the obvious
(news, weather, sports) to the ethereal ("Opportunity for Local Self-
Expression").[76] The FCC further stated that if a broadcaster was
responsive to the "tastes, needs and desires" of his community, "he
has met his responsibility."[77] Thus, the FCC's pursuit of diverse
programming began with platitudes that seemed difficult for a

---

74. 114 S. Ct. at 2457 (footnote omitted).
75. Network Programming Inquiry, Report and Statement of Policy, 25 FED. REG.
7291 (1960) [hereinafter *1960 Programming Statement*]. The definitive analysis of that
regulatory policy is KRATTENMAKER & POWE, *supra* note 22, at 76–81.
76. 25 FED. REG. at 7295.
77. *Id.*

conscientious broadcaster to avoid fulfilling in his ordinary self-interested pursuit of profit. Nevertheless, the FCC required broadcasters to explain in renewal proceedings their failures to achieve sufficiently diverse programming.[78] Naturally, the threat of being denied renewal of one's license was a sword of Damocles over broadcasters' heads. By 1984, however, the FCC realized that expanded markets virtually guaranteed that broadcasters would meet requirements for diverse programming, so the guidelines no longer served any useful purpose.[79]

Diversity of Opinion and the Fairness Doctrine

The "Fairness Doctrine" required broadcast licensees "to provide coverage of vitally important controversial issues of interest in the community served by the licensees" and "to provide a reasonable opportunity for the presentation of contrasting viewpoints on such issues."[80] The FCC began its entanglement with fairness in its 1949 *Report on Editorializing by Broadcast Licensees*, requiring broadcasters to provide reply time for opposing viewpoints on controversial issues.[81] The FCC codified the so-called personal attack and political editorializing rules in 1967.[82] The personal attack rule required broadcasters to give an individual or group personally attacked during a discussion of a matter of public importance time to reply. The political editorial rules required a broadcaster that presented an editorial policy favoring one political candidate to give reply time to

---

78. Revision of Programming and Commercialization Policies, Ascertainment Requirements, and Program Log Requirements for Commercial Television Stations, Report and Order, MM Dkt. No. 83-670, 98 F.C.C.2d 1076, 1078 n.3 (1984) [hereinafter *Television Deregulation*].

79. *Id.* at 1080–85.

80. Concerning General Fairness Doctrine Obligations of Broadcast Licensees, Report, GEN Dkt. No. 84-282, 102 F.C.C.2d 143, 146 (1985) [hereinafter *1985 Fairness Report.*]

81. Editorializing by Broadcast Licensees, 13 F.C.C. 1246 (1949); *see also* Great Lakes Broadcasting Co., 3 FRC ANN. REP. 32 (1929), *rev'd on other grounds*, 37 F.2d 993 (D.C. Cir. 1930), *cert. dismissed*, 281 U.S. 706 (1930).

82. Amendment of Part 73 of the Rules to Provide Procedures in the Event of a Personal Attack or Where a Station Editorializes as to Political Candidates, Mem. Op. and Order, Dkt. No. 16574, 8 F.C.C.2d 721 (1967).

the other. The Supreme Court upheld the constitutionality of those rules in *Red Lion*. Also in 1967 the FCC extended the Fairness Doctrine to cigarette advertising,[83] but it abandoned that position in 1974, when it began to appear that the agency would have to apply the doctrine to all advertising.[84] Still, the FCC insisted that the doctrine was constitutional. The commission explained that "the First Amendment impels, rather than prohibits, government promotion of a system which will ensure that the public will be informed of the important issues which confront it. . . . The purpose and foundation of the Fairness Doctrine is therefore that of the First Amendment itself."[85]

By 1987 the FCC, staffed with Reagan administration deregulators, had changed its mind. In 1985 the FCC explained that it was "firmly convinced that the fairness doctrine, as a matter of policy, disserves the public interest."[86] The FCC found that the growth in the number of broadcast stations reduced the need for the doctrine, that it discouraged broadcasters from addressing controversial subjects, and that it required the government to evaluate broadcast program content. Nonetheless, concerned that the doctrine might be statutorily mandated, the FCC declined to eliminate it.[87] The FCC also declined to address the argument that the doctrine was unconstitutional; the D.C. Circuit returned the case to the FCC to decide that issue.[88] Meanwhile, in another case the D.C. Circuit had concluded that the Communications Act did *not* mandate the doctrine.[89] On remand, the FCC reiterated its conclusion that the doctrine no longer served the public interest.[90] The FCC also concluded that the Fairness Doctrine was unconstitutional because it "chills speech and is not

---

83. Complaint Directed to Station WCBS-TV, New York, N.Y., Concerning Fairness Doctrine, 8 F.C.C.2d 381 (1967).

84. Handling of Public Issues Under the Fairness Doctrine, Fairness Report, Dkt. No. 19260, 48 F.C.C.2d 1, 26 (1974); *see also* Friends of the Earth v. FCC, 449 F.2d 1164 (1971) (requiring fairness doctrine to be applied to advertisements for automobiles and gasoline).

85. 48 F.C.C.2d at 5–6.

86. *1985 Fairness Report*, 102 F.C.C.2d at 148.

87. *Id.* at 148.

88. Meredith Corp. v. FCC, 809 F.2d 863, 872 (D.C. Cir. 1987).

89. *TRAC*, 801 F.2d 501.

90. Syracuse Peace Council, 2 F.C.C. Rcd. 5043, 5066 n.120 (1987), *recons. denied*, 3 F.C.C.2d 2035 (1988).

narrowly tailored to achieve a substantial government interest."[91] The FCC ruled that "under existing Supreme Court precedent, as set forth in *Red Lion* and its progeny, . . . the Fairness Doctrine contravenes the First Amendment and thereby disserves the public interest."[92] The FCC declared the Fairness Doctrine to be repealed. Without address- ing the constitutional issue, the D.C. Circuit affirmed the FCC's decision to abandon the Fairness Doctrine on the grounds that it no longer served the public interest under the Communications Act.[93]

Diversity of Ownership

The rationale for ownership limits in telecommunications is that changing the identity of a programmer will broaden the menu of programs. Thus, the FCC reasons that programs might be more diverse if owners of program outlets are more diverse. Those ostensi- bly content-neutral regulations can present serious First Amendment concerns. Moreover, those regulations can have important ramifica- tions for the anomalous restrictions on foreign ownership that the FCC imposes when it is supposedly attempting to promote diverse speech and media ownership.

Five broadcast ownership regulations directly affect diversity of programming content.[94] First, the FCC treats ownership by a racial minority as a "plus" in considering which applicant shall be awarded a broadcasting license. Second, the FCC limits the number of outlets within the same broadcast service and in the same local market that a licensee may control. Third, the FCC places a limit on the number of outlets nationwide that a licensee may own within the same broadcast service. Fourth, the FCC restricts the conditions that a television network may impose on its affiliated stations. Fifth, the FCC prohibits cross-ownership of a television station and a daily newspaper in the same market.

Minority Preferences. In an attempt to broaden minority ownership of radio and television stations, the FCC adopted racial preferences in

---

91. 2 F.C.C. Rcd. at 5057.
92. *Id.*
93. Syracuse Peace Council *v.* FCC, 867 F.2d 654 (D.C. Cir. 1989).
94. Some of those regulations are discussed in KRATTENMAKER & POWE, *supra* note 22, at 89.

licensing.[95] Initially, the FCC gave a minority applicant a preference only if he showed that his minority background would influence programming.[96] The D.C. Circuit overruled that policy in 1973 in *TV-9, Inc.* v. *FCC*, however, and held that the FCC had to grant minority preferences on the *assumption* that an applicant's minority background would influence programming.[97] The D.C. Circuit in effect forced the FCC into assuming that diverse ownership would create diverse programming.

The FCC implemented *TV-9* in 1978. First, the FCC accorded a preference to minority applicants in comparative licensing hearings where the minority owner intended to participate in the day-to-day management of the broadcast outlet.[98] Second, the FCC permitted broadcasters who faced hearings and license revocation to avoid both by selling the station at a discount to a minority-controlled group.[99] Third, the agency offered capital gain tax deferral to broadcasters who sold to minority groups.[100] After much controversy in Congress and the D.C. Circuit, the Supreme Court upheld the FCC's minority preference and distress sale policies in *Metro Broadcasting*.[101]

Paradoxically, during the same period that the FCC was encouraging minority ownership to promote diverse programming, the agency also was aggressively probing foreign ownership in television stations that proposed to provide foreign-language programming to minority audiences. Thus, in one case the FCC disqualified an applicant with Taiwanese investors who proposed to provide programming in Mandarin to viewers in the greater San Francisco market.[102]

---

95. Commission Policy Regarding the Advancement of Minority Ownership in Broadcasting, 92 F.C.C.2d 849 (1982).

96. Mid-Florida Television Corp., 33 F.C.C.2d 1, 17–18 (Rev. Bd.), *review denied*, 37 F.C.C.2d 559 (1972).

97. 495 F.2d 929 (D.C. Cir. 1973), *cert. denied*, 419 U.S. 986 (1974).

98. "Minority" was defined as "Black, Hispanic Surnamed, American Eskimo, Aleut, American Indian, and Asiatic American extraction." Statement of Policy on Minority Ownership of Broadcast Facilities, 68 F.C.C.2d 979, 980 n.8 (1978).

99. *Id.* at 983.

100. *Id.* at 982.

101. 497 U.S. 547 (1990). In reliance on *Metro Broadcasting*, the D.C. Circuit subsequently struck down a similar gender preference used in FCC comparative licensing hearings. Lamprecht v. FCC, 958 F.2d 382, 386–88 (D.C. Cir. 1992) (Thomas, J.). That court ruled that the FCC had shown no evidence that women were likely to program differently from men.

102. Pan Pacific Television, Inc., 3 F.C.C. Rcd. 6629 (1988).

In another series of cases, the FCC forced a Mexican citizen to divest his interest in Spanish-language television stations throughout the southwestern United States.[103] The FCC's notion of diversity ended abruptly at the nation's boundaries without regard to the "paramount" right of the listener to hear what he liked, even if it had a foreign accent.[104]

Local Ownership Limits. The FCC has also established ownership limits in local markets. Specifically, the FCC proscribed common ownership of two or more AM, FM, or TV stations in the same market, as well as common ownership of a VHF station and a radio station in any local market.[105] In recognition of the need for radio stations to cut costs, the FCC in 1992 loosened those rules in large markets. The agency allowed a single licensee to control up to two AM and two FM stations in any market with fifteen or more stations, so long as their combined market share does not exceed 25 percent. In smaller markets, a single licensee was allowed to own up to three stations, no more than two of which may be AM or FM, so long as the jointly owned stations constitute less than 50 percent of the stations in the market.[106] The Telecommunications Act of 1996 substantially liberalized those remaining local ownership rules.[107]

National Ownership Limits. The FCC formerly limited the number of broadcast stations that any licensee could control nationally.[108] The rules, however, left intact affiliation agreements into which networks may enter. As a result, large owners could circumvent the limits by

---

103. *Seven Hills*, 2 F.C.C. Rcd. at 6876 ¶ 33.

104. *Red Lion*, 396 U.S. at 389.

105. FCC, Sixth Annual Report 68 (1940) (prohibiting "duopoly" in FM radio and TV); Multiple Ownership of Standard Broadcast Stations, 8 Fed. Reg. 16,065 (1943) (forbidding operation within same market of two AM radio or TV stations); Amendment of §§ 73.35, 73.240, and 73.646 of the Commission Rules Relating to Multiple Ownership of Standard, FM, and Television Broadcast Stations, First Report and Order, Dkt. No. 18110, 22 F.C.C.2d 306 (1970), *modified*, 28 F.C.C.2d 662 (1971) (permitting existing combinations to continue and proposing to treat UHF-radio combinations on a case-by-case basis).

106. *See* Revision of Radio Rules and Policies, Report and Order, MM Dkt. No. 91–140, 7 F.C.C. Rcd. 2755 (1992).

107. 47 U.S.C. §§ 202(b), (c)(2).

108. *See* Krattenmaker & Powe, *supra* note 22, at 95–96.

forming networks of affiliates.[109] FCC rules also capped television station ownership at twelve stations or any lesser number of stations that reached 25 percent of the national audience.[110] The Telecommunications Act of 1996 raised or repealed those national limits.[111]

Regulation of Network Affiliations. The FCC regulates the relations between television networks and their affiliates and program suppliers. The FCC adopted network affiliate rules to limit the ability of dominant networks to extract supposedly onerous contract terms from their affiliates. Thus, network-affiliate agreements may not prevent affiliates from broadcasting programs of another network,[112] may not confer exclusive territories on affiliates,[113] may not grant networks "options" on affiliates' time,[114] and may not prevent or hinder an affiliate from altering its rates for the sale of nonnetwork broadcast time.[115] Those rules increase diversity only in the sense that substituted programs will differ from those that might otherwise have been carried if there were no regulations. Although section 310(b) does not apply to television networks, the network affiliation rules could become a device for asserting that a foreign-owned network was impermissibly controlling its affiliates, which would, of course, be radio licensees subject to the restrictions on foreign ownership or control.

Limits on Newspaper-Television Cross-Ownership. The FCC limits the ability of a single entity to own both a television station and a daily newspaper in the same city. The agency, however, generously grandfathered existing cross-ownership situations at the time of the regulation's promulgation. Thereafter, the FCC has allowed waivers of that rule on a case-by-case basis.

---

109. *Id.* at 97.

110. Amendment of § 73.3555 of the Commission's Rules Relating to Multiple Ownership of AM, FM, and Television Broadcast Stations, Mem. Op. and Order, GEN Dkt. No. 83-1009, 100 F.C.C.2d 17 (1984), *on recons.*, 100 F.C.C.2d 74 (1984).

111. 47 U.S.C. §§ 202(a), (c)(1).

112. 47 C.F.R. § 73.658(a).

113. *Id.* § 73.658(b).

114. *Id.* § 73.658(d).

115. *Id.* § 73.658(h).

The rule is significant to the foreign ownership restrictions in several respects. First, foreigners face no legal impediment to owning newspapers in the United States. Second, the newspaper-television cross-ownership rule is an ostensibly content-neutral regulation of industry structure that the D.C. Circuit has recognized can be enforced by the FCC in a content-based manner. Third, the FCC's application of the rule to Rupert Murdoch's News America was found to violate the First Amendment because it was intended to silence Murdoch based on his political views or national origin.

In *News America Publishing, Inc.* v. *FCC*,[116] the agency argued that *Syracuse Peace Council* rested narrowly on the "conclusion . . . that scarcity did not justify content regulation," and that the decision was therefore irrelevant to "structural regulation of ownership requirements,"[117] such as the newspaper-television cross-ownership rule invoked against Murdoch.[118] Writing for the D.C. Circuit, Judge Stephen Williams intellectually devastated the FCC's claim that structural broadcast regulation should automatically receive a less intense standard of judicial review than content regulation. Even content-neutral FCC regulations that purport to address solely matters of market structure must be scrutinized "under a test more stringent than the 'minimum rationality' criterion typically used for conventional economic legislation under equal protection analysis."[119] Judge Williams characterized broadcast regulation as a continuum, such that ostensibly structural regulations can have the practical effect of restricting broadcasters' freedom of speech: "Clearly one can array possible rules on a spectrum from the purely content-based (e.g., 'No one shall criticize the President') to the purely structural (e.g., the cross-ownership rules themselves)."[120] Along that continuum, a structural prohibition may be "structural only in form," revealing "well recognized ambiguities in the content/structure dichotomy."[121] *News America*, therefore, repudiated the FCC's assertion that

---

116. 844 F.2d 800 (D.C. Cir. 1988).

117. Brief for the Federal Communications Commission at 20, News Am. Publishing, Inc. *v.* FCC, 844 F.2d 800 (D.C. Cir. 1988) (No. 88-1037).

118. 47 C.F.R. § 73.3555(c).

119. *News America*, 844 F.2d at 802; *see also id.* at 814.

120. *Id.* at 812.

121. *Id.* (citing Geoffrey R. Stone, *Restrictions of Speech Because of Its Content: The Peculiar Case of Subject-Matter Restrictions*, 46 U. CHI. L. REV. 81 (1978)).

structural regulation is qualitatively different from content regulation. Instead, the decision implied what some economists long had argued: economic freedom and freedom of speech are inextricably linked.[122]

That more demanding standard of judicial review under the First Amendment eventually will topple the fallacy of spectrum scarcity and, with it, the many statutes and FCC regulations artificially constraining the structure of the telecommunications industry in the name of promoting diversity of expression. The foreign ownership restrictions exemplify the numerous regulatory policies that rest ostensibly on the rationale that, to promote "diversity of expression," government must allocate spectrum and regulate the industrial organization of telecommunications markets in a manner that is not neutral with respect to the identity and message of the person licensed to speak.

Summary. The FCC's regulations attempt to promote diversity of programming through rules concerning actual program content, ownership of media outlets, and program acquisition. As the *News America* case makes clear, those regulations are susceptible to enforcement against foreigners (and even naturalized Americans, such as Rupert Murdoch) in a manner antithetical to the First Amendment.

CABLE TELEVISION

Proponents of regulating cable television have tried to justify reduced levels of First Amendment scrutiny for such regulations by analogizing cable to broadcast television. The analogy is immediately problematic because cable television differs technologically from traditional broadcast television. Terrestrial broadcasting transmits electromagnetic signals to all persons within the broadcaster's service contour who have television receivers and antennae. Cable television operators transmit signals through fiber-optic or coaxial cables laid to the subscribers' premises. As discussed above, the federal government regulates broadcasting far more intrusively than print. The arguments

---

122. *See* OWEN, *supra* note 59, at 21–24, 26–28; R. H. Coase, *The Market for Goods and the Market for Ideas*, 64 AM. ECON. REV. PAPERS & PROC. 384 (1974); Aaron Director, *The Parity of the Economic Market Place*, 7 J.L. & ECON. 1, 3–7 (1964); *see generally* Thomas G. Moore, *An Economic Analysis of the Concept of Freedom*, 77 J. POL. ECON. 532 (1969).

for regulation of cable television revolve around scarcity and notions that cable occupies government property.

Scarcity

We have already considered and rejected the scarcity rationale for broadcast regulation. Different versions of the scarcity argument have been advanced to justify cable regulation as well, but they are inapplicable for technological reasons and for the general theoretical reasons already discussed in connection with broadcasting.

Scarcity Rationales. One can envision five different versions of "scarcity."[123] The first form is "static technological scarcity," which refers to the problem of interference when multiple broadcasters transmit on the same frequency. The argument from static technological scarcity is that regulation prevents overlapping transmissions and thus prevents chaos on the air.[124] That type of scarcity is irrelevant to cable television because cable companies use specific, shielded cables that prevent overlap between systems. Competing entities could only interfere with each other's programming by sharing a single cable, which never happens because the cables are private property.

The second form of scarcity, "dynamic technological scarcity," is the notion that the broadcast spectrum will eventually be exhausted because it is inherently finite—one cannot produce more spectrum. That form of scarcity also fails when applied to cable, however, because one can always manufacture more cables. The resources required to manufacture fiber-optic or coaxial cable are no more limited than are those used to produce paper. Also, the use of digital compression will permit existing cables to have far greater capacity.

"Excess demand scarcity" is a third scarcity concept connoting that more people wish to have spectrum rights than there are possible rights to distribute.[125] There is no excess demand for cable television

---

123. The scarcity nomenclature used in this section is derived from THOMAS W. HAZLETT & MATTHEW L. SPITZER, PUBLIC POLICY TOWARD CABLE TELEVISION: REGULATION AND THE FIRST AMENDMENT ch. 4 (AEI Working Paper in Telecommunications Deregulation 1996).

124. *NBC*, 319 U.S. at 212; *Red Lion*, 395 U.S. at 375–78.

125. *Red Lion*, 395 U.S. at 398–99. The FCC calls this "allocational scarcity." Syracuse Peace Council, 2 F.C.C. Rcd. at 5048–49 ¶¶ 37–39.

systems or for the cable equipment with which to build them, however, because both are allocated through the market at market-clearing prices.

The fourth form of scarcity, "entry scarcity," refers to the argument that broadcasting should be licensed because entry into broadcasting is more difficult than entry into the print media.[126] That argument is circular: The primary obstacles to entry in broadcasting are the burdensome regulations and restrictions on entry themselves. Even so, that form of scarcity barely applies to cable. The productive elements of a cable television system are distributed through the market, without onerous FCC restrictions on transfers. Cities do license large cable systems, but that fact cannot serve as a justification for regulation: it is the *product* of regulation. The only similarity for scarcity purposes between broadcasting and cable television is that large amounts of capital are required for both. Yet the same is true for a large print institution, such as the *New York Times*.

The last scarcity rationale, "relative scarcity," maintains that broadcast spectrum is substantially more scarce than paper. The argument, however, overlooks that one cannot quantify "units" of spectrum in comparison with units of paper, or cable, without considering the ultimate purpose that those units will serve. Often, only one medium is suitable for a particular job: one cannot broadcast music through paper. That scarcity rationale attempts to compare the incomparable.

Scarcity in the Courts. The federal courts have generally refused to accept scarcity as a justification for limiting the speech of cable television operators.[127] So did the major case involving First Amendment protection of cable television, the Supreme Court's 1994 decision in *Turner Broadcasting System, Inc.* v. *FCC.*[128]

---

126. Berkshire Cablevision *v.* Burke, 571 F. Supp. 976, 987 n.10 (D.R.I. 1983), *aff'd*, 773 F.2d 382 (1st Cir. 1985).

127. Preferred Communications, Inc. *v.* City of Los Angeles, 754 F.2d 1396, 1402–05 (9th Cir. 1985), *aff'd on other grounds*, 476 U.S. 488 (1986); Omega Satellite Prods. Co. *v.* City of Indianapolis, 694 F.2d 119, 127 (7th Cir. 1982); Home Box Office, Inc. *v.* FCC, 567 F.2d 9, 44–46 (D.C. Cir.), *cert. denied*, 434 U.S. 829 (1977).

128. 114 S. Ct. 2445 (1994).

In *Turner*, the Court considered the constitutionality of the "must-carry" rules in the Cable Television Consumer Protection and Competition Act of 1992.[129] Those rules required cable systems to set aside a portion of their channels for retransmission of local broadcast programming. The D.C. Circuit had already struck down must-carry rules under the First Amendment in two earlier cases. In *Quincy Communications Corporation* v. *FCC*,[130] the court struck down must-carry rules because the FCC failed to justify its position that permitting cable operators to refuse local programming would threaten over-the-air broadcasting. In response to *Quincy*, the FCC formulated temporary must-carry rules that, in part, required cable operators to provide consumers with A/B switches, which would allow consumers to choose between using the cable feed and a rooftop antenna. The FCC deemed that the interim rules should stay in effect for five years—for consumers to learn how to use a switch. The court struck down the interim rules in *Century Communications Corporation* v. *FCC*[131] because it rejected the FCC's conclusion that it would take five years for consumers to grasp the use of an A/B switch.

The Court in *Turner* set forth some unanimous propositions before discussing scarcity. First, the Court held that cable television is protected by the First Amendment[132] and that the must-carry rules restricted speech in two ways: "The rules reduce the number of channels over which cable operators exercise unfettered control, and they render it more difficult for cable programmers to compete for carriage on the limited channels remaining."[133] Second, the Court agreed that cable, which often has a local monopoly, controls a type of transmission bottleneck—cable has the ability to prevent entry of noncable signals to the market comprising cable subscribers.[134] Third, the Court found that Congress passed the must-carry provisions to curb cable operators' potential monopoly power to reduce the supply of programming sources.[135]

---

129. Pub. L. No. 102-385, 106 Stat. 1460 § 4 (1992) (codified at 47 U.S.C. §§ 534–35).
130. 768 F.2d 1434 (D.C. Cir. 1985).
131. 837 F.2d 517 (D.C. Cir. 1988).
132. *Turner*, 114 S. Ct. at 2456.
133. *Id.*
134. *Id.* at 2466.
135. *Id.* at 2454–55, 2467.

The Court then rejected the government's argument that, because cable raised scarcity problems just as broadcasting did, the Court should review the must-carry rules under intermediate scrutiny. The majority asserted:

> The broadcast cases are inapposite in the present context because cable television does not suffer from the inherent limitations that characterize the broadcast media. Indeed, given the rapid advances in fiber optics and digital compression technology, soon there may be no practical limitation on the number of speakers who may use the cable medium. Nor is there any danger of physical interference between two cable speakers attempting to share the same channel. In light of these fundamental technological differences between broadcast and cable transmission, application of the more relaxed standard of scrutiny adopted in *Red Lion* and other broadcast cases is inapt when determining the First Amendment validity of cable regulation.[136]

The Court nonetheless applied intermediate scrutiny, but on the different rationale that the must-carry rules were not motivated by an intent to discriminate on the basis of programming content.[137] Concluding that the government's asserted justification—protection of broadcasting for those too poor to afford cable—was substantial enough to satisfy intermediate scrutiny,[138] the Court remanded the case for a determination of how narrowly tailored the statute was to achieving that purpose.

The same reasoning by which the Court distinguished cable television from broadcasting also would distinguish broadcasting from wireless point-to-point communications and even broadcasting as it is likely to be conducted in the near future. Digital compression will make all wireless services (including broadcasting) more spectrally efficient. Spread spectrum techniques, moreover, will make existing frequencies allocated to cellular telephony far more capacious and will greatly reduce the likelihood of physical interference. Thus, while purporting to distinguish *Red Lion* from *Turner*, the Court was actually sowing the seeds of *Red Lion*'s demise. When *Red Lion* is

---

136. *Id.* at 2457.
137. *Id.* at 2462.
138. *Id.* at 2461.

inevitably overruled or distinguished away into oblivion, the Court will cite the preceding passage from *Turner* as support for its holding.

Public Property

As with the argument that the broadcast spectrum is public property, and therefore may be regulated by the government, the question arises whether cable television's use of public property justifies regulation as well. A cable company that wishes to distribute its services must use and disrupt streets to some extent. The use and disruption are most severe when the company digs up the street to lay its cable. To the extent that local governments permit cable companies to do so, the argument goes, those governments should be permitted to subject cable operators to reasonable regulations. Thus, for example, government municipalities may franchise cable operators, subject them to fees and access obligations, and impose public service requirements.[139]

Those issues concerning use of public property return us to the public forum doctrine, mentioned earlier. As with all arguments in favor of reduced First Amendment protection, however, the use of public property fails for cable television as well.

First, most cable is not placed on government property. Most cable is attached to telephone poles, which are usually owned jointly by telephone and electric companies.[140] The ground on which the poles rest may be owned as easements granted by private landowners or the government, or in fee by the electric or telephone company. In none of those arrangements can one say that the cable operator uses government property.

There is a stronger argument from public property when cable lines are laid underground. The cable lines may be laid before a residential subdivision is completed, in which case no streets are disrupted. If streets already exist, the cable operator may either rent space in telephone company conduits and pull its wire through them, or cut trenches in the street to lay a new line. That option raises the

---

139. Community Communications Co. *v.* City of Boulder, 660 F.2d 1370, 1377–78 (10th Cir. 1981), *cert. dismissed*, 456 U.S. 1001 (1982); Omega Satellite Prods. Co. *v.* City of Indianapolis, 694 F.2d 119, 127 (7th Cir. 1982).

140. *See* Stuart N. Brotman, *Communications Policy Making at the FCC: Past Practices, Future Direction*, 7 CARDOZO ARTS & ENTERTAINMENT L.J. 55 (1985).

question of who owns the street and the land beneath it. That question, as Thomas Hazlett and Matthew Spitzer have shown, is unclear, as the municipality may or may not hold title, and various utilities generally hold easements throughout the land (for pipes, sewers, cables, and so forth).[141]

Even if one assumes that the local government owns the streets, that fact would not justify reduced First Amendment protection for cable operators. The mere fact that government owns the street does not imply that it may control what is said there.[142] Under the public forum doctrines, government-owned land is either a traditional public forum, a designated public forum, or a nonpublic forum. In general, streets are considered traditional public fora.[143] Regulation of the content of what is expressed in a traditional public forum must be narrowly drawn to serve a compelling government interest.[144] Time, place, and manner restrictions in the forum must be narrowly tailored to serve a significant government interest while leaving open alternative channels of communication.[145]

There are two possible scenarios for applying the public forum doctrines to cable regulations. First, a cable company may hang its cables from poles. As discussed above, the poles on which cables are hung are seldom public property, and thus fail to provide an adequate argument for government regulation on that ground. But cable companies would have to use the streets to hang and service their cables. Although streets are the quintessential public forum, their traditional First Amendment uses include marching and leafleting—not hanging cable. The cable company's use might seem more like conduct than speech. Yet the processes of communication between marchers and leafleteers involves much intermediate conduct—such as marching and handing out leaflets—that resembles hanging cable. Both are necessary antecedents to the communication of ideas. One hands out leaflets before they are read by a recipient, yet the courts consider leafleting to be speech even though the leaflets themselves are literally the only speech involved. So too, hanging cable is a

---

141. *See* HAZLETT & SPITZER, *supra* note 123.

142. Forsyth County *v.* Nationalist Movement, 505 U.S. 123 (1992); Hurley *v.* Irish-American Gay, Lesbian & Bisexual Group of Boston, 115 S. Ct. 2338 (1995).

143. Perry Educ. Ass'n *v.* Perry Local Educators' Ass'n, 460 U.S. 37, 45 (1983).

144. *Id.* at 45.

145. Police Dep't of Chicago *v.* Mosley, 408 U.S. 92, 95–96 (1972).

predicate to electronic speech, even though the only literal speech involved is what subsequently flows over the wires.

If a court were to rule that the use of the street involves a public forum, then the access regulations would have to be time, place, and manner restrictions that were content-neutral and reasonable. But nothing resembling the cable regulation challenged in *Turner* can be justified by time, place, and manner restrictions on when cables may be strung. If a court were to rule that use of the street did not involve a public forum, the regulation might still fail, even under nonpublic forum analysis. That is so since a municipality's attempt to prevent excess traffic congestion and damage to roads—the inevitable justifications for limits on cable-hanging activity—have nothing to do with franchise fees, access rules, or public service requirements. The practical necessity of regulating street access, in short, has no rational extension to control over the content or distribution of speech that flows through the cable company's wires.

A similar analysis would apply if cable were laid underground. Although the question of street disruption would be identical to that which would arise when cable operators use streets to hang cable on poles, the issue of laying cable in subterranean space would be somewhat different. It is likely that this space, if owned by government at all, would not be deemed a public forum: traditional uses of public fora never involved going *beneath* the street. The underground layer would probably be deemed a nonpublic forum. But, as discussed above, even if that conclusion were true, it would not justify extensive regulation of the content or distribution of cable signals. Monopoly franchising might survive, because it arguably would limit street disruption to a single cable. Thus, certain ancillary regulations on the monopolist might be upheld, such as leased access and public service requirements. But, absent a monopolist, regulations going to the heart of cable transmission (such as those that exist for radio and television) would still not be justified.

In short, the government property rationale fails almost entirely to justify any cable regulation. Much of the property that the cable operator would use is not government property. Most of what is government property constitutes a public forum. Only where the government is found to "own" the ground under the street might cable regulation be upheld on a government property rationale. But even then, a municipality would be hard-pressed to justify regulations that are not rationally related to issues of public convenience and

safety, which would be the putative basis for regulating cable placement in the first instance.

In any event, the distinction between the street's surface and
subsurface is too flimsy to support the entire body of cable regulation.
If the First Amendment is to remain flexible, and thus capable of
accommodating unanticipated technologies such as cable television,
courts must see that for cable transmission purposes the subsurface is
the electronic equivalent of the street's surface. Regulation of cable
television should rest on a rationale more substantial than the cable
operator's use of government property. That rationale certainly does
not support restricting a cable operator's First Amendment rights.

TELEPHONY

The provision of telephone service differs from broadcasting and
cable television in the obvious sense that there is no single source of
speech. Although broadcasters and cable operators provide programming to listeners, telephone companies traditionally route calls between individuals on a point-to-point basis. Telephone companies are
common carriers providing a pathway for communication rather than
the communication itself. Furthermore, with the exception of the
newer wireless services such as cellular telephony, paging, and
mobile data transmission, telephone communication does not require
use of the radio spectrum, but instead is provided through terrestrial
networks of wires and switches. Therefore, all the arguments discussed above support the conclusion that, of all telecommunications
media, telephone service should be the most protected from regulations restricting the First Amendment rights of providers.

In fact, the notion that telephone service providers have First
Amendment rights in the first place is new. The question arose in the
landmark case of *Chesapeake and Potomac Telephone Company of
Virginia* v. *United States (CPT)*.[146] The case involved a challenge to
the constitutionality of the "cable–telco entry ban" in the Cable
Communications Policy Act of 1984,[147] which essentially prohibited
local telephone companies from offering, with editorial control, cable

---

146. 42 F.3d 181 (4th Cir.), *cert. granted*, 115 S. Ct. 2608 (1995), *vacated*, 116
S. Ct. 1036 (1996).
147. Pub. L. No. 98–549, 98 Stat. 2779 (codified at 47 U.S.C. §§ 521 *et seq.*).

television services to their common carrier subscribers.[148] Curiously, the legislative history of that provision was silent, but for a boilerplate assertion that the provision was intended to codify existing FCC regulations.[149]

Clearly, the new ability of telephone companies to provide cable programming raised the possibility of new competition in the cable television market. What public interest benefit could therefore justify suppressing such competition and compromising the free speech of telephone companies in the process? The government raised two justifications for section 533(b) of Title 47: preventing telephone companies from engaging in monopolistic practices against the cable industry (principally through the misallocation of common fixed costs) and maintaining diversity of ownership of communications outlets.[150]

The U.S. Court of Appeals for the Fourth Circuit began by recognizing that, in light of *Turner Broadcasting*, the provision of cable television service is protected speech under the First Amendment.[151] The question before the court, then, was whether section 533(b) violated the First Amendment. The court first had to decide what level of scrutiny to employ. It rejected arguments for minimal scrutiny based on scarcity, diversity of ownership, regulation as the quid pro quo for the grant of local monopoly status, and construction of section 533(b) as a generally applicable antitrust law.[152]

The court then rejected strict scrutiny. It first found that section 533(b) was not content-based.[153] The regulation did not discriminate on the basis of the content of speech, but only distinguished speech on the basis of its mode of delivery—in this case, as video programming. The government did not evaluate the content of the speech transmitted, and so the court reasoned that the government did not burden or benefit speech on the basis of its content.[154] The court then found that section 533(b) was not rooted in a discriminatory intent.[155] Rather, the court's review of the legislative history revealed that

---

148. 47 U.S.C. §§ 533(b)(1), (2).
149. *CPT*, 42 F.3d at 187.
150. *Id.* at 190.
151. *Id.*
152. *Id.* at 191–92.
153. *Id.* at 192–95.
154. *Id.* at 194–95.
155. *Id.* at 195–98.

Congress's motivation was to prevent monopolistic practices and preserve diverse ownership.[156] In addition, the court found that section 533(b) did not invidiously target a particular group of speakers, nor grossly diminish the quantity of speech available.[157]

The Fourth Circuit therefore applied intermediate scrutiny.[158] It first found "no question" that the interests to be served by section 533(b) were "significant," thus satisfying the first prong of intermediate-scrutiny review.[159] The court next ruled, however, that the statute failed the second prong because it was not narrowly tailored to meet those significant government interests.[160] The court was particularly disturbed that Congress failed to buttress section 533(b) with any factual findings.[161] Thus, the court held that the government failed to demonstrate why section 533(b) does not burden more speech than is necessary.[162] Last, the Fourth Circuit found that section 533(b) failed the third prong of intermediate scrutiny—the statute did not leave telephone companies with ample alternative channels for communication.[163] Although the First Amendment may tolerate regulations that ban a particular manner or type of expression at a given time or place, it does not accommodate regulations that ban a particular manner of expression altogether.[164] Because the statute could not satisfy the three requirements of intermediate scrutiny, the Fourth Circuit concluded that section 533(b) violated the First Amendment. In June 1995 the Supreme Court granted certiorari to hear the case.[165] After oral argument, the Court vacated the case in light of the enactment of the Telecommunications Act of 1996, which repealed the cable–telco entry ban, and remanded the case to the Fourth Circuit to determine whether the case had been mooted.[166]

CPT is a powerful precedent in support of First Amendment rights for telephone companies. It also symbolizes a consensus among

---

156. *Id.* at 195.
157. *Id.* at 196–98.
158. *Id.* at 198.
159. *Id.* at 199.
160. *Id.* at 199–202.
161. *Id.* at 201.
162. *Id.* at 202.
163. *Id.* at 202–3.
164. *Id.* at 203.
165. 115 S. Ct. 2608 (1995).
166. 116 S. Ct. 1036 (1996).

the lower federal courts. *Every* challenge to section 533(b) prevailed on First Amendment grounds.[167] That body of case law opened the door for greater competition in multichannel video by enabling telephone companies to exercise editorial discretion rather than function solely as passive common carriers.

CPT is highly pertinent to the constitutionality of the foreign ownership restrictions. Although the Fourth Circuit acknowledged the substantiality of the government's objectives underlying the statute, the court found the means employed to achieve that objective to be too loosely woven, and the alternative channels of expression too limited, to satisfy the First Amendment. As we shall now see, those same infirmities plague the foreign ownership restrictions.

FOREIGN OWNERSHIP AND
FREEDOM OF SPEECH

The preceding pages have reviewed the major regulations affecting broadcasting, cable television, and telephony that can restrict freedom of speech. We have seen that print receives far greater protection than broadcasting and cable television for reasons that are largely historical, factually incorrect, and intellectually specious. We have also seen that video programming over telephone lines is now receiving far more solicitous consideration under the First Amendment. We shall now apply that body of law to the foreign ownership restrictions in section 310(b) of the Communications Act to determine whether those restrictions violate the First Amendment.

It is first necessary to determine what level of scrutiny to employ. We shall then analyze the interests purportedly served by the foreign ownership restrictions and the fit between those ends and the means by which Congress and the FCC have chosen to achieve them. Under several different lines of analysis, the existing case law would

167. U S West, Inc. *v.* United States, 48 F.3d 1092 (9th Cir. 1994); Southern New England Tel. Co. *v.* United States, 886 F. Supp. 211 (D. Conn. 1995); Southwestern Bell Corp. *v.* United States, No. 3:94-CV-0193-D (N.D. Tex. Mar. 27, 1995); United States Tel. Ass'n *v.* United States, No. 1:94-CV-0196-1 (D.D.C. Jan. 27, 1995); GTE South, Inc. *v.* United States, No. 94-1588-A (E.D. Va. Jan. 13, 1995); NYNEX Corp. *v.* United States, No. 92-323-P-C, 1994 U.S. Dist. LEXIS 20414 (D. Me. Dec. 8, 1994); BellSouth Corp. *v.* United States, 868 F. Supp. 1335 (N.D. Ala. 1994); Ameritech Corp. *v.* United States, 867 F. Supp. 721 (N.D. Ill. 1994).

support the conclusion that various applications of the foreign ownership restrictions violate the First Amendment.

THE APPLICABLE LEVEL
OF JUDICIAL SCRUTINY

The restrictions in section 310 limit foreign investment in American
telecommunications firms. The section applies only to radio-based
technologies. First, our choice of a level of scrutiny must reference
the standards of review that courts have used to analyze First Amendment challenges to regulations governing the various modes of
electronic speech, particularly radio.

But that is only a first step. The activities to which section 310
applies include broadcasting and common carriage, and, under some
circumstances, private carriage. Some activities to which section 310
applies might be considered nonspeech. Furthermore, defenders of the
constitutionality of section 310 might be able to advance a content-
neutral rationale for regulating some types of activities, but not
others.

Furthermore, individuals to which section 310 restrictions apply
might themselves be alien natural persons, or corporations organized
under the laws of a foreign government, or domestic corporations
with foreign investors, officers, or directors. In some circumstances,
the identity of the plaintiff might affect the standard of review.

Thus, it is unlikely that a court would apply the same standard
of review to all applications of section 310. The sections below will
show, however, that at least intermediate levels of scrutiny will apply
to many applications of section 310.

Red Lion or Turner Broadcasting?

Traditionally, the electronic media have received less constitutional
protection than the print media. The early radio cases involving prior
restraints, *Shuler* and *Brinkley*, point toward a minimal scrutiny, or
rational relation, standard of review.[168] That is true of the cable

---

168. Trinity Methodist Church *v.* FRC, 62 F.2d 850 (D.C. Cir. 1932), *cert.
denied*, 288 U.S. 599 (1933) (*Shuler*); KFKB Broadcasting *v.* FRC, 47 F.2d 670 (D.C.
Cir. 1931) (*Brinkley*).

television taxation case, *Leathers* v. *Medlock*, as well.[169] In both situations, the courts deferred to Congress's commerce power on the one hand and the states' general power to tax on the other. The Court's recent opinion in *Turner Broadcasting* suggests that First Amendment jurisprudence has advanced beyond that point, at least for cable.

But section 310 involves radio, not cable. Therefore, constitutional analysis of section 310 will, for a broad category of cases, depend on how a modern court would treat broadcasting. In light of the advances in First Amendment jurisprudence since the 1930s, no court today would rely on the discredited logic in *Shuler* or *Brinkley* in assessing the constitutionality of section 310. But one must still contend with *Red Lion* and *Pacifica*.

The first part of this chapter shows that the scarcity reasoning behind *Red Lion* has been amply rebutted. The decision is discredited and embarrassing—but not yet overruled. What standard of review, then, will apply so long as *Red Lion* remains the law?

The answer is complicated. For all its warts, *Red Lion* at least recognized that broadcasters did have First Amendment rights. But the exact "test" that the Court used to determine whether those rights had been violated was not clear. At issue was the legitimacy of a regulation that required a broadcaster to offer reply time to persons (usually candidates for office) whom the broadcaster (or even an unrelated third party) had personally attacked during a broadcast. In upholding that regulation, the Court spoke of various factors relevant to its analysis, chiefly concerning monopoly and the rights of listeners, but the Justices did not formulate a general test. Although the right of reply was clearly content-related, the Court did not apply the now familiar constitutional framework, under which the government must show a compelling interest for content-related regulations.

Nonetheless, the Court did apply something more than a rational basis test and gave the FCC's proffered rationales for the right of reply something more than cursory scrutiny. That aspect of *Red Lion* implies that the Court might not approve of some regulations of broadcasting:

> There is no question here of the Commission's refusal to permit the broadcaster to carry a particular program or to publish his

---

169. 499 U.S. 439 (1991).

own views; of a discriminatory refusal to require the licensees to
broadcast certain views which have been denied access to the
airwaves; of government censorship of a particular program
contrary to § 326; or of the official government view dominating
public broadcasting. Such questions would raise more serious First
Amendment issues. But we do hold that the Congress and the
Commission do not violate the First Amendment when they
require a radio or television station to give reply time to answer
personal attacks and political editorials.[170]

That passage suggests that the standard in *Red Lion* is a form of
intermediate scrutiny.

Until 1984 the Court's subsequent decisions failed to clarify
where, along the continuum from rational basis to strict scrutiny, one
would find the appropriate standard of review for broadcast regulation. That year, in *FCC* v. *League of Women Voters*, the Court held
that a provision of the Public Broadcasting Act that forbade public
television stations to "engage in editorializing" violated the First
Amendment.[171] The Court explained that in earlier broadcast cases,
such as *Red Lion*, broadcast regulations

have been upheld only when we were satisfied that the restriction
is narrowly tailored to further a substantial government interest,
such as ensuring adequate and balanced coverage of public issues.
Making that judgment requires a critical examination of the
interests of the public and broadcasters in light of the particular
circumstances of each case.[172]

The Court then identified the regulation in question as being "defined
solely on the basis of the content of the suppressed speech."[173] *League
of Women Voters* thus established that content-related broadcast
regulations would get intermediate, not strict, scrutiny.

Perhaps it would follow that content-neutral regulation of
broadcast speech would get only rational basis review. But the D.C.
Circuit rejected that argument in *News America Publishing*, noting
that "the Supreme Court has for the regulation of speech insisted on

---

170. *Red Lion*, 395 U.S. at 396.
171. 468 U.S. 364 (1984).
172. *Id.* at 380.
173. *Id.* at 383.

a closer fit between a law and its apparent purpose than for other legislation."[174] Judge Williams wrote that the court would look for a fit that snug even in broadcast regulation.[175] The court appears to have applied something between a rational basis test and intermediate scrutiny.[176]

Thus, if *Red Lion* is not immediately overruled, section 310 would probably receive intermediate scrutiny; as we shall see below, it would be hard to argue that section 310 is content-neutral. Intermediate scrutiny requires the government to show that the restrictions serve an important governmental objective, are narrowly tailored to that objective, and preserve ample alternative channels for communication. Applications of section 310 that were determined to be content-neutral would receive something less than intermediate scrutiny, but more than minimum rationality. Finally, *Red Lion* has been applied to other uses of radio besides ordinary broadcasting; private carriers or common carriers challenging section 310 would be unlikely to escape *Red Lion* by noting that they were not broadcasters.[177]

If the Court were willing to overrule *Red Lion* and bring broadcasting under the precedents used for other media, then section 310 would get strict scrutiny if found to be content-related, and intermediate scrutiny if content-neutral, as in *Turner Broadcasting*.[178] There, the majority applied an intermediate level of scrutiny to the must-carry rules; the dissenters applied strict scrutiny, believing the rules not to be content-neutral.[179] The Fourth Circuit also used intermediate scrutiny in striking down the cable–telco entry ban in *CPT* and would have used strict scrutiny had it found the rules to be content-related.[180] The same standard would apply if the radio spectrum were treated as a public forum, as discussed earlier in this chapter. As the analysis below will make clear, the demise of *Red Lion* would also mean the demise of most applications of section 310.

---

174. *News America Publishing*, 844 F.2d at 805.
175. *Id.* at 805.
176. *Id.* at 814.
177. *See TRAC*, 801 F.2d 501.
178. 114 S. Ct. 2445.
179. *Id.* at 2462.
180. *CPT*, 42 F.3d at 198.

Content-Specific or Content-Neutral?

Almost all regulations subjected to strict scrutiny fail the test. Almost all regulations subjected to a rational basis test pass. The outcome of constitutional analysis of section 310 therefore would likely turn on which standard of review the Court applied. That analysis in turn would depend on whether section 310 was content-related or content-neutral. Content-related restrictions are those that "suppress, disadvantage, or impose differential burdens upon speech because of its content."[181] In determining whether a statute is content-related, courts ask first whether the statute on its face discriminates against certain content.[182] But that analysis does not end the inquiry. A law that is content-neutral on its face will be deemed content-related if there is evidence that the statute was intended to suppress certain content.[183] "Our cases have recognized," said the Court in *Turner*, "that even a regulation neutral on its face may be content-based if its manifest purpose is to regulate speech because of the message it conveys."[184]

Applying that basic framework to the various provisions of section 310 raises an additional problem. The statute does not make any direct reference to content. The restrictions do not distinguish between types of foreign speech. According to that view, the limits on foreign ownership apply regardless of what the foreigner wishes to say. Technically, the section is on its face content-neutral. But the statute does single out a class of speakers—foreigners, among all others—for differential treatment. The restrictions can be enforced in an invidiously discriminatory manner, presume the speech of foreigners to be inherently suspect, and limit speech solely on the basis of its source.

Legislation that singles out certain speakers for deferential treatment has been treated with suspicion by the Court.[185] Current

---

181. *Turner Broadcasting*, 114 S. Ct. at 2459.

182. Simon & Schuster, Inc. *v.* New York Crime Victims Bd., 502 U.S. 105, 115–16 (1991).

183. Texas *v.* Johnson, 491 U.S. 397, 402 (1989).

184. *Turner Broadcasting*, 114 S. Ct. at 2461.

185. Minneapolis Star & Trib. Co. *v.* Minnesota Comm'r of Revenue, 460 U.S. 575, 584, 591–92 (1983). Justice O'Connor's dissent in *Turner* explained:

Laws that treat all speakers equally are relatively poor tools for controlling public debate, and their very generality creates a substantial political check

precedents, however, do not quite establish the principle that speaker-specific laws will be always be treated as if they are on their face content-discriminatory. *Turner* rejected "the broad assertion that all speaker-partial laws are presumed invalid,"[186] explaining that, "speaker-based laws demand strict scrutiny when they reflect the Government's preference for the substance of what the favored speakers have to say (or aversion to what the disfavored speakers have to say)."[187] The Court added that "laws favoring some speakers over others demand strict scrutiny when the legislature's speaker preference reflects a content preference."[188] The description of suspicious speaker-specific laws in *Turner* corresponds to the type of speaker-specific classification that section 310 creates. Section 310 singles out foreigners because Congress did not like what some of them might say. Section 310 therefore is likely to be treated as a statute that is content-related on its face.

The fact that the statute will be treated as content-related on its face has an important impact on the inquiry into the statute's legislative history. If content discrimination appears on the face of a statute, a court will often disregard content-neutral justifications for the law contained in the legislative history.[189] If the statute is on its face neutral, a court will look to the legislative history to see whether the law's purpose is content-related. But, the court is less likely to overrule the law on the basis of content-related commentary in the legislative history if it also finds that the main purpose of the law is not content-related. In *Turner*, for example, the Court stated: "Our review . . . persuades us that Congress' overriding objective in enacting must-carry was not to favor programming of a particular subject matter, viewpoint, or format, but rather to preserve access to free television programming for the 40 percent of Americans without cable."[190] In other words, just because some legislators had a content-

---

that prevents them from being unduly burdensome. Laws that single out particular speakers are substantially more dangerous, even when they do not draw explicit content distinctions.

114 S. Ct. at 2476.
186. *Turner Broadcasting*, 114 S. Ct. at 2467.
187. *Id.*
188. *Id.*
189. Carey *v.* Brown, 447 U.S. 455, 466–68 (1980).
190. *Turner Broadcasting*, 114 S. Ct. at 2461.

related purpose in mind when voting for the statute does not mean that a court would necessarily strike down the statute.

What would the outcome be if those precedents were applied to section 310? The legislation should be treated as if it is content-related on its face. And, the legislative history makes clear that the primary purpose of the foreign ownership restrictions is content-related. The target was foreign speech that Congress thought might be a threat to U.S. interests, especially during war.[191] The overwhelming legislative history shows that section 310 is content-related.

Depending upon the extent to which the reviewing court had rejected *Red Lion*, the foreign ownership restrictions would properly be reviewed under strict scrutiny, or at least intermediate scrutiny. A court applying strict scrutiny to section 310 would strike down the statute unless the government showed that the statute serves a compelling government interest and is narrowly tailored to meet that end in the manner that least restricts speech.

As always, however, defenders of the constitutionality of the section could raise one final argument: Although the legislative history is replete with evidence that the law was content-related, one can offer content-neutral justifications for section 310(b) *not* found in the legislative history, such as its value as a tool of trade policy. Any court applying more than a rational basis test, however, would not take seriously that stratagem of *post hoc* justification. Furthermore, because section 310(b) is content-related on its face, a court may disregard proffered content-neutral rationales. Those canons of interpretation, of course, in no way limit the possibility that Congress could reenact the foreign ownership restrictions without referring to any forbidden purpose the second time around.

If the reviewing court had rejected *Red Lion* and found the foreign ownership restrictions to be content-neutral, it would analyze them under intermediate scrutiny. If, alternatively, a reviewing court clung to *Red Lion*, the restriction would get a level of scrutiny in between the intermediate test and the rational basis test.

Given the legislative history, the proposition that section 310 is content-neutral lacks any support. We may therefore assume that,

---

191. Ian M. Rose, Note, *Barring Foreigners from Our Airwaves: An Anachronistic Pothole on the Global Information Highway*, 95 COLUM. L. REV. 1188, 1211 (1995).

other things being equal, a First Amendment attack on section 310 would be subject at least to intermediate scrutiny.

Speakers, Would-Be Speakers, and Carriers

So far, we have assumed that a plaintiff who challenged section 310 would be in a position to claim that the government had interfered with some aspect of his rights to speak. Not every activity, however, is considered speech. Challenges to section 310 brought by some hypothetical plaintiffs might not come under the protection of the First Amendment at all.

Clearly, some activities with which section 310 would interfere with *are* covered by the First Amendment. Suppose, for example, that a domestic broadcasting corporation that already held a broadcast license was found to be controlled by another corporation of which foreigners held 26 percent of the stock. That condition would violate the FCC's interpretation of section 310(b)(4), and the domestic corporation's broadcast license could be revoked, which would effectively silence its speech. That type of entity—a company that is already broadcasting—is the type of complainant whose First Amendment rights the Court recognized in *Red Lion*. Likewise, a company whose license to provide itself with aeronautical radio service (private carriage) was revoked would be treated as a speaker under the First Amendment. So would a common carrier that provided some sort of information service.

But what about a company or individual who did not yet hold a radio license of any kind? One might argue that this type of plaintiff is too remote from actually speaking to count under the First Amendment. That reasoning, however, conflicts with many First Amendment cases that recognize applicants for permits of every kind (would-be cable franchisees[192] or applicants for public assembly or parade permits[193]) as having First Amendment standing.

---

192. *See, e.g.,* Preferred Communications, Inc. *v.* City of Los Angeles, 13 F.3d 1327 (9th Cir. 1994).

193. Forsyth County *v.* Nationalist Movement, 505 U.S. 123 (1992) (striking down a permit scheme giving local government discretion to adjust parade fees to the level of policing that it thought was necessary).

Red Lion, however, suggests an additional argument for denying an applicant for a license standing under the First Amendment—the fear that to recognize an applicant's First Amendment rights would require the FCC to license everyone everywhere, which would result in chaos. That fear, in combination with the premise that the FCC could deny an applicant a license for perfectly legitimate reasons—if the spectrum that the applicant wished to use was already occupied, for instance—led the FCC, quoting *Red Lion* in the agency's *Seven Hills* decision, into folly:

> Seven Hills' constitutional claims are faulty, for its compact rhetoric fatally confuses (1) the First Amendment rights of free speech accruing to those federally licensed to broadcast with (2) the conditional privilege of a broadcast license. "No one has a First Amendment right to a license . . . ." The Supreme Court has decreed [that] ["w]here there are substantially more individuals who want to broadcast than there are frequencies to allocate, it is idle to posit an unabridgeable First Amendment right to broadcast comparable to the right of every individual to speak, write, or publish.["][194]

The assertion that no one has a First Amendment right to a broadcast license is beside the point, and the rest of the argument is fallacious.

First, the FCC's logic goes too far. Consistent with the Supreme Court's assertion in *Red Lion* that no one has a right to a license, one might claim that *no* broadcaster, licensed or not, has any First Amendment rights at all. There is no way to distinguish the would-be licensee from the extant licensee, as the FCC tried to do in *Seven Hills*. Indeed, the *Red Lion* Court itself said: "By the same token, as far as the First Amendment is concerned those who are licensed stand no better than those to whom licenses are refused."[195]

Furthermore, it does not follow from the fact that the FCC should be able to refuse to license a station for some reason unrelated to its content that the FCC should be able to refuse to license a station for any reason at all. It does not cause the whole fabric of FCC spectrum allocation to unravel to recognize the First Amendment

---

194. *Seven Hills*, 2 F.C.C. Rcd. at 6876 ¶ 34 (quoting *Red Lion*, 395 U.S. at 388, citation omitted).

195. *Red Lion*, 395 U.S. at 389.

claims of an applicant for a radio license. One need not hold that the FCC must grant him a license, only that the agency must not refuse to consider his application because Congress has directed it to censor a certain group. In short, a constitutional challenge to section 310 brought by a would-be radio licensee should not fail because he has not yet obtained a license.

Suppose instead that the plaintiff was a common carrier. That case would present not a would-be speaker but a carrier of others' speech. The pure common carrier is more analogous to Federal Express than to a leafleteer who is also a speaker, or to a telephone company that wants to use its own network to provide cable programming of its own selection. But the First Amendment does protect some distributors who do not themselves necessarily speak—newsracks[196] and bookstores,[197] for example. And any common carrier that provided interactive broadband and enhanced services would be a speaker as well. As the convergence of technology daily erodes the distinction between common carriers and private carriers, the First Amendment claim here will grow stronger.

So far, the Court has clung to the distinction between conduit and editor, most recently in *Hurley* v. *Irish-American Gay, Lesbian and Bisexual Group of Boston*.[198] An association of veterans' groups, the sponsors of Boston's St. Patrick's Day parade, refused to allow an association of gays, lesbians, and bisexuals to march in the parade. The Court held that applying a state public accommodations statute to require parade organizers to include a group imparting a message that the organizers did not wish to convey violated the First Amendment.[199] The Court analogized the organizers' control of the parade to a editor's control over a newspaper.[200] But the Court was hard-

---

196. City of Cincinnati v. Discovery Network, Inc., 507 U.S. 410 (1993) (ban on newsracks' dispensing "commercial handbills" turns on content of handbills and thus is content-based); Miami Herald Publishing Co. v. City of Hallandale, 734 F.2d 666 (11th Cir. 1984) (invalidating municipal ordinance giving city council discretion to deny newsrack license when applicant did not comply with certain regulations).

197. *Ex parte* Jackson, 96 U.S. 727, 733 (1877); Lovell v. Griffin, 303 U.S. 444 (1938); Board of Educ. v. Pico, 457 U.S. 853, 867 (1982); Smith v. California, 361 U.S. 147, 150 (1959).

198. Hurley v. Irish-American Gay, Lesbian & Bisexual Group of Boston, 115 S. Ct. 2338 (1995).

199. *Id.* at 2341.

200. *Id.* at 2345–46.

pressed to distinguish its opinion in *Turner*, in which it was willing
to approve, if the government could show on remand that the fit
between means and ends was tight enough, regulations requiring cable
systems to set aside capacity for broadcasters. To distinguish *Turner*,
the Court emphasized its reliance on the idea that cable systems were
a mere conduit for broadcasting.[201]

Media and Nonmedia
Domestic Corporations

The various provisions of section 310 may be applied to either media
or nonmedia domestic corporations. That dichotomy further compli-
cates the First Amendment analysis. Some cases suggest that
corporations have full First Amendment rights; others suggest that the
Court will allow those rights to be abridged in circumstances in which
it would not allow an individual's rights to be abridged.

In *Grosjean*, the Court easily concluded that corporations
enjoyed First Amendment rights.[202] Although that case, of course,
involved a First Amendment challenge brought by a media corporation, a newspaper, the Court did not attach any significance to that
fact. Denying corporations full First Amendment protection would
effectively undermine the media. Since the Court decided *Grosjean*
in 1936, it has not hinted in any case that a media corporation could
be given less than full First Amendment protection because it was a
corporation.[203]

The treatment of nonmedia corporations, however, has been less
consistent. In *First National Bank* v. *Bellotti*, the Court struck down
restrictions on a corporation's political action committee expenditures.
The Court found that such expenditures were a form of speech and
that a corporation did indeed enjoy First Amendment protection.[204]
The Court stated broadly that the "inherent worth of the speech in
terms of its capacity for informing the public does not depend upon
the identity of its source, whether corporation, association, union, or
individual."[205] But the Court refrained from deciding whether in all

---

201. *Id.* at 2348.
202. 297 U.S. at 244.
203. *See* Philadelphia Newspaper, Inc. *v.* Hepps, 475 U.S. 767 (1986).
204. 435 U.S. 765, 777 (1978).
205. *Id.*

cases corporations would be as protected as individuals: "In deciding whether this novel and restrictive gloss on the First Amendment comports with the Constitution and the precedents of this Court, we need not survey the outer boundaries of the Amendment's protection of corporate speech, or address the abstract question whether corporations have the full measure of rights that individuals enjoy under the First Amendment."[206] Later, in *Austin* v. *Michigan Chamber of Commerce*, the Court upheld a statute restricting independent corporate expenditures in election campaigns. The Court found that the state had a compelling interest in preventing the appearance of corruption.[207] In *Buckley* v. *Valeo*, however, the Court had struck down application of such a restriction to individuals. The Court found that the same asserted interest was not compelling.[208] The *Austin* Court attempted to justify its unequal treatment of corporations by explaining that "[s]tate law grants [them] special advantages."[209] The Court did not alter the standard of review—only the balance it struck between government interests and First Amendment rights.

The Court's reasoning in *Austin* would also imply that media corporations could be given less than full First Amendment protection. Justice Scalia, dissenting, objected strongly to the Court's reasoning, noting that it proves too much: "[O]ther associations and private individuals [are] given all sorts of special advantages that the State need not confer, ranging from tax breaks to contract awards to public employment to outright cash subsidies. It is rudimentary that the State cannot exact as the price of those special advantages the forfeiture of First Amendment rights."[210] Justice Scalia thus joined former Justice Brennan in opposing the idea that nonmedia corporations should be given less than full First Amendment protection.[211] Other cases involving the First Amendment rights of corporations do not contain the fallacies of *Austin*; it seems safe to say that, outside the campaign expenditure cases, corporations do enjoy full First Amendment rights.[212] Under those precedents, whether an individual,

---

206. *Id.* at 777–78.
207. 494 U.S. 652 (1990).
208. 424 U.S. 1 (1976).
209. *Austin*, 494 U.S. at 658.
210. *Id.* at 680 (Scalia, J., dissenting).
211. *Hepps*, 475 U.S. at 779–80 (Brennan, J., concurring).
212. Pacific Gas & Elec. Co. *v.* Public Util. Comm'n of Cal., 475 U.S. 1 (1986);

a media corporation (such as a broadcaster), or a nonmedia corporation brought a challenge to section 310 would not affect the chances of its success.

*Foreigners, Foreign Corporations,
and Foreign Governments*

Under many circumstances, section 310 will apply to foreign natural persons, to foreign corporations, or to foreign governments. A brief outline of cases involving the constitutional rights of those potential plaintiffs is necessary before analyzing the constitutionality of section 310.

First, are foreigners less entitled to the protections of the Constitution than U.S. citizens? The Supreme Court does not generally support the idea that nonresident aliens have a First Amendment right to speak in America.[213] Several cases concerning the First Amendment rights of foreigners—even resident aliens—suggest that those rights are curtailed whenever they conflict with the federal government's plenary power over immigration.[214] Other cases suggest that resident aliens are under the full protection of the Constitution, including the First Amendment.[215] The extent to which resident aliens may assert First Amendment rights against the government outside the immigration context is not resolved.

It would be surprising if the federal government could shut down a resident alien who owned a bookstore or a newspaper—other than in the exercise of the immigration power—under circumstances where the First Amendment would protect a U.S. citizen. The availability

---

Consolidated Edison Co. *v.* Public Serv. Comm'n of N.Y., 447 U.S. 530, 544 (1980); *see also* HENRY N. BUTLER & LARRY E. RIBSTEIN, THE CORPORATION AND THE CONSTITUTION 59–106 (AEI Press 1995).

213. Rose, *supra* note 191, at 1207.

214. *See* Note, *Silencing the Speech of Strangers: Constitutional Values and the First Amendment Rights of Aliens*, 81 GEO. L.J. 2073, 2074 (1993).

215. Bridges *v.* Wixon, 326 U.S. 135, 148 (1945); Kwong Hai Chew *v.* Colding, 344 U.S. 590, 596 n.5 (1953) ("[O]nce an alien lawfully enters and resides in this country he becomes invested with the rights guaranteed by the Constitution to all people within our borders. Such rights include those protected by the First and the Fifth Amendment and by the Due Process Clause of the Fourteenth Amendment. None of these provisions acknowledges any distinction between citizens and resident aliens"); *see* Rose, *supra* note 191, at 1208-9.

of the immigration power to the federal government in such a case is a strong argument that the government does not need even broader general powers over resident aliens, particularly where exercise of those powers would erode the protections of the First Amendment. That consideration should incline a court to treat a constitutional challenge to section 310(b) brought by foreigners as comparable to a challenge brought by a U.S. citizen; at least intermediate scrutiny would still apply, rather than the rational basis test employed in immigration cases.

An additional argument in favor of that view is that section 310(b) as applied to foreign broadcasters (at least) implicates, not only the foreigner's rights, but also the rights of U.S. citizens to hear foreign speech.[216] Where there has been no immigration issue, the Court has recognized those rights. In *Lamont* v. *Postmaster General*, the Court declared unconstitutional a statute that required addressees of "communist political propaganda" to send a reply card to the post office to receive such mail.[217] The "propaganda" in that case was a Chinese newspaper; the Court did not address the sender's rights, but found that to require the recipient to return a reply card to the post office violated the addressee's First Amendment rights.[218] It would break new ground for a court to announce that it would not protect the rights of citizens to hear the speech of a resident (or even nonresident) foreigner because the First Amendment must always take a back seat to the power of the federal government to regulate trade or national security. It is not clear, however, whether a foreign plaintiff could assert a U.S. citizen's rights in that context; a citizen with a persuasive standing argument certainly could.[219]

It has not yet been settled to what extent a foreign corporation's First Amendment rights would be recognized. Cases involving long-arm jurisdiction statutes as applied to foreign corporations do establish that they have rights under the Due Process Clause.[220]

---

216. Rose, *supra* note 191, at 1205–6; J. Gregory Sidak, *Don't Stifle Global Merger Mania*, WALL ST. J., July 6, 1994, at A18.

217. Lamont *v.* Postmaster Gen., 381 U.S. 301 (1965); Rose, *supra* note 191, at 1205–7.

218. 381 U.S. at 307.

219. Meese *v.* Keene, 481 U.S. 465 (1987).

220. *E.g.*, Asahi Metal Indus. Co. *v.* Superior Ct. of Cal., 480 U.S. 102, 108 (1987).

By contrast, the First Amendment does not protect foreign governments and their official representatives.[221] In *Mendelsohn* v. *Meese*, a district court rejected a First Amendment challenge to the Anti-Terrorism Act of 1987 (ATA), which was intended to curb the operation of the Palestine Liberation Organization.[222] The challenge was brought by U.S. citizens who claimed that the act, by barring them from obtaining funds from the PLO, prevented them from speaking on behalf of the PLO in an official or unofficial capacity. The court accepted the government's argument that *official* agents of a foreign government have no constitutional rights:

> A "foreign state lies outside the structure of the Union." The same is true of the PLO, an organization whose status, while uncertain, lies outside the constitutional system. It has never undertaken to abide by United States law or to "accept the constitutional plan." No foreign entity of its nature could be expected to do so . . . . It would make no sense to allow American citizens to invoke their constitutional rights in an effort to act as official representatives of foreign powers upon which the political branches have placed limits. Doing so would severely hamper the ability of the political branches to conduct foreign affairs. Any action harming the interests of a foreign power could otherwise be challenged in court as a violation of Americans' due process or First Amendment rights. Diplomatic relations could not be severed, for the foreign government could enlist American citizens to act as its representatives.[223]

The *Mendelsohn* court, however, refused to extend such reasoning to those plaintiffs who claimed to be acting in an *unofficial* capacity. The Court cited the Supreme Court's opinion protecting such an unofficial representative in *Communist Party* v. *Subversive Activities Control Board*.[224] Nonetheless, the court in *Mendelsohn* ultimately rejected the unofficial representative's First Amendment claims because it found

---

221. *See* LOUIS HENKIN, FOREIGN AFFAIRS AND THE CONSTITUTION 254 (Columbia University Press 1972).

222. 695 F. Supp. 1474 (S.D.N.Y. 1988).

223. *Id.* at 1481 (quoting Principality of Monaco *v.* Mississippi, 292 U.S. 313, 330 (1934)); *see also* Lori Fisler Damrosch, *Foreign States and the Constitution,* 73 VA. L. REV. 483, 518 (1987).

224. 367 U.S. 1, 95–96 (1961).

the statute to be content-neutral and the government's asserted interest sufficiently important.[225]

THE NATIONAL SECURITY RATIONALE

In *United States* v. *Robel*,[226] the Supreme Court declared a provision of the Subversive Activities Control Act[227] to be unconstitutional. In violation of the statute, a member of the Communist Party had remained in his job as a machinist at a shipyard after it had been designated a defense facility. Writing for the majority, Chief Justice Warren explained that "the phrase 'war power' cannot be invoked as a talismanic incantation to support any exercise of congressional power which can be brought within its ambit."[228] Despite the concerns of Congress and the executive branch over "internal subversion," the chief justice said, such reasoning would defeat the purpose of creating a constitutional democracy in the first place:

> [T]his concept of "national defense" cannot be deemed an end in itself, justifying any exercise of legislative power designed to promote such a goal. Implicit in the term "national defense" is the notion of defending those values and ideals which set this Nation apart. For almost two centuries, our country has taken singular pride in its Constitution, and the most cherished of those ideals have found expression in the First Amendment. It would indeed be ironic if, in the name of national defense, we would sanction the subversion of one of those liberties . . . which makes the defense of the Nation worthwhile.[229]

The Court was dismayed to find that the statute made no attempt to exempt individuals who might be "passive or inactive member[s] of a designated organization," or who might be unaware of or disagree with the organization's unlawful aims, or who might "occupy a nonsensitive position in a defense facility."[230] The Court held that "because [the statute] sweeps indiscriminately across all types of

---

225. 695 F. Supp. at 1483.
226. 389 U.S. 258 (1967).
227. 64 Stat. 992 (1950) (codified at 50 U.S.C. § 784 (a)(1)(D)).
228. 389 U.S. at 263–64.
229. *Id.* at 264.
230. *Id.* at 266.

association with Communist-action groups, without regard to the quality and degree of membership . . . it runs afoul of the First Amendment."[231] The Court could have lodged virtually the same criticisms against section 310 of the Communications Act.

The Effect of Bellicosity or Peace

National security has long been the asserted rationale for section 310. As chapter 2 documented, section 310 serves the national security objective of protecting the United States from foreign threats during time of war or national emergency.[232] The legislation principally reflected fears that radio stations within the United States could be used for point-to-point communications with the enemy. Before America's entry into World War I, German-controlled wireless stations on the Atlantic coast communicated with German vessels. That concern was foremost in the minds of senior Navy officers who lobbied Congress for the foreign ownership restrictions and established RCA as a government-controlled wireless company owned and managed by Americans. A second national security objective, which does not find nearly so much historical support for it as does the first, is the prevention of the dissemination of enemy propaganda during wartime. Despite the relatively slim historical support for such an objective, the proposition that Congress intended section 310(b) to prevent wartime propaganda is firmly fixed in the conventional wisdom surrounding the foreign ownership restrictions and is routinely cited by courts and the FCC.[233]

Some might argue also that section 310 addresses the danger of sabotage, as do other restrictions on foreign direct investment.[234] That argument, however, is conspicuously absent from the legislative history of section 310 and its predecessor provisions in the Radio Acts of 1912 and 1927. Indeed, if alien sabotage of the telecommunications network, rather than foreign control of the wireless portion of it, had been a concern to Congress, then the foreign ownership

---

231. *Id.* at 262.

232. *See, e.g., Hearings on S. 2910 Before the Senate Comm. on Interstate Commerce*, 73d Cong., 2d Sess. 170 (1934).

233. *E.g.,* Noe *v.* FCC, 260 F.2d 739, 741–42 (D.C. Cir. 1958).

234. Elliot L. Richardson, *United States Policy Toward Foreign Investment: We Can't Have It Both Ways,* 4 AM. U.J. INT'L L. & POL'Y 281, 307–8 (1989).

restrictions would have covered wireline telephony and telegraphy, which are far more vulnerable than wireless to destruction of physical infrastructure, such as wires and switches. To the contrary, Congress confined section 310 to wireless communications. Also, one must ask why foreigners would want to invest in communications assets of any kind with the intention of destroying them in wartime.

In short, the national interest in restricting the electronic speech of aliens has diminished with the end of the cold war. The level of international security in the late 1990s, despite regional conflicts in Africa and the Balkans, is far removed from a condition of world war in which a formidable enemy with superior weaponry is consistently attacking civilian shipping and transportation. New risks to national security, of course, may arise and justify relatively burdensome regulation of foreign ownership of wireless. A handful of rogue nations—Iran, Iraq, Libya, and North Korea—remains hostile to the United States. But that condition does not justify turning away foreign investment from the many nations that are America's allies. Indeed, to apply the foreign ownership restrictions to investment from a friendly nation is not even consistent with the rationale routinely imputed to Congress for the statute's enactment.

General Strength of Interest
in National Security

If national security is the asserted government purpose of section 310, the next question is whether that interest is content-related or content-neutral. Clearly, concern with propaganda is content-related. But concern with opportunities for espionage appears to be content-neutral. Depending on the circumstances, then, the asserted government interest must be either "important" or "compelling."

Such goals are important and have even been deemed compelling during wartime.[235] But they lose their urgency during times of peace. The justification for the foreign ownership restrictions is wartime xenophobia and fear of subversion. With the end of the cold war, there is less basis than in 1934 to suspect foreigners when they speak to Americans; and, with the growth of international communication over the Internet, there is less reason to believe that Americans would

---

235. Korematsu *v.* United States, 323 U.S. 214 (1944).

uncritically accept propagandistic ideas. Furthermore, it would prove too much to convert the national security interest to one of "preparedness" on the rationale that the outbreak of war cannot be predicted, for that logic would justify any manner of infringement on civil liberties. In short, a court could reasonably conclude that the national security rationale for section 310(b) fails both intermediate and strict scrutiny when it is applied in undifferentiated form during times of peace.

The Fit between National Security Interests and Section 310 Generally

If national security is nonetheless deemed an important or even compelling interest, we must next determine whether the foreign ownership restrictions are narrowly tailored to meet national security concerns. In particular, are there alternative means available that would restrict speech to a lesser extent? Clearly, yes.

The foreign ownership restrictions are both overinclusive and underinclusive. They apply with equal force to *all* foreigners, whether citizens of friendly or hostile nations. As a result, the United States limits the investments of Britons to the same extent that it limits the investments of North Koreans and Iraqis. If the foreign ownership restrictions were narrowly tailored, they would distinguish among nations and restrict only citizens of those states that pose a realistic threat to U.S. national security. If a nation such as North Korea or Iraq is a security threat to the United States, then it would not be in the public interest to allow one of its citizens to own *any* fraction of the stock in a U.S. radio license. Yet, section 310(b) would freely allow such a person (through direct investment and indirect investment in a corporate subsidiary) to acquire 40 percent of the U.S. firm, as chapter 4 demonstrates. Moreover, since 1996 the statute would have allowed all of the licensee's officers and directors to be North Koreans or Iraqis. Section 310(b) is thus underinclusive. At the same time, the statute is overinclusive. Narrowly tailored foreign ownership restrictions would exempt, or at least treat more favorably, nations with which the United States is allied in security treaties such as NATO. Section 310(b) makes no attempt to do so. It is not reasonable—it is not rational—to characterize a nation in whose defense the United States is bound by treaty to commit military forces as a nation whose citizens, for purposes of section 310(b), should be restricted on national security grounds in their ability to make direct investments in U.S. radio licenses.

The foreign ownership restrictions are also overinclusive and underinclusive in the sense that they do not take into account the convergence of telecommunications technologies that has occurred and will continue to occur. The concern over foreign propaganda was directed at broadcasting as it existed at the time of FDR's fireside chats in the 1930s: an omnidirectional, point-to-multipoint radio transmission. Today, however, multichannel video is (or soon will be) available over wires from cable systems and telephone companies. Yet those wireline media are exempt from the foreign ownership restrictions. At the same time, cellular telephony—which was invented in the 1940s and is a radio-based substitute for wireline telephony—poses no threat of being used for mass propaganda yet *is* subject to section 310(b). The blanket restrictions in section 310(b) are hopelessly outdated and fail to differentiate adequately among the various uses for radio technologies that have emerged since 1934.

The problem of overinclusiveness and underinclusiveness is entirely preventable and remediable. The statutory means exist to protect the national security in a manner less restrictive of free speech than the foreign ownership restrictions. Section 606 of the Communications Act empowers the president to seize communications facilities during wartime.[236] Woodrow Wilson exercised such power, conferred on him by section 2 of the Radio Act of 1912,[237] immediately upon America's entry into World War I.[238] That safeguard alone should be sufficient to protect the nation from propaganda and improper political influence.

If his power of confiscation were not enough, the president also has the unconditional power, under the Alien Enemy Act of 1798, to order summarily the arrest, internment, and removal of any enemy alien during a declared war.[239] He also may impose conditions on an enemy alien's continued stay in the United States. During the world

---

236. 47 U.S.C. § 606.

237. 37 Stat. 302, § 2 (1912).

238. Exec. Order (Apr. 6, 1917), *reprinted in* 17 A COMPILATION OF THE MESSAGES AND PAPERS OF THE PRESIDENTS 8241; Exec. Order (Apr. 30, 1917), *reprinted in id.* at 8254.

239. Act of July 6, 1798, ch. 66, § 1, 1 Stat. 577 (current version at 50 U.S.C. § 21).

wars, Presidents Wilson and Roosevelt aggressively exercised those powers.[240] By June 30, 1942, the Department of Justice had supervised administrative hearings for 9,121 cases, which resulted in the internment of 4,132 of the 900,000 persons the department had identified as enemy aliens on December 7, 1941.[241]

Finally, the national security goals of section 310 could be achieved with equal or greater efficacy by an FCC regulation requiring the transfer of any foreign investment exceeding the current benchmarks to an American trustee during wartime or national emergency. The FCC already has a similar trustee mechanism to accommodate the transfer-of-control issues raised by hostile tender offers for FCC licenses.[242] As part of a similar policy to address foreign ownership, the FCC could simply require the foreign investor to designate in advance a person who could immediately serve as trustee of the foreigner's investment in the event of a national emergency.

Alternative Channels of Communications

If the foreign ownership restrictions pass the first two prongs of intermediate scrutiny, the third prong asks whether the restrictions preserve ample alternative channels of communication for foreigners. On their face, the restrictions preserve *no* alternatives. Further, to the extent that ownership of a radio licensee confers editorial ability, the restrictions close to foreigners the major outlet of electronic speech.

---

240. *See* J. Gregory Sidak, *War, Liberty, and Enemy Aliens*, 67 N.Y.U. L. REV. 1402, 1413–19 (1992).

241. *Id.* (citing 1942 ATT'Y GEN. ANN. REP. 14; 1943 ATT'Y GEN. ANN. REP. 9).

242. Tender Offers and Proxy Contests, Policy Statement, MM Dkt. No. 85-218, 59 Rad. Reg. 2d (P&F) 1536 (1986), *appeal dismissed sub nom.* Office of Communication of the United Church of Christ *v.* FCC, 826 F.2d 101 (D.C. Cir. 1987). Ordinarily, 47 U.S.C. § 309(b) governs applications for approval of substantial transfers of control of FCC licensees. The FCC's policy statement on tender offers fashioned a different, two-step procedure for tender offers. The first step involves transferring control from existing shareholders to a voting trustee pursuant to a short-form application subject to 47 U.S.C. § 309(f). The second step, which is planned to follow consummation of the transfer of control from the voting trustee, involves transferring control from the voting trustee to the tender offeror (which has designated the voting trustee for the first step) pursuant to a long-form application subject to 47 U.S.C. § 309(b). *See, e.g.*, CNCA Acquisition Corp., 3 F.C.C. Rcd. 6088 (1988).

TRADE RECIPROCITY AS AN
UNENUNCIATED GOVERNMENT INTEREST

Section 310's limits on foreign control of radio licenses also potential-ly present a collision between the First Amendment and the power of the national government over trade policy. The most likely form that the argument would take is a claim that section 310 imposes a crude version of reciprocity. A number of nations have enacted rules that restrict access of U.S. citizens and corporations to the nations' media outlets. When the FCC considered imposing a foreign ownership restriction on cable systems, proponents of the limit reminded the FCC that Canada had adopted strict limits on foreign ownership of its own cable systems.[243] In 1989 the European Communities (EC) adopted an *EC Directive Concerning the Pursuit of Television Broadcasting Activities.*[244] The directive obligates member states to reserve a majority of broadcast time for "European works."[245] Representatives of the United States decried the agreement as a form of protectionism and threatened to retaliate.[246]

Trade Policy and the Constitution

The Constitution grants Congress the power to "regulate Commerce with foreign Nations."[247] Defenders of section 310 could argue that the foreign ownership limits represent a valid exercise of the national power over trade policy. As the appendix to chapter 3 documents, the federal government restricts foreign investment in banking, air transportation, and other industries.

But telecommunications is different. Speech is protected by the First Amendment. In the immigration context, however, the federal government's plenary powers over immigration policy have been held

---

243. Amendment of Parts 76 and 78 of the Commission's Rules to Adopt General Citizenship Requirements for Operation of Cable Television Systems and for Grant of Station Licenses in the Cable Television Relay Service, Mem. Op. and Order, 77 F.C.C.2d 73, 76 ¶ 6 (1980) [hereinafter *Foreign Ownership of CATV Systems*].

244. *See* Fred H. Cate, *The First Amendment and the International "Free Flow" of Information*, 30 VA. J. INT'L L. 371, 402–3 (1990).

245. *Id.*

246. *Id.* at 415.

247. U.S. CONST. art. I, § 8.

to override the normal operation of the First Amendment. One might argue that the same result should hold for the federal government's power to regulate trade by the specific means employed by section 310.

That argument proves too much. It would enable the government to enact broad, sweeping restraints on speech—restrictions on the importation of foreign books, for example—in the name of trade policy. And, unlike in the immigration context, portions of section 310 would impair the rights of domestic corporations as well as the rights of foreigners and foreign corporations. Finally, the congressional power to regulate trade with foreign nations is in the same sentence of the Constitution as the power to regulate trade among the several states. If Congress cannot use its power to regulate commerce among the several states to restrict speech, it would be arbitrary to argue that Congress has the power to do so in the name of regulating commerce with foreign nations.

The better analysis would treat trade policy as just another government interest. Precedent supports that view. When foreign affairs laws or executive orders conflict with the rights of U.S. citizens, the citizen's free speech rights remain intact.[248]

Strength of the Government Interest

Unlike the argument for promoting national security, the government's asserted interest in trade policy is content-neutral. At most, then, the trade policy argument might face intermediate scrutiny. That does not mean, however, that the government could convincingly defend section 310 as a legitimate exercise of trade policy. First, there is no evidence in the legislative history through 1934, or from 1934 until 1995, that section 310 was ever intended to be an instrument of trade policy. When Congress enacted sweeping reforms of telecommunications law in 1996, it considered and declined to enact a reciprocity test for section 310(b). To the contrary, the repeal in 1996 of the restrictions on foreign officers and directors unilaterally reduced U.S. trade barriers, albeit slightly. It would be unconvincing, therefore, for the government to argue *post hoc* that section 310

---

248. Boos *v.* Barry, 485 U.S. 312, 321–29 (1988); New York Times Co. *v.* United States, 403 U.S. 713, 714 (1971) (*per curiam*).

in fact furthers a substantial government interest in trade policy. That kind of after-the-fact reasoning would be unlikely to pass even a rational basis test.

Second, it is doubtful that the government's interest in using section 310(b) as a bargaining chip in trade negotiations could amount to a substantial interest. If that proposition is not immediately clear in the context of the electronic media, consider banning the importation of foreign books. Or suppose that the U.S. trade representative determined that denying American dealers of Lexus and Infiniti automobiles the opportunity to advertise in newspapers and over radio and television would be even more effective than threatening a 100 percent import duty to force the Japanese government to open its markets to U.S. automobile and parts manufacturers. A restriction on speech does not become more permissible under the First Amendment simply because the government's interest has shifted from national security to trade policy.

Third, as a matter of trade policy, the arguments in favor of liberalizing the current foreign ownership provisions are stronger than tightening them. Historically, the United States has adopted a policy of defending freedom of speech worldwide.[249] As a practical matter, allowing more foreign investment strengthens competition in domestic markets and thus provides consumers with substantial benefits. For example, BT's investment of capital in MCI was intended to enable MCI to spend $20 billion to upgrade its networks to provide expanded voice, video, and data communications.[250] Competition for ABC, NBC, and CBS came from Australia in the form of News Corp.'s investment in the creation of Fox Broadcasting Company.[251]

Means Employed

Some of the arguments considered earlier also show the poor fit between the restrictions in section 310 and trade policy. Section 310 does not sort friend from foe in the context of a trade war. Alternative instruments of trade policy are available, including means that do not restrict speech at all, such as imposing tariffs on Montrachet and

---

249. Cate, *supra* note 244, at 371–88.
250. MCI Comm. Corp., 9 F.C.C. Rcd. 3960, 3964 n.45 (1994).
251. Rose, *supra* note 191, at 1227.

BMWs or arguing eternally in the face of European stubbornness that American cultural imperialism is not to be feared. Although those alternatives might not seem so effective as those available to the government during wartime, they are probably at least as effective as section 310 would be.

In its current form, section 310 would be a very clumsy instrument for prying open foreign telecommunications markets to U.S. direct investment. Reciprocity essentially means restricting the access of foreigners to a nation's market—and thus ignoring the economic welfare of its own citizens in the present—in the hope that the foreigners will turn on their own government, which has shown its willingness to ignore the economic welfare of its own citizens. So roundabout a means of solving the problem is likely to backfire.

When considering adopting such a policy for foreign investment in cable television in 1980, the FCC itself emphasized that reciprocity is just as likely to result in more restrictions all around than in a mutual lessening of them. The FCC also noted that the most certain effect of adopting reciprocity would be to insulate domestic cable systems from competition:

> We do not believe a desire for reciprocity in international investment policies by itself provides an adequate basis for action on our part. Nor are we, in any case, in a position to know if such a policy on our part would in fact have the result intended or if, to the contrary, it would lead to increasing trade barriers in other areas. . . . There is no showing in this proceeding that a reciprocal agreement would improve communications service available in the United States. To the contrary, it seems likely that reciprocal treatment between the U.S. and Canada would merely reduce competition to provide cable television service in the U.S. . . . At this time it is difficult for us to perceive how the television viewing public would benefit in any way from the regulation requested. Rather it would appear that such a restriction would merely promote the self-interests of the domestic cable television industry at the expense of additional competitive alternatives for the public in the franchising process.[252]

---

252. *Foreign Ownership of CATV Systems*, 77 F.C.C.2d 73, 79 ¶ 13, 80 ¶ 15, 80 ¶ 18 (1980).

Although not offered in the context of First Amendment litigation, the FCC's reasoning in 1980 nonetheless sheds light today on the dubious constitutionality of using section 310(b) as tool of trade policy. The possibility of showing a good fit between the intended result (the opening of markets overseas to direct investment by U.S. telecommunications firms) and the means chosen (the FCC's conditional authorization of foreign direct investment in U.S. radio licensees) is negligible.

APPLICATION OF THE
STANDARD OF REVIEW TO THE
FOREIGN OWNERSHIP RESTRICTIONS

Consider now various hypothetical applications of the standard of review to the individual parts of section 310. We shall first examine the constitutionality of section 310(a) and then proceed to each subsection of section 310(b).

Section 310(a)

Section 310(a) imposes a blanket ban on all radio licensing of foreign governments and their representatives: "The station license required under this Act shall not be granted to or held by any foreign government or the representative thereof."[253] Because foreign governments and their official representatives enjoy no First Amendment rights, section 310(a) is constitutional as applied to those entities.

A U.S. citizen who speaks on behalf of a foreign entity in an *unofficial* capacity retains his First Amendment rights.[254] But section 310(a) probably would not be applied to the unofficial representative of a foreign government. For instance, the FCC has ruled that it did not violate section 310(a) to grant a license to an honorary counsel of Bolivia, who received no compensation for his services from the Bolivian government.[255]

---

253. 47 U.S.C. § 310(a).
254. *Mendelsohn*, 695 F. Supp. at 1481; *Communist Party*, 367 U.S. at 96.
255. Russell G. Simpson, 2 F.C.C.2d 640 (1966).

Should the FCC alter that interpretation, a court would almost certainly find the application of 310(a) to an unofficial representative to be unconstitutional under either intermediate or strict scrutiny. Even if such a representative were considered a security risk, the fit between the statute and the goal of preserving national security is poor. The legislation makes no attempt to sort friends from foes. Legislation that requires a person who speaks on behalf of foreign governments (officially or not) to register as an agent of a foreign government has been upheld, but courts upholding those requirements emphasize that the laws do not restrict speech but only require disclosing the agent's identity.[256] Section 310(a) is different. It operates as a complete ban on such speech.

We may thus conclude that section 310(a) is constitutional as applied to foreign governments and their official representatives. On the other hand, application of the bar to an unofficial representative, particularly a U.S. citizen, would be unconstitutional.

Section 310(b)(1) and 310(b)(2)

Section 310(b) generally prevents any of the entities that the statute covers from holding a "broadcast or common carrier or aeronautical en route or aeronautical fixed radio station license."[257] Section 310(b)(1) applies the bar to "any alien or the representative of any alien," and section 310(b)(2) applies it to "any corporation organized under the laws of any foreign government."[258] First, consider application of section 310(b)(1) to a foreign private citizen, lawfully living in the United States, who wishes to provide broadcast services, such as a Spanish-language radio station. Assuming that a court would treat the foreign corporation as it would a foreign natural person, that hypothetical also illustrates application of the standard of review to

---

256. *Communist Party*, 367 U.S. at 97 (Subversive Activities Control Act does not prohibit speech); Attorney Gen. *v.* Covington & Burling, 411 F. Supp. 371, 376 (D.D.C. 1976) (FARA registration provisions require disclosure without burdening speech unnecessarily); United States *v.* Peace Info. Center, 97 F. Supp. 255, 262–63 (D.D.C. 1951) (FARA withstands First Amendment scrutiny because, although it requires registration, it does not regulate speech); United States *v.* Auhagen, 39 F. Supp. 590, 591 (D.D.C. 1941) (FARA does not prohibit distribution of propaganda but only requires disclosure of distributor).

257. 47 U.S.C. § 310(b).

258. *Id.* §§ 310(b)(1), (2).

section 310(b)(2), which applies the ban on licensing to corporations organized under the laws of a foreign government.

The absurd rationales for distinguishing broadcast from the print media aside, the FCC's denial of a license to broadcast is analogous to content-based denial of permission to engage in leafleting. Because of the uncertainty surrounding the extent to which foreigners and foreign corporations are entitled to the full protections of the First Amendment, however, we shall not assume that sections 310(b)(1) and 310(b)(2) would face strict scrutiny. Nonetheless, those restrictions would fail even intermediate scrutiny.

If the asserted government interest is national security, how could it be an "important" interest in that hypothetical? *Even if* one accepts the view that the broadcaster's audience would believe whatever is transmitted (a bit of paternalism one would think could not attain the status of constitutional law, if listeners have any rights at all), the national security interest in this case is implausibly weak in the absence of evidence that the foreigner has any hostile intention. For the same reason, the effect of the ban is grossly overbroad; the means is not at all tailored to the end. True, one could argue that there are alternative channels. The foreigner could buy a cable television system. But pointing out the availability of that alternative merely makes the total ban on the foreign broadcaster seem more foolish and unnecessary if the objective of section 310(b) is to prevent foreign ideas from reaching a mass audience in the United States in the first place. The repeal in 1996 of the restrictions on foreign officers and directors in sections 310(b)(3) and 310(b)(4) can only accentuate the looseness of fit between means and ends in sections 310(b)(1) and 310(b)(2): A Canadian citizen or Bell Canada would be forbidden from being a U.S. radio licensee, while Saddam Hussein would be allowed to be a director and the chief executive officer of the largest broadcasting company in the United States. In short, application of sections 310(b)(1) and 310(b)(2) to a foreign broadcaster would almost certainly be found unconstitutional.

Suppose instead that the foreigner wished to provide common carriage services. Again, the national security interest would be weak, as the FCC itself has admitted when approving waivers to provide common carrier service under section 310(b)(4).[259] If, however, the

---

259. Upsouth Corp., 9 F.C.C. Rcd. 2130, 2131 ¶ 13 (1994); *MCI*, 9 F.C.C. Rcd.

asserted government interest were the prevention of sabotage, the interest would be content-neutral, and the restrictions would probably face only a rational basis test. Still remaining as open questions, however, would be whether it is rational to presume (1) that only foreigners are responsible for sabotage committed in the United States, (2) that a foreigner will pay good money to invest in radio facilities so that he may destroy them in time of war or international crisis, and (3) that investment in a U.S. radio licensee gives a foreigner any greater opportunity to sabotage American telecommunications than he would have without making that investment. Experience and common sense give good cause to reject all three presumptions as irrational. The FCC, for example, quietly prosecuted at least one U.S. citizen in the late 1980s for intentionally interfering with air traffic control communications.[260] And, the bombing of the federal building in Oklahoma City in 1995 appears to be one in a long line of acts of terrorism or sabotage (including the Haymarket riot of 1886, the Wall Street bombing of 1924, and the Weathermen bombing at the University of Wisconsin during the Vietnam War) that were committed on American soil by Americans.

Finally, a common carrier may have difficulty persuading a court to recognize its First Amendment standing at all. The same reasoning would apply to an entity that sought to provide private aeronautical radio service for hire (not for its own use). The outcome of that analysis is uncertain.

On the other hand, suppose that the foreigner wanted to establish its own private carriage network for its own communications by obtaining the sort of aeronautical fixed or mobile radio station denied it by section 310(b). Here, the foreigner wants to speak as well as to provide carriage facilities. As a constitutional matter, point-to-point carriage is at least as well protected as cable television, perhaps more so. In *Sable*, the Supreme Court concluded that restrictions on

---

at 3964 ¶ 23.

260. I became aware of this case while deputy general counsel of FCC, but I have been unable to find any reported decision. The agency's reason for minimizing publicity at the time was to prevent copycat offenses. *See also Va. Man Accused of Interfering with Air Traffic Control*, WASH. POST, Sept. 25, 1993, at D2; Jeffrey A. Perlman, *Voice "Pirate" over Airport Is Investigated*, L.A. TIMES, Apr. 7, 1992, at B1 (Orange County ed.); Joseph W. Queen, *Warnings of Radio Phantom; Pilots Told to Beware of False Instructions*, NEWSDAY, Sept. 23, 1989, at 7.

indecent dial-a-porn (as opposed to outright obscene dial-a-porn) were unconstitutional. The Court employed a strict scrutiny test because the regulations were not content-neutral.[261] Analyzing sections 310(b)(1) and 310(b)(2) under even an intermediate level of scrutiny, there seems to be no substantial national security argument for denying all foreigners a radio common carrier license during peacetime. Again, the fit of the legislation is bad. Pointing out that the foreigner has a ready alternative of hiring private carriage from a third party again highlights the weakness of the argument that those restrictions are necessary national security measures. The third party is not likely to be monitoring conversations on its wavelengths to be certain that they comport with the national interest. In short, the application of section 310(b)(1) or 310(b)(2) in that circumstance is unconstitutional.

Finally, section 310(b)(1) might be applied to a U.S. citizen who is considered the representative of a foreigner. Such a provision would almost certainly receive strict scrutiny. Again, the application of the section 310 ban in that case would also be unconstitutional, for the reasons indicated above.

Section 310(b)(3)—Seven Hills Revisited

Now consider section 310(b)(3), which covers corporations arguably under the direct control of foreigners in their capacity as investors (rather than officers or directors). Its scope is "any corporation of which more than one-fifth of the capital stock is owned of record or voted by aliens or their representatives or by a foreign government or representatives thereof or by any corporation organized under the laws of a foreign country."[262] If such a corporation were organized under the laws of a foreign government, section 310(b)(2) would apply as well—and if application of section 310(b)(2) were found to be unconstitutional, section 310(b)(3) would be as well.

Section 310(b)(3) could apply to a corporation organized under the laws of the United States if more than 20 percent of the stock of that corporation were owned by foreigners. Again, suppose that such a corporation sought to become a broadcaster. Application of section 310(b)(3) to that case would be unconstitutional. None of the cases

---

261. *Sable*, 492 U.S. at 124.
262. 47 U.S.C. § 310(b)(3).

suggesting that foreigners are entitled to lesser First Amendment protection would apply for the simple reason that the plaintiff would *not* be foreign. Again, the national security interest would be weak. The hopelessly overbroad legislation makes no effort to sort friend from foe. Again, pointing out the alternative channels of communication that the foreigner's investment in a cable system would provide would merely make section 310(b)(3) look sillier. The law would fail intermediate scrutiny. In light of the repeal of the prohibition on direct participation of foreigners in management as officers and directors, section 310(b)(3) might even fail a rational basis test.

A stronger case might be made for the constitutionality of the application of section 310(b)(3) to a corporation of which more than one-fifth of the capital stock is owned or voted by a foreign government. Here the national security interest seems stronger. The fit of the legislation remains poor, however, with still no attempt made to sort friend from foe. In that case, application of section 310(b)(3) could be constitutional only if the corporation were somehow considered an official representative of the foreign government.

Suppose that a corporation covered by section 310(b)(3) sought to become a common carrier or a private carrier for hire. As with sections 310(b)(1) and 310(b)(2), providing that sort of distribution facility might not yet count as protected speech. The case law must catch up with the technology.

Finally, as with sections 310(b)(1) and 310(b)(2), that provision is probably unconstitutional as applied to a corporation covered by section 310(b)(3) that seeks to become a private aeronautical carrier on its own behalf.

Section 310(b)(4)

Section 310(b)(4) applies to "any corporation directly or indirectly controlled by any other corporation of which more than one-fourth of the capital stock is owned of record or voted by aliens, their representatives, or by a federal government or representative thereof, or by any corporation organized under the laws of a foreign country, if the Commission finds that the public interest will be served by the refusal or revocation of such license."[263] As discussed in chapter 3,

---

263. *Id.* § 310(b)(4).

the plain language of the statute does not operate as an absolute bar to a U.S. holding company of a foreign investor owning more than a 25 percent interest in a radio licensee subject to section 310(b). Indeed, the provision should not operate as a bar at all, unless and until the FCC affirmatively shows that the foreign presence in the holding company poses a genuine danger to the public interest.

As written, section 310(b)(4) is probably constitutional as applied to *any* of the entities it covers. Because the statute requires the FCC to make affirmative findings before the provision operates as a bar to foreign investment, section 310(b)(4) should not operate as a bar in any case where the national security interest is weak or where there is no proper fit. The problem, of course, is that the FCC has not applied the statute as it is written. As chapter 3 shows, the FCC has created a presumption that the public interest will not be served by any company whose level of foreign ownership exceeds the limits of section 310(b)(4); any company that does exceed those limits must apply for a waiver. And, as chapter 4 documents, the waiver process is a sticky, expensive affair. In the case of a broadcaster, forcing a company to jump through those hoops is akin to requiring special licensing for a publisher or leafleteer. Although the waiver might ultimately be granted, there is always a chill on speech. The tax on advertising revenues invalidated in *Grosjean* was invalidated as an attack on a newspaper's ability to raise funds.[264] The FCC's waiver process is properly viewed as a tax on media corporations that wish to raise investment capital abroad.[265]

Assuming that national security was the asserted interest, the FCC's interpretation of section 310(b)(4) would face at least intermediate scrutiny. Because of the weakness of the national security interest and the bad fit between means and ends, the FCC's requirements would fail intermediate scrutiny.

But the FCC might be wily and attempt to justify its interpretation of section 310(b)(4) as an exercise in trade policy. Here, however, the FCC has already refuted its own argument. In its 1980

---

264. The Court held that a tax on the advertising revenues of newspapers with a circulation of more than 20,000 violated the First Amendment. Justice Sutherland, writing for the majority, described the law as a double restraint on the press: "First, its effect is to curtail the revenues realized from advertising; and, second, its direct tendency is to restrict circulation." *Grosjean*, 297 U.S. at 244–55.

265. Rose, *supra* note 191, at 1209–10.

decision on foreign ownership in cable television, the FCC explained that it had no jurisdiction under the Communications Act of 1934 (or any other statute, for that matter) to determine trade policy:

> The Commission's responsibilities relate to "interstate and foreign communications," that is to telecommunications within the United States and between the United States and foreign countries. This does not imply, however, any responsibility for investment policy with respect to communications systems in foreign countries. We do not believe a desire for reciprocity in international investment policies by itself provides an adequate basis for action on our part .... It is a matter which we believe is appropriately considered by other branches of the government.[266]

Thus, the FCC would be unable to defend the constitutionality of its interpretation of section 310(b)(4) as applied to a broadcast licensee.

Again, however, for reasons explained above, a corporation covered by section 310(b)(4) might not succeed under current case law in making a First Amendment claim for the right to be a common carrier or a private carrier for hire. If it sought to provide some content in the form of enhanced services, the corporation would have a better argument.

Finally, such a corporation should be able to establish successfully a constitutional right to obtain an aeronautical radio license for private use. Again, the national security interest is virtually nonexistent.

RELATED CLAIMS BASED
ON EQUAL PROTECTION

The guarantee of equal protection of the laws is an additional ground on which section 310(b) may be legally suspect. The claim that a person has been denied the equal protection—perhaps the dominant theory in modern litigation over constitutional rights—arises when a law classifies persons differently who ought to be treated the same or, conversely, does not distinguish persons who ought to be treated differently. The federal courts generally permit disparate treatment of

---

266. *Foreign Ownership of CATV Systems*, 77 F.C.C.2d at 78–79 ¶ 13 (citation omitted).

foreigners if the federal government supplies a minimally rational justification. Not surprisingly, therefore, only one equal protection challenge has been made in federal court to the constitutionality of section 310(b), and it failed.

In *Moving Phones Partnership L.P.* v. *FCC*, the FCC denied, under section 310(b)(3), an application for a license to operate a cellular telephone system because the applicant had foreigners among its general partners.[267] The FCC had rejected the contentions that dismissal of the application violated the applicant's Fifth Amendment right to equal protection regardless of alienage.[268] The D.C. Circuit applied a rational basis test and ruled that section 310(b) does not abridge the equal protection of foreigners. Relying on Supreme Court precedent, the court stated that "classifications based on alienage in federal statutes are permissible so long as the challenged statute is not a 'wholly irrational' means of effectuating a legitimate government purpose."[269] The court elaborated that application of strict scrutiny to foreigners as a class "has been limited to 'exclusions which struck at the noncitizens' ability to exist in the community.'"[270] Stating that the opportunity to own a broadcast or common carrier radio station "is hardly a prerequisite to existence in a community," the court applied a weak rational basis test to section 310(b)(3).[271] The court determined that the policy to "'safeguard the U.S. from foreign influence'" bore a rational relationship to the classification in question.[272] It therefore upheld section 310(b)(3).

The U.S. Court of Appeals for the Seventh Circuit has considered an equal protection challenge to section 303(l),[273] which prohibits the FCC from granting commercial radio operator licenses to

---

267. 998 F.2d 1051, 1053 (D.C. Cir. 1993).

268. *Id.* at 1054. Equal protection claims against the federal government are brought under the Due Process Clause of the Fifth Amendment, U.S. CONST. amend. 5, because the Fourteenth Amendment addresses only the states. The substance of those provisions, however, is identical. Adarand Constructors, Inc. v. Pena, 115 S. Ct. 2097 (1995); Bolling v. Sharpe, 347 U.S. 497 (1954).

269. *Moving Phones*, 998 F.2d at 1056 (quoting Mathews v. Diaz, 426 U.S. 67, 83 (1965)).

270. *Id.* (quoting Foley v. Connelie, 435 U.S. 291, 295 (1978)).

271. *Id.*

272. *Id.* (quoting Kansas City Broadcasting Co., 5 Rad. Reg. (P & F) 1057, 1093 (1952)).

273. 47 U.S.C. § 303(l).

foreigners. In *Campos* v. *FCC*, the FCC denied lawful permanent resident foreigners the chance to take the qualifying examination to secure a radio operator license.[274] The Seventh Circuit denied the foreigners' claim that section 303(l) violated their Fifth Amendment right to equal protection.[275] The court stated that "where, as here, no substantive constitutional right is impaired, federal regulation of foreigners must be upheld unless wholly irrational."[276] Relying on the Supreme Court's decision in *Mathews* v. *Diaz*, the court held that the national interest in providing an incentive for foreigners to become naturalized citizens was rationally related to the classification at issue.[277] The court upheld section 303(l) because it was not a "wholly irrational" means of serving the interest to be advanced.[278]

Although one can make better arguments than those advanced in *Moving Phones* that the foreign ownership restrictions violate the equal protection component of due process under the Fifth Amendment, the necessary legal arguments add little to what a foreigner could argue, with greater forcefulness, under the First Amendment. If a court were sympathetic to a constitutional challenge to the foreign ownership restrictions, it would more likely base its decision on a finding that the restrictions violated the freedom of speech rather than on a finding that they impermissibly discriminated against foreigners as a class.

CONCLUSION

The premise of any constitution is that a nation may formulate general rules to govern the conduct of its affairs and that those rules will remain valid over time. The specific premise of the First Amendment is that Congress may not be trusted with the power to control speech. Between technological revolutions, such as the development of broadcasting, and political revolutions, such as World Wars I and II and the rise of communism, the early twentieth century threw those premises into doubt. The Supreme Court, declining to halt the experiment in rationing and centralized control that Congress had initiated

---

274. 650 F.2d 890 (7th Cir. 1981).
275. *Id.* at 892.
276. *Id.* at 893 (citing Mathews *v.* Diaz, 426 U.S. 67, 83 (1965)).
277. *Id.* at 894 (citing Hampton *v.* Mow Sun Wong, 426 U.S. 88, 105 (1965)).
278. *Id.*

with the Radio Act of 1927 and continued with the Communications Act of 1934, decided to loosen the constitutional constraints on Congress. The result was a line of decisions culminating in *Red Lion*.

Red Lion was intended to be a modern doctrine for the modern age. It is ironic, then, that two decades after *Red Lion* was decided, its faith in government and centralized control seemed more medieval than modern. The scarcity logic of *Red Lion* belongs in the dustbin, beside the command-and-control economic policies that collapsed with the Berlin Wall. The Court's opinion in *Turner*, in rejecting the application of scarcity logic to another new media, cable, represents a return to the premise of the First Amendment. It promises, though perhaps does not quite deliver, a victory of rules supported by reason over fear.

Section 310 epitomizes the mood of the *ancien régime*. It embodies fear and the exercise of power without understanding. It restricts foreign ownership of most radio licensees without undertaking any inquiry into whether such ownership would genuinely pose any danger to national security. Indeed, the national security rationale became entirely incredible in 1996, when Congress authorized foreigners to manage the very radio licensees in which the FCC will permit them to be only minority shareholders. What remains of section 310(b) inhibits certain forms of electronic speech and appears to be motivated by little else than Congress's desire to censor certain content. As such, section 310(b) is too medieval even for *Red Lion*. Under current jurisprudence, the statute is plainly unconstitutional in most familiar circumstances. It is certainly unconstitutional in virtually any application to a U.S. citizen or domestic corporation.

8

The 1997 World Trade Organization Agreement—and Beyond

THE SUCCESSFUL COMPLETION of the World Trade Organization talks in February 1997 made clear that global competition in telecommunications is the next frontier after the privatizations and domestic deregulation of the 1990s. Indeed, the demand for seamless international telecommunications services is a propellant of all three of those phenomena. To be sure, reform of the foreign ownership restrictions in the Communications Act is only one component of the set of policy initiatives that will be necessary to unleash the potential of telecommunications technologies on a global scale. But it is a good starting point, in part because it challenges the United States to accept the opportunity to do what it does well—to lead by example. The U.S. trade representative did so in Geneva in 1997, with the result that other nations agreed to reduce, among other impediments to global competition, their restrictions on foreign direct investment in telecommunications services. That commendable result, however, will surely focus the attention of American policymakers on the question of whether section 310(b) must or should be amended or repealed to comply with the WTO agreement.

THE 1997 WTO AGREEMENT

On February 15, 1997, seventy countries working within the framework of the World Trade Organization agreed on a multilateral reduction of regulatory barriers to competition in international telecommunications services.[1] The agreement will be annexed to the Fourth Protocol of the General Agreement on Trade in Services and has an official ratification date of November 30, 1997, and an effective date of January 1, 1998. The accord was the historic result of a process that had begun nearly three years earlier but collapsed in April 1996, when the United States walked out of the negotiations. If the 1997 talks also had collapsed, it would have been unlikely that the WTO would have revisited trade liberalization for telecommunications services until after 2000.[2] In addition, the failure of a second round of WTO talks on telecommunications services could have damaged the WTO's ability to reach an agreement on liberalization of financial services, which was an important objective of the U.S. trade representative.[3]

---

1. In alphabetical order, the following sixty-nine countries joined in the agreement: Antigua and Barbuda, Argentina, Australia, Bangladesh, Belize, Bolivia, Brazil, Brunei Darussalam, Bulgaria, Canada, Chile, Colombia, Côte d'Ivoire, Czech Republic, Dominica, Dominican Republic, Ecuador, El Salvador, European Union (Austria, Belgium, Britain, Denmark, Finland, France, Germany, Greece, Ireland, Italy, Luxembourg, Netherlands, Portugal, Spain, Sweden), Ghana, Grenada, Guatemala, Hong Kong, Hungary, Iceland, India, Indonesia, Israel, Jamaica, Japan, Malaysia, Mauritius, Mexico, Morocco, New Zealand, Norway, Pakistan, Papua New Guinea, Peru, Philippines, Poland, Romania, Senegal, Singapore, Slovak Republic, South Africa, South Korea, Sri Lanka, Switzerland, Thailand, Trinidad and Tobago, Tunisia, Turkey, United States, and Venezuela. World Trade Organization, The WTO Negotiations on Basic Telecommunications (Feb. 17, 1997) (unofficial briefing document) [hereinafter *WTO Negotiations*]; Reuters Financial Service, *Trade-Telecoms-Countries (Scheduled)*, Feb. 16, 1997. The European Union negotiated on behalf of all of its fifteen members in the WTO and was officially considered a single signatory. *Id.* But because the EU's individual market-opening offers varied, the member countries were counted individually. *Id.* "The number of governments annexing their commitments to the Protocol is 69, rather than 71, because two governments (St. Vincent and the Grenadines and the Bahamas) were not in a position to finalize their offers as of 15 February [1997]." *WTO Negotiations, supra*, at n.1. "Other commitments may emerge later in the 1997 under late submission procedures laid out in the Decision on Commitments in Basic Telecommunications; the governments of Guyana, Honduras and Tanzania have formally stated intentions to try to pursue this option and others have made informal inquiries about doing likewise." *Id.* n.3. By February 19, 1997, Cyprus reportedly was ready to sign the WTO agreement also. Reuters World Service, *Press Digest—Cyprus*, Feb. 19, 1997.

2. Alan Cane, *Getting Through: Why Telecoms Talks Matter*, Fin. Times, Feb. 14, 1997, at 6.

3. *Id.*

In the 1997 round, as in the 1996 round, the United States played a game of brinkmanship, assuming the role of "the world's lone naysayer,"[4] in the hope of securing greater concessions from nations reluctant to lift restrictions advantageous to their state-controlled telephone monopolies. The American leverage came from the fact that the U.S. market for telecommunications services is the world's largest. In the end, of course, the United States went along. Signatory nations to the WTO agreement, representing markets generating 95 percent of the $600 billion in global telecommunications revenues,[5] are now legally bound to open their state-controlled telephone monopolies to competition.

The WTO agreement was announced with predictions of steep price reductions in many parts of the worldwide communications market.[6] Reed E. Hundt, chairman of the Federal Communications Commission, estimated that the average cost of international telephone calls would drop by 80 percent—from $1 per minute on average to twenty cents per minute.[7] In addition to inducing price reductions, the WTO agreement was predicted to have at least two other salutary economic effects. By encouraging foreign direct investment and promoting the reform of national regulatory regimes, the agreement was expected to hasten the modernization of public switched telephone networks around the world.[8] Further, the agreement would

---

4. Edmund L. Andrews, *U.S. Appears to Ease Its Stance in Global Talks on Phone Deal*, N.Y. TIMES, Feb. 15, 1997, at 35 [hereinafter *U.S. Appears to Ease Its Stance*]; *see also* Edmund L. Andrews, *68 Nations Agree to Widen Markets in Communications*, N.Y. TIMES, Feb. 16, 1997, at 1 [hereinafter *68 Nations Agree*].

5. Edmund L. Andrews, *In Global Push for Phone Deal, U.S. Is Odd Man Out*, N.Y. TIMES, Feb. 14, 1997, at D1 [hereinafter *Global Push*]. According to acting U.S. Trade Representative Charlene Barshefsky: "The United States [in] April [1996] put a halt to the talks because we felt that there simply was not enough reciprocity for us to bind open our market. . . . At that time, we had offers from 47 countries. This agreement covers 70. The previous offers covered 60 percent of global telecommunications revenues, of which we were half. Now we are covering 95 percent of telecommunications revenues." Anne Swardson & Paul Blustein, *WTO Reaches Phone Pact*, WASH. POST, Feb. 16, 1997, at A33.

6. Edmund L. Andrews, *60 Nations Sign Telecom Pact*, N.Y. TIMES, Feb. 17, 1997, at 1.

7. Statement of FCC Chairman Reed Hundt Concerning WTO Agreement on Telecom Services, Feb. 15, 1997 (released Feb. 18, 1997) [hereinafter *Hundt Statement*].

8. Cane, *supra* note 2.

stimulate international trade in other goods and services that use telecommunications services as principal inputs.[9] Acting U.S. Trade Representative Charlene Barshefsky noted, "The world's businesses today spend more money on telecommunications than on oil."[10] WTO Director-General Renato Ruggiero predicted that liberalization of international barriers to trade in telecommunications services could increase income by $1 trillion in the decade following the agreement. That figure is approximately 4 percent of world GDP at 1997 prices.[11]

Elements of the Agreement

The WTO agreement has three elements: market access, investment, and procompetitive regulatory principles. Fifty-three countries offered to open their markets to international services and facilities. Those countries account for 99 percent of all 1995 revenues from telecommunications services worldwide.[12] Almost every offer made to provide market access for facilities-based competition also included the opportunity to resell service and to interconnect with existing networks at reasonable rates, terms, and conditions.[13] Additionally, forty-two countries guaranteed greater market access for satellite services, up from twenty-eight a year earlier. The forty-two countries represented 80 percent of the WTO member countries' total satellite services revenues.[14]

With respect to investment, fifty-six of the sixty-nine offers tendered on February 15, 1997, allowed foreign ownership of *all* telecommunications facilities, compared with only twenty-seven offers made in the 1996 negotiations that would have done so.[15] Of those fifty-six offers, seventeen allowed 100 percent ownership starting

---

9. *Id.*

10. Office of U.S. Trade Representative, Statement of Ambassador Charlene Barshefsky, Basic Telecom Negotiations (Feb. 15, 1997) [hereinafter *Barshefsky Statement*].

11. Renato Ruggiero, Remarks Before the WTO (Feb. 17, 1997) [hereinafter *Ruggiero Remarks*]; *see also* BEN PETRAZINNI, GLOBAL TELECOM TALKS: A TRILLION DOLLAR DEAL (Institute for International Economics 1996).

12. *Barshefsky Statement, supra* note 10.

13. *Id.*

14. *Id.*

15. *Ruggiero Remarks, supra* note 11; *WTO: Global Telecoms*, AFP-Extel News, Feb. 17, 1997.

January 1, 1998, and thirteen others lifted ownership restrictions by 2004.

The WTO's April 1996 Reference Paper, which is reproduced in the appendix, outlined the third element—"procompetitive regulatory principles"—that the organization sought in the negotiations on telecommunications services. Among other obligations, the principles in the Reference Paper commit foreign countries to establish independent regulatory bodies, guarantee that U.S. companies may interconnect with foreign networks at fair prices, forbid anticompetitive cross-subsidization, and mandate transparency of government regulations and licensing.[16] Those principles received nearly universal endorsement in the 1997 Geneva talks:

> By the February 1997 deadline, 61 of the 69 governments submitting schedules included commitments on regulatory disciplines, with 55 of these committing to the Reference Paper in whole or with only minor modifications. This compares favourably with the April 1996 results, when 44 out of the 48 governments submitting offers included commitments on regulatory disciplines and only 31 of these inscribed the Reference Paper.[17]

Chairman Hundt of the FCC characterized the acceptance of the Reference Paper as nothing less than the exportation and wholesale adoption of enlightened American regulatory policies:

> By this agreement, the Telecommunications Act enacted a year ago by Congress has become the world's gold standard for pro-competitive deregulation. Sixty-five countries have bound themselves to the Reference Paper embodying the Congressional vision of free competition, fair rules, and effective enforcement.
> In Buenos Aires three years ago, at the first International Telecommunications Union development conference, Vice President Gore challenged the nations of the world to build a network around the globe linking all human knowledge and creating global opportunities. One year ago, Congress delivered a clear and compelling blueprint for the competition that will build

---

16. World Trade Organization, Negotiating Group on Basic Telecommunications, Reference Paper (Apr. 24, 1996).

17. *WTO Negotiations, supra* note 1.

this network. Today, the nations of the world endorsed that blueprint.[18]

There is reason, however, to question whether the Reference Paper, as interpreted by American telecommunications regulators, will indeed translate into *deregulatory* principles. In the week following the successful completion of the WTO agreement on telecommunications services, Deputy U.S. Trade Representative Jeffrey Lang, commenting to a Washington, D.C., audience on the principles contained in the Reference Paper, observed that "to move from what was regarded for 100 years as not just a monopoly but a natural monopoly . . . to a system of enforced competition means not deregulation but *reregulation*. And that is what the pro-competitive principles embody."[19] In other words, the promotion of competition requires reregulation. At the same event Chairman Hundt said that, just as "the laws of physics are everywhere the same, . . . it may well be that the laws of economics can be demonstrated to everywhere be the same," such that there would be no need to have "different ways to resolve issues such as forward-looking pricing."[20] It is true that microeconomic principles are the same everywhere. But the danger inherent in Chairman Hundt's view is that if the FCC were to produce misguided policies—such as its 1996 *First Report and Order* on the pricing of unbundled network elements under the Telecommunications Act of 1996[21]—then the American gloss placed on the WTO's Reference Paper would cause other nations to copy a defective public policy.

---

18. *Hundt Statement, supra* note 7. Acting U.S. Trade Representative Barshefsky similarly described the Reference Paper as "an extraordinary testimony to the compelling nature of Congress' vision in this area." *Barshefsky Statement, supra* note 10.

19. Remarks by Jeffrey Lang, Deputy U.S. Trade Representative, to the Center for Strategy and International Studies, Washington, D.C. (Feb. 21, 1997) [hereinafter *Lang Remarks*] (emphasis added).

20. *Id.* (comments of Reed E. Hundt).

21. Implementation of the Local Competition Provisions in the Telecommunications Act of 1996 and Interconnection Between Local Exchange Carriers and Commercial Mobile Radio Service Providers, First Report and Order, CC Dkt. Nos. 96-98, 95-185, 11 F.C.C. Rcd. 15,499 (1996).

Foreign Opposition to
the U.S. Negotiating Position

The success of the WTO talks at the same time underscored the continuing importance and controversy of restrictions on foreign direct investment in telecommunications, such as section 310(b). By the start of 1997, it had become clear that the so-called critical mass existed for an agreement at the WTO talks.[22] The most contentious issue remaining on the table, however, and the issue having the greatest potential to upset American participation in an agreement was the elimination of restrictions on foreign ownership of radio licensees. In the final week before the February 15 deadline, the U.S. delegation threatened not to sign the agreement (and thus doom it) unless Canada, Japan, Mexico, and South Korea made concessions on their foreign ownership restrictions.[23] The United States proposed to allow foreign investors to own 100 percent of American telecommunications companies (that is, radio common carrier licensees) and demanded that other nations allow a foreign company to own at least controlling stakes (51 percent) of their domestic carriers. Mexico, which had proposed a 40 percent foreign ownership limit, and South Korea, which had proposed a 33 percent limit, compromised with the U.S. negotiators by allowing 49 percent foreign ownership.[24] In addition, South Korea raised the foreign equity limit in Korea Telecom, the dominant operator, from 20 percent to 33 percent and agreed to accept entirely

---

22. "The WTO negotiations on market access for basic telecommunications had resulted in 34 offers (covering 48 governments) by the time they were originally scheduled to end in April 1996. Mid-November 1996 marked the first tangible signs of renewed progress in the negotiations when the European Union, the United States and the Slovak Republic became the first to formally submit revisions of the offers contained in the package achieved in April. In January 1997 momentum began to build as seven more governments submitted revised offers and six governments added completely new offers to the package of results. In February, 17 more new offers were submitted and another 22 governments submitted first-time revisions of their April 1996 offers. This raised to 23 the total number of new offers submitted and to 32 the total number of revisions (covering 46 governments) of the 34 offers tabled in April." *WTO Negotiations, supra* note 1.

23. Andrews, *Global Push, supra* note 5.

24. *Id.*; Andrews, *U.S. Appears to Ease Its Stance, supra* note 4; Frances Williams, *Telecoms: South Korea Puts in New Offer*, FIN. TIMES, Feb. 14, 1997, at 6.

the WTO's Reference Paper on regulatory rules intended to protect competition.[25]

Canada and Japan, however, resisted the American pressure. Canada offered to cap foreign ownership of a land-based telecommunication carrier at 46.7 percent, but it refused to raise that limit to 49 percent, let alone 51 percent.[26] Although the proposed liberalization of foreign ownership limits being negotiated at the WTO talks did not extend to the mass media, Canada was nonetheless concerned that, as telephone companies increasingly ventured into the delivery of video programming, the foreign ownership limits proposed by the United States would provoke the perennial resentment over American domination of popular culture and the mass media in Canada.[27] As a consequence, Canada refused to open its market to the delivery of television programming by foreign satellite operators. Canada did, however, ease certain "restrictions in its satellite communications market and brought forward the date for ending the Telesat monopoly on fixed satellite services from 2002 to 2000."[28]

Despite a personal letter from President William J. Clinton seeking higher permissible limits on foreign investment,[29] Japan did not budge from its offer to cap foreign ownership at 20 percent in Nippon Telegraph & Telephone (NTT), the dominant provider of domestic telephone services, and Kokusai Denshin Denwa (KDD), the dominant provider of international telecommunications services. The *Financial Times* reported that Prime Minister Ryutaro Hashimoto, in an apparent reference to section 310(b), responded that "Japan's lifting of its restriction on foreign equity participation in NTT and KDD depended on similar deregulation" in the United States.[30] Nonetheless, as a concession to the United States, Japan offered to allow up to 100 percent foreign ownership of any *new* companies formed

---

25. *Id.*

26. Andrews, *68 Nations Agree, supra* note 4; Edmund L. Andrews, *U.S. May Block Pact Opening Worldwide Phone Markets*, N.Y. TIMES, Feb. 3, 1997, at D1 ("For the United States, the biggest obstacle to a deal may turn out to be Canada," which "has refused to agree to scale back tough restrictions on foreign ownership of telephone companies.").

27. Andrews, *U.S. Appears to Ease Its Stance, supra* note 4.

28. Frances Williams & Michiyo Nakamoto, *Telecoms Pact Waits for US Verdict*, FIN. TIMES, Feb. 15, 1997, at 3.

29. *Id.*

30. *Id.*

to compete against NTT or KDD in the Japanese telecommunications market.[31]

Congressional Opposition to
the U.S. Negotiating Position

While the U.S. delegation was dickering with Japan and Canada, Senator Ernest F. Hollings of South Carolina, the ranking Democrat on the Senate Commerce Committee, and Representative Edward J. Markey of Massachusetts, the ranking Democrat on the Telecommunications Subcommittee of the House Commerce Committee, threatened, in separate letters to President Clinton, to upset U.S. participation in a WTO agreement.[32] Senator Hollings and Representative Markey argued that, in the absence of prior legislation from Congress, USTR could not negotiate an agreement liberalizing foreign ownership of radio licensees in the United States without violating section 310(b), as well as section 310(a).

In his letter dated February 4, 1997, Senator Hollings argued that USTR was basing its negotiating position on an incorrect reading of the Communications Act that would produce an unconstitutional result of excluding Congress's consideration of the WTO agreement:

> During consideration of the landmark Telecommunications Act of 1996, the United States Trade Representative (USTR) requested changes to the Communications Act of 1934 which limits certain foreign investment in communications licenses. Congress failed to achieve consensus on this contentious issue. Now it is our understanding that after two years of asserting that changes must be made to U.S. law, USTR has crafted a novel interpretation of the statute that stands current practice on its head and would commit the U.S. to an agreement that is inconsistent with the law.
>
> Furthermore, USTR should not commit the U.S. to a trade agreement that limits the scope of the public interest test administered by the Federal Communications Commission (FCC). Any agreement that results in changes to current U.S. law and FCC

---

31. Andrews, *U.S. Appears to Ease Its Stance, supra* note 4.

32. Letter from Senators Ernest F. Hollings, Daniel K. Inouye, Robert C. Byrd, and Byron L. Dorgan to President William J. Clinton, Feb. 4, 1997 [hereinafter *Hollings Letter*]; Letter from Representative Edward J. Markey to President William J. Clinton, Feb. 5, 1997 [hereinafter *Markey Letter*].

practices should be done only with the approval of the Congress in accordance with our Constitutional obligation to regulate foreign commerce.

We take strong exception to this action by USTR which we believe is specifically designed to circumvent the constitutional role delegated to the Congress. While implementing legislation may be viewed by some as inconvenient, the U.S. offer is inconsistent with the law and therefore it is our view that implementing legislation is necessary.[33]

Two days later, on February 6, 1997, Senator Hollings introduced S. 287, the Approval of Trade Agreements Act of 1997, which would "require congressional approval before any trade agreement is entered into under the auspices of the World Trade Organization."[34] In its entirety, the bill provided:

No international trade agreement negotiated under the auspices of the World Trade Organization the application of which to the United States would require a change in United States law or practice may be implemented by or in the United States until the agreement is approved by the Congress by law. For purposes of this Act, the term "change in United States law or practice" means a change in any activity, procedure, limitation on exports or imports, duty, requirement, or practice that is contrary to any law, regulation, rule, or practice of the United States in effect on the date on which the change would, but for this Act, be implemented.[35]

By its terms, of course, Senator Hollings's bill would affect *all* WTO negotiations, not merely the agreement on telecommunications services. In floor remarks, Senator Hollings emphasized what he regarded to be the constitutional need for his bill:

---

33. *Hollings Letter, supra* note 32. Senator Hollings incorrectly stated that the Communications Act limits foreign investment in communications *licenses*, as opposed to *licensees*. As we shall see momentarily, Representative Markey made the same mistake when purporting to educate President Clinton about the meaning of section 310(b).

34. S. 287, 105th Cong., 1st Sess. (1997).

35. *Id.*

. . . I rise today to restore the constitutional balance to our trade policy and preserve the Congress' constitutional obligation to regulate foreign commerce. The bill I introduce requires that before a trade agreement negotiated under the auspices of the World Trade Organization is accorded the force of law, it must be ratified by the Congress. It is a simple bill, but I believe it protects a fundamental principle of our democracy, the separation of powers.[36]

As a matter of constitutional law, Senator Hollings's argument was backward. His bill would not "restore the constitutional balance," if by that phrase he meant the original meaning of the Constitution. Article II, section 2, of the Constitution does not specify, as Senator Hollings implied, that treaties "be ratified by the Congress" as a whole. Rather, the president "shall have Power, by and with the Advice and Consent *of the Senate*, to make Treaties, provided two thirds of the Senators present concur."[37] The Framers, in other words, did not intend to obligate the president to secure the consent of the House of Representatives to treaties that he had negotiated. Rather, the Framers specifically required treaty ratification by a supermajority vote of that house of Congress that they expected to have greater institutional inertia and continuity because of the staggered six-year terms of its members and the original method of having state legislatures elect U.S. senators.[38] Notwithstanding some fashionable legal scholarship to the contrary,[39] it is plain that Senator Hollings's desired approval "by the Congress by law"—that is, approval by means of an ordinary statute enacted by simple majority votes in the House and Senate and presented to the president—would not be a satisfactory substitute for securing two-thirds of the votes in the Senate. In short, it is paradoxical that a senator would argue that the defense of the Constitution requires enacting a statute that would

---

36. 143 CONG. REC. S1108 (daily ed. Feb. 6, 1997) (remarks of Sen. Hollings).

37. U.S. CONST. art. II. § 2 (emphasis added).

38. *Id.*, art. I, § 3; *see also* THE FEDERALIST NO. 66, at 445, 450 (Alexander Hamilton) (Jacob E. Cooke ed., Wesleyan University Press 1961). Direct election of senators did not occur, of course, until ratification of the Seventeenth Amendment in 1913. U.S. CONST. amend. XVII.

39. RESTATEMENT (THIRD) OF THE FOREIGN RELATIONS LAW OF THE UNITED STATES § 303 comment e (1987); Harold Hongju Koh, *The Coase Theorem and the War Power: A Response*, 41 DUKE L.J. 122, 126 (1991).

have the conspicuous constitutional defect of disabling one-third plus one members of the Senate from vetoing a treaty that they disfavored.

Alternatively, Senator Hollings may have meant that the WTO agreement on telecommunications services was an executive agreement that would be inconsistent with the existing domestic law of the United States. Under that theory, Congress would have to enact legislation, through the usual process encompassing bicameralism and presentment to the president, that would amend the Communications Act to the extent that it and the WTO agreement were in conflict. But for two reasons that argument is not persuasive either. First, there is no conflict between the WTO agreement and section 310 of the Communications Act. A foreign government, investing through a U.S. holding company, could indirectly own 100 percent of a U.S. radio licensee, unless the FCC found that the public interest would be harmed by the foreign government's ownership of more than 25 percent of the holding company. In other words, the existence of section 310(b)(4) would permit an American court to read the WTO agreement in a way that would not create any conflict with existing American statutory law.[40] Plainly, a court would construe the statute to avoid creating a conflict with the trade agreement.

Second, suppose counterfactually that the WTO agreement could be read only in a manner that created a conflict between it and section 310(a) of the Communications Act. Would the WTO agreement implicitly repeal section 310(a)? No. It could do so only if the WTO agreement on telecommunications services were itself considered to be a treaty and subsequently were ratified by two-thirds of the Senate. If the WTO agreement were instead viewed as an executive agreement, and if the president or the FCC were to seek to enforce that agreement, then a court would find the WTO agreement to be invalid and unenforceable to the extent that it contradicted section 310(a).[41]

---

40. *See* 143 CONG. REC. S1965 (daily ed. Mar. 5, 1997) (remarks of Sen. Rockefeller) ("As I understand it, the U.S. offer in the telecommunications agreement tracks U.S. law, meaning this dispute is really over the interpretation of current U.S. law by the FCC, which the ranking member of the Commerce Committee [Senator Hollings] does not like, not the trade agreement reached by USTR.").

41. *See id*. at S1956 (remarks of Sen. Roth) ("The [Hollings] amendment gives the erroneous impression that the President is currently able to implement international trade agreements calling for changes in U.S. statutory law without the passage of implementing legislation by Congress. . . . Trade agreements are executive agreements. And the simple fact is that if there is an inconsistency between an executive agreement

The law is clear on that score. The WTO telecommunications agreement is a trade agreement reached under the auspices of the Uruguay Round of trade negotiations, which culminated in the Uruguay Round Agreements. In the Uruguay Round Agreements Act, Congress expressly provided: "No provision of any of the Uruguay Round Agreements, nor the application of any such provision to any person or circumstance, that is inconsistent with any law of the United States shall have effect."[42] Moreover, the Uruguay Round Agreements Act itself "shall not be construed to amend or modify any law of the United States" or "to limit any authority conferred under any law of the United States . . . , unless specifically provided for" in that trade statute.[43] Consequently, if the WTO agreement on telecommunications services were to conflict with domestic U.S. law, the latter would trump the former unless the requisite changes in U.S. law were specified in separate implementing legislation, which would require bicameral approval in Congress and presentment to the president.

In short, contrary to Senator Hollings's fear, the WTO agreement could not have the effect of automatically amending or repealing conflicting portions of section 310. To reiterate, however, the wording of section 310(b)(4) already in effect at the time of the WTO agreement permitted the levels of foreign investment to which USTR committed the United States at the negotiations in Geneva.

Notwithstanding the multiple weaknesses of his legal argument, Senator Hollings sought to increase his leverage by vowing to attach his bill as an amendment to legislation necessary to confirm Charlene Barshefsky as U.S. trade representative.[44] (During the WTO negotiations Ms. Barshefsky had been *acting* U.S. Trade Representative, pending a Senate vote on her nomination by President Clinton.[45]) In

---

and a statute, the statute prevails. In other words, a law passed by Congress remains on the books in full force and effect and cannot somehow be trumped by an executive agreement or any other action by the President.").

42. 19 U.S.C. § 3512(a)(1).

43. *Id.* § 3512(a)(2).

44. Lorraine Woellert, *Pact Will Initiate Huge Savings on Overseas Calling*, WASH. TIMES, Feb. 16, 1997, at A1; John Maggs, *Congress Threatens to Unravel Telecom Pact; Sentiment Grows That Accord Should Be OK'd on Capitol Hill*, J. COMMERCE, Feb. 10, 1997, at 1A.

45. At the time of the WTO negotiations in Geneva, Ms. Barshefsky's nomination for U.S. trade representative had "been held up for months because of a rule that US

response to concerns raised by the Clinton administration, Senator Hollings narrowed the wording of his amendment on February 13, 1997, so that it would not address "practices or rules or regulations."[46] As revised, the relevant wording of his amendment became:

> No international trade agreement which would in effect amend or repeal United States law may be implemented by or in the United States until the agreement is approved by the Congress.[47]

Even that modification was unacceptable to USTR, for the U.S. negotiators believed (correctly, as chapter 3 indicates) that the WTO agreement necessitated no amendment to section 310(b)(4) of the Communications Act to allow more liberal levels of foreign investment in American radio licensees through a holding company structure.[48]

Like Senator Hollings, Representative Markey warned President Clinton on February 5, 1997, that he was "compelled to seek implementing legislation to effectuate any deal to which USTR agrees."[49] Representative Markey based that admonition on the following legal reasoning:

> As you may recall, Congress grappled with the contentious issue of foreign investment in U.S. communications licenses as part of last year's deliberation on the Telecommunications Act of 1996. Although the House-passed version would have permitted increased foreign ownership of common carrier licenses, Congress ultimately decided not to change current law restricting foreign investment of common carrier or broadcast licenses. It has come to my attention that USTR now espouses a reading of Section 310 of the Communications Act of 1934 that effectively renders last

---

trade representatives cannot have had a foreign government as a client. As a trade lawyer, Ms Barshefsky provided advice to the Canadian government. Negotiations with the Senate finance committee resulted in a waiver for Ms Barshefsky, leaving the nomination on course for approval." Nancy Dunne, *Senator Puts Up Obstacles over Fast-Track Procedure on Pacts*, FIN. TIMES, Feb. 13, 1997, at 5. It was that waiver that Senator Hollings was seeking to amend.

46. *Hollings Alters His Amendment to Barshefsky Waiver*, NAT'L JOURNAL'S CONGRESSDAILY, Feb. 14, 1997 (quoting Sen. Hollings).

47. *Id.*

48. *Id.*

49. *Markey Letter, supra* note 32.

year's congressional deliberation (or any future ones) meaningless. USTR apparently believes that it can consummate a trade deal that would allow 100 percent foreign ownership of U.S. common carrier licenses, limit the Federal Communications Commission (FCC) in its interpretation of the public interest test, and that such a deal can simply be implemented administratively.

I strongly disagree with this interpretation of the statute. USTR's position, as it pertains to foreign ownership of common carrier licenses, also implicates foreign ownership of *broadcast* licenses since such licenses fall under the same statutory prohibitions. The logical conclusion of USTR's interpretation is that foreign ownership of U.S. television and radio licenses today hinges solely on the FCC's willingness to block such ownership on public interest grounds. Such a distorted interpretation contradicts long-standing practice and established law. It also plainly flies in the face of common sense.[50]

Like Senator Hollings's reading of section 310(b), Representative Markey's reading of the statute ignored the FCC's long-standing statutory misinterpretation of section 310(b)(4). His reading contained another glaring error. Section 310(b) addresses the foreign ownership of radio *licensees*, not radio licenses. Section 310(b) does not speak of a radio license's being *owned* by any entity, foreign or domestic. Rather, the statute speaks of a radio license's being "granted to or held by" an entity, which in turn may be a corporation, some or all of whose *stock* is owned by foreigners.[51] Surely, the more plausible interpretation of the WTO agreement was that it would increase the ability of foreigners to invest in U.S. radio licensees, not that it would authorize foreigners to do something that American citizens may not under current law—to "own" radio frequencies.

Not surprisingly, the Clinton administration asserted that the U.S. offer made at the WTO talks would not require congressional approval of the sort that Senator Hollings proposed. In a written response to questions from Senator Trent Lott, acting U.S. Trade Representative Barshefsky explained:

> Based on Section 310(b)(4) of the Act, the offer places no new restrictions on indirect foreign ownership of a U.S. corporation

---

50. *Id.* (emphasis in original).
51. 47 U.S.C. § 310(b)(3), (4).

holding a radio license. Section 310(b)(4) allows such indirect foreign ownership unless the Federal Communications Commission finds that the public interest will be served by the refusal to grant such a license. The U.S. offer is to allow indirect foreign ownership, up to 100%, under this provision.

The U.S. offer permits a foreign government indirectly to own a radio license, unless the FCC finds that such ownership is not in the public interest. Under the public interest test, the FCC looks at many factors, such as financial and technical ability of the applicant, international agreements, national security concerns, foreign policy concerns, law enforcement concerns and the effect of entry on competition in the U.S. market. In the event of a successful conclusion to these negotiations, the U.S. offer will allow the FCC to continue to apply these public interest criteria, as long as they do not distinguish among applicants on the basis of nationality or reciprocity, consistent with the obligations of the General Agreement on Trade in Services.[52]

Nonetheless, USTR attempted to smooth ruffled feathers. "The administration believes we have regulatory authority to implement the agreement," said acting U.S. Trade Representative Barshefsky after the successful completion of the February 15, 1997, agreement, "but we have told members of Congress previously and will continue to tell them that we would be very pleased to work with [them] to see if appropriate implementing legislation can be formulated."[53]

Paradoxically, the Hollings-Markey initiative prompted Republicans in Congress to come to the rescue of the Clinton administration. Key members of the House and Senate—including Senator John McCain of Arizona, chairman of the Senate Commerce Committee, and Representative Thomas Bliley of Virginia, chairman of the House Commerce Committee—sent a letter of support to the U.S. trade representative.[54] Senator McCain and Representative Bliley argued that section 310(b)(4) already empowered the FCC to approve

52. Written Responses of Acting U.S. Trade Representative Charlene Barshefsky to Questions from Senator Lott (undated), *reprinted in* 143 CONG. REC. S1962 (daily ed. Mar. 5, 1997).

53. Woellert, *supra* note 44.

54. Letter from Senators John McCain, Conrad R. Burns, John Ashcroft, and Sam Brownback, and Representatives Thomas J. Bliley, Jr., and Michael G. Oxley to President William J. Clinton, Feb. 6, 1997.

levels of foreign ownership in radio licensees that exceed 25 percent and that, consequently, Senator Hollings's bill was unnecessary.[55] In a February 11, 1997, letter to President Clinton, Senator McCain, joined by Senator Conrad Burn, chairman of the Senate Subcommittee on Communications, and Representative Michael G. Oxley of Ohio, vice chairman of the House Telecommunications Subcommittee, reiterated that the U.S. trade representative's interpretation of section 310(b)(4) corrected the long-standing misreading of the statute discussed in chapter 3:

> It has been stated that USTR sought amendments to the Telecommunications Act of 1996 to clarify legal limits on foreign investment in U.S. telecommunications firms. This is incorrect. As the authors of the Senate and House foreign ownership provisions, we wish to state for the record that we were acting on our own initiative and that no Administration official requested that we legislate in this area. Any discussions we had with the Administration on these issues came at our request.
>
> We firmly believe that the Administration possesses the authority to negotiate an agreement without implementing legislation. Indeed, the correct legal interpretation of the relevant statute is that private foreign firms are free to invest in American firms without restriction unless "the [Federal Communications] Commission finds that the public interest will be served by the refusal or revocation" of a telecommunications license. To allege that implementing legislation is necessary is to misinterpret the law. Indeed, it is the very prevalence of such misreadings that caused us to attempt to reform the ownership rules.[56]

On March 5, 1997, Representative Oxley introduced the International Telecommunications Investment Clarification Act of 1997, a bill that would extend the FCC's authority to waive foreign ownership limits on direct investment in U.S. telecommunications companies—foreign investment that would *not* be structured in the holding company fashion envisioned by section 310(b)(4).[57] He noted that "[t]he FCC

---

55. *Id.*

56. Letter from Senators John McCain and Conrad Burns and Representative Michael G. Oxley to President William J. Clinton, Feb. 11, 1997 (quoting 47 U.S.C. § 310(b)(4)), *reprinted in* 143 CONG. REC. S1957 (daily ed. Mar. 5, 1997).

57. H.R. 954, 105th Cong., 1st Sess. (1997). As introduced, the bill would amend

already possesses such authority in the case of indirect investment, which is the tool the Administration used to get the agreement" in Geneva.[58] The Oxley legislation would (1) repeal section 310(b)(4)'s

---

section 310(b) to read as follows:

(b)(1) No broadcast or common carrier or aeronautical en route or aeronautical fixed radio station license shall be granted to or held by—

(A) any alien or the representative of any alien;

(B) any corporation organized under the laws of any foreign government; or

(C) any corporation of which more than one-fifth of the capital stock is owned of record or voted by a foreign government or representative thereof.

(2) No broadcast radio station license shall be granted to or held by—

(A) any corporation of which more than one-fifth of the capital stock is owned of record or voted by aliens or their representatives or by any corporation organized under the laws of a foreign country; or

(B) any corporation directly or indirectly controlled by any other corporation of which more than one-fourth of the capital stock is owned of record or voted by aliens, their representatives, or by a foreign government or representative thereof, or by any corporation organized under the laws of a foreign country, if the Commission finds that the public interest will be served by the refusal or revocation of such license.

(3) No common carrier or aeronautical en route or aeronautical fixed radio station license shall be granted to or held by any corporation of which more than one-fifth of the capital stock is owned of record or voted by aliens or their representatives or by any corporation organized under the laws of a foreign country, if the Commission finds that the public interest will be served by the refusal or revocation of such license.

(4) Nothing in this subsection or subsection (a) prohibits a common carrier or aeronautical en route or aeronautical fixed radio station license from being granted to or held by any corporation directly or indirectly controlled by any other corporation of which more than one-fourth of the capital stock is owned of record or voted by aliens, their representatives, or by a foreign government or representative thereof, or by any corporation organized under the laws of a foreign country.

Id.

58. *Top Lawmaker Tells CompTel Annual Convention & Trade Expo Attendees He Plans Legislation on FCC Authority over Foreign Ownership Limits on Direct Investment in U.S. Telecom Firms*, PR NEWSWIRE, Feb. 18, 1997. "While my bill will represent a modest change in the way ownership limits are interpreted, in that it merely harmonizes limitations, it is important in that it gives the FCC and the Office of the U.S. Trade Representative additional leverage in their dealings with our trading

statutory benchmark on foreign indirect investment in U.S. corporations holding common carrier or aeronautical radio licenses, although not broadcast licenses; (2) allow foreign direct investment greater than 20 percent in U.S. corporations holding common carrier or aeronautical radio licenses, if the FCC were to find that such investment served the public interest; and (3) explicitly prohibit any corporation with more than 20 percent direct foreign government ownership from holding common carrier, aeronautical, or broadcast licenses.

Chairman Hundt, a staunch supporter of USTR, also prepared for Democratic opposition in Congress to the WTO agreement. He told reporters on February 19, 1997, that the FCC would drop its effective competitive opportunities test and instead consider under the public interest standard how a foreign firm's entry into the United States would affect competition in that market.[59] Two days later, in public remarks that presumably pleased Senator Hollings, the chairman said that he opposed *any* foreign ownership of American broadcast licensees:

> On the foreign ownership issue, . . . let me state a couple of things that are obvious, but that ought to be stated. One is: Broadcast is not part of this deal. Second, the consensus view in this country—and a view that I share—is that we should not allow foreign ownership of broadcast properties. It is in fact a good idea to rewrite Section 310(b) to make it very clear that we should not allow foreign ownership of broadcast. And I believe that Charlene Barshefsky has stated that and stated that the administration would very much like to work with Congress to make sure that 310(b)—which does in fact have the capability of being read to permit waivers of the foreign ownerships vis-à-vis broadcast—in fact should not be interpreted that way in practice by the FCC. But since I can't guarantee to stay there forever—and there is an issue as to whether you want an FCC to be empowered to exercise such a waiver—in fact it would be a very, very good idea to write 310(b) so as to reflect the will of the country, which is that we don't want foreign ownership of broadcast, period.[60]

---

partners, especially in Europe, helping U.S. companies penetrate foreign markets." *Id.*

59. *FCC to Drop ECO Test, Use Public Interest Review for Foreign Applicants*, COMM. DAILY, Feb. 20, 1997.

60. *Lang Remarks, supra* note 19 (comments of Reed E. Hundt).

Those remarks may have been a sop to Senator Hollings and Representative Markey to secure easier congressional approval of the WTO agreement. If taken seriously, however, Chairman Hundt's remarks were schizophrenic. At an event venerating the reduction of barriers to foreign investment in telecommunications, the chairman of the FCC, without any awareness of cognitive dissonance, advocated a *complete ban* on foreign investment in American radio and television broadcasters—because *he* perceived such a change in existing law to be "the will of the country." Apart from being uninhibited by the historical fact that Congress has repeatedly and incrementally liberalized section 310(b), Chairman Hundt's proposed course of action would unconstitutionally infringe the freedom of speech for the reasons discussed in chapter 7.

On March 5, 1997, the day of Ms. Barshefsky's confirmation vote, Senator Hollings sought support for his amendment with the rallying cry that "with respect to foreign trade, with respect to global competition, we are in the hands of the Philistines, we are in the hands of the multinationals."[61] In addition to his earlier arguments about section 310(b)(4), Senator Hollings argued that the WTO agreement would violate section 310(a), which forbids any foreign government from holding or being granted an applicable radio license.[62] That argument, however, was plainly wrong. As noted earlier, a court would strive to read the WTO agreement and section 310 in a manner that would not create a conflict with domestic U.S. law, and indeed such a reading could readily be found through the proper interpretation of section 310(b)(4).

In his floor speech, Senator Hollings also objected to the commitments that the U.S. negotiators secured from Japan, Canada, and Mexico by the February 15, 1997, deadline:

> They just gave away 100 percent in violation of 310(a). They didn't just do the 25 percent in 310(b). They go in, as naive as get out, I can tell you that. I want to build a bridge back to the old-fashioned Yankee trader. Come in and say, look, we have the largest and the richest market; what can you come up with? Let's see what you propose and we will work with it. Instead, like goody-goody two shoes, this touchy-feely crowd that we have up

---

61. 143 Cong. Rec. S1945 (daily ed. Mar. 5, 1997).
62. *Id.*

here in Washington says, "We will give you 100 percent and let's see what you come up with." Nippon Telephone & Telegraph [*sic*] says, "Thank you for the 100 percent, bug off, you get nothing from us." And you go down the list. No country gave us any kind of 50-percent ownership. Our best of allies and friends in international trade, Canada and Mexico, in NAFTA, said, "No, you can't get a 50-percent." Under 50 percent. So you can see what a spurious approach they used, in violation of the law.[63]

Senator Hollings's argument was flawed. It does not at all follow logically that, because the Japanese, Canadians, and Mexicans did not fully open their telecommunications markets to foreign investment, the acting U.S. trade representative violated section 310(b)(4) by committing the United States to open its market fully. There has never been a statutory requirement in the Communications Act that indirect foreign investment in radio licensees exceeding 25 percent be conditioned on mirror reciprocity with respect to the treatment of U.S. investment in the foreigner's home market.

By a vote of 84 to 16, the Senate rejected Senator Hollings's amendment to Ms. Barshefsky's nomination.[64] Bearing no grudge, the senator then proceeded to vote in favor of her confirmation.[65]

It may remain a mystery why Senator Hollings and Representative Markey opposed the WTO agreement taking shape in early February 1997. Their objections may have sincerely reflected a specific concern over the proper scope of the Senate's treaty ratification powers or Congress's power to enact domestic legislation to implement an executive agreement on international trade. Or their objections may have manifested a general aversion to free trade—reminiscent of Senator Hollings's proposal in 1995 for a "snapback" provision for section 310(b) and Representative Markey's unsuccessful attempts in earlier years to extend section 310(b) to cable television. But in Washington, D.C., a city where little is left to spontaneity, it is equally plausible that the opposition of those two prominent Democrats in Congress was entirely acceptable to the White House, USTR, the FCC, and the Republicans in Congress. The timing of Chairman Hundt's proposal that Congress ban all foreign

---

63. *Id.*
64. *Id.* at S1970.
65. *Id.* at S1973. The vote was 99 to 1.

investment in U.S. radio and television broadcasters was particularly suspicious in that regard. So was Senator Hollings's ultimate vote in favor of Ms. Barshefsky's nomination. In short, the threats to derail the WTO agreement could have been an instance of "good cop, bad cop"—a contrived fit of breast beating intended to give other nations the impression that the U.S. delegation in Geneva was under serious domestic political pressure to walk away from any agreement that was not fully satisfactory to American interests and to continue to negotiate aggressively with Canada, Japan, and Mexico in bilateral talks before the November 30, 1997, ratification of the WTO agreement.[66] In any event, the Senate's resounding rejection of the Hollings amendment represented a vindication of sorts of the argument that section 310(b)(4) should be read in the plain manner in which Congress originally drafted the statute, rather than in the manner in which the FCC had applied the statute until the agency's late conversion to textual fidelity in support of the WTO agreement in 1997.

Resolution of the Disagreement
with Canada and Japan

Despite the resolute positions Canada and Japan took in the week before the February 15, 1997, deadline in Geneva, the U.S. negotiators agreed to join in the WTO agreement. The United States, however, took retaliatory measures against both countries. With respect to Canada, explained acting U.S. Trade Representative Barshefsky, the United States "took action multilaterally" against Canadian satellite broadcasters to prevent them from beaming television images into the United States, although she predicted that it was "inevitable" that on a bilateral basis Canada and the United States would have "continued discussion" concerning the matter.[67] In an

---

66. After completion of the agreement, an aide to Senator Hollings grumbled: "There is just not that much there. . . . We're disappointed with the Canadians, Mexicans and Japanese." Swardson & Blustein, *supra* note 5.

67. *US Hails Telecoms Deal in Geneva*, AGENCE FRANCE-PRESSE, Feb. 15, 1997. The German press noted that the United States itself had compromised its free-trade ideals at moments in the talks: "The Americans diluted part of an offer to reduce their own restrictions on audiovisual markets such as satellite radio and television, which are increasingly linked with the phone business." *WTO States Approve World Telecoms*

attempt to pressure Canada to make a better offer before ratification of the telecommunications agreement on November 30, 1997, USTR negotiators filed an exemption to deny Canada most-favored-nation status.[68] As summarized by the WTO, the relation of MFN exemptions to the multilateral WTO agreement on telecommunications services is as follows:

> On the multilateral level, the results of the telecommunications negotiations are to be extended to all WTO members on a non-discriminatory basis through m.f.n. treatment. However, the legal basis for the negotiations made it possible for each WTO Member to decide individually whether or not to file an m.f.n. exemption on a measure affecting trade in basic telecommunications services. On 15 February [1997], nine governments submitted m.f.n. exemption lists to be annexed to the Protocol. The exemption by the United States relates to one-way satellite transmission of DTH and DBS television services and digital audio services. . . . While m.f.n. exemptions can sometimes be required by legal technicalities, a decision to file one can also depend on whether a participant is satisfied with the quality of commitments made. Without an m.f.n. exemption, a Member must treat the services or service suppliers of every other Member as favourably as those of any other country, Member or not. But even if it files an exemption, a Member may only apply it to unscheduled services or to grant special preferences over and above the market access restrictions indicated in its schedule.[69]

If the United States were to strip Canada of its MFN status, Canada might be forced back to the negotiating table, and congressional opponents of free trade would therefore have less reason to criticize the U.S. delegation for being soft on Canada.[70] But resort to that trade weapon could just as easily backfire and provoke an undesirable and counterproductive response from Canada.

With respect to Japan's 20 percent ceiling on foreign investment in NTT and KDD, the U.S. delegation joined in the WTO agreement subject to the proviso that the United States would reserve the option,

---

Liberalization Agreement, DEUTSCHE PRESSE-AGENTUR, Feb. 15, 1997.

68. *WTO: Global Telecoms*, AFP-EXTEL NEWS, Feb. 17, 1997.

69. *WTO Negotiations, supra* note 1.

70. *WTO: Global Telecoms*, AFP-EXTEL NEWS, Feb. 17, 1997.

described by Edmund Andrews of the *New York Times* as a mere "face-saving measure," to impose a corresponding 20 percent limitation on Japanese investment in U.S. telecommunications firms.[71] Although one cannot quarrel with the assessment of U.S. Trade Representative Barshefsky that "Japan keeping those restrictions is a terrible signal to many countries around the world who hide behind Japan's restriction as justification for their own,"[72] it nonetheless remains the case that, for the reasons discussed earlier in chapter 6, it would harm the interests of American consumers for the United States actually to impose a bilateral sanction of that sort on Japanese direct investment in American telecommunications firms. As in the case of Canada, U.S. negotiators concluded in February 1997 that the November 30, 1997, deadline for ratification would give them time to urge Japan to drop its 20 percent caps on foreign ownership of NTT and KDD. There was some indication at the time of the February 15, 1997, agreement that Japan would work over the following nine months to improve its offer in that respect.[73]

U.S. TELECOMMUNICATIONS POLICY AFTER THE WTO AGREEMENT

The 1997 WTO agreement ensured that the controversy over foreign investment in American telecommunications would continue. More than six decades have elapsed since the enactment of section 310(b), and more than eight since Congress placed the first U.S. restrictions on foreign investment in wireless. The original and foremost justification for those restrictions has been national security. Yet we have known since at least Pearl Harbor that encryption technology, and not mere access to wireless communications, is the real threat to national security. Denying foreigners the full opportunity to invest in the U.S. wireless industry on the grounds that they might send harmful messages is like forbidding the sale of ink and paper to foreigners on the grounds that they might use them to write secret notes. For more than a half century, the national security justification for section 310(b) has been untenable. It is ironic, perhaps fitting, that on

---

71. Andrews, *U.S. Appears to Ease Its Stance, supra* note 4.
72. *US Hails Telecoms Deal in Geneva*, AGENCE FRANCE-PRESSE, Feb. 15, 1997.
73. *WTO: Global Telecoms*, AFP-EXTEL NEWS, Feb. 17, 1997.

February 15, 1997, the same day that the *New York Times* and the *Washington Post* reported that the United States would support a WTO agreement on telecommunications liberalization, the front pages of both newspapers ran stories of possible illegal contributions to an American political party by foreign governments[74]—a vivid reminder that there are more direct ways for a foreign power to influence American policy and public sentiment that for foreign corporations to invest in U.S. wireless licensees.

Moreover, if national security *were* a compelling justification for section 310(b), then Congress would be remiss in not rewriting the statute to close the multitude of loopholes that it—and the FCC, through its patchwork enforcement of the statute—have allowed to develop. Since 1934 Congress has, to the contrary, repeatedly amended section 310(b) to narrow its scope. In 1996 Congress pulled whatever rickety support remained out from under the national security rationale when it repealed the restrictions on foreigners' participation in the management of U.S. radio licensees as officers and directors. Today the foreign ownership restrictions are applied in a way that makes arbitrary distinctions between different radio services that cannot plausibly be justified on the grounds that foreign ownership of one constitutes a larger security threat than foreign ownership of the other. Similarly, the statute irrationally regulates passive foreign investment while permitting active foreign management. Meanwhile, behind that foreground of utterly whimsical rules is a landscape of alternative statutory powers conferred on the president and the FCC that are far better suited to thwarting spies and provocateurs than is section 310(b).

The arbitrary distinctions that the FCC has lent in its administrative decisions to the already arbitrary statutory contours of section 310(b) have produced a body of law and agency folklore as intricate as a Persian rug. The only beneficiaries of that state of affairs are Washington communications lawyers, whom clients must retain to contort straightforward international business transactions. There are obvious transactions costs associated with that regulatory burden. But the greater costs are the agency costs that arise when parties cannot

---

74. Stephen Labaton, *Clinton and Gore Received Warnings on Asian Donors*, N.Y. TIMES, Feb. 15, 1997, at 1; Sharon LaFranier & Susan Schmidt, *NSC Gave Warnings About Asian Donors*, WASH. POST, Feb. 15, 1997, at A1.

freely arrange the ownership and control of a firm in a manner that optimally allocates risk among willing parties and protects the firm's owners against the possibility that management will deviate from profit-maximizing behavior. The FCC's administration of the foreign ownership restrictions has been oblivious to the drag that it has imposed on productive economic activity. The public interest, it would seem, could not possibly concern such mundane matters as preoccupy Nobel laureates.

In light of the costs and risks that the foreign ownership restrictions create, it no surprise that, until BT's proposal in 1996 to acquire all of MCI, foreign investors had made relatively few billion-dollar investments in U.S. radio licensees. Admittedly, the small number of large transactions also reflected that the most likely invest-ors—large foreign telecommunications carriers—were, until their recent privatizations, state-owned monopolies. Consequently, they were completely barred from being U.S. radio licensees. That state of affairs is changing, however, as even state-owned telephone monopolies that have not yet been fully or partially privatized, such as France Télécom and Deutsche Telekom, are investing to fashion global networks to compete with those offered by AT&T and BT-MCI. The harm to consumers of America's traditional inhospitality to foreign direct investment in telecommunications will become more apparent as the regional Bell operating companies are eventually allowed under the Telecommunications Act of 1996 to compete in the interLATA market. It would seem inevitable that one or more of those companies will combine with one or more of the major foreign carriers to form at least a fourth "supercarrier." The immediate benefits to American consumers from such a combination will be commensurate with the degree to which the various segments of the long-distance market are currently less than perfectly competitive. Those prospective consumer benefits in telephony, meanwhile, must be added to the benefits that American consumers have already received because News Corp. endured the gauntlet of foreign owner-ship challenges by incumbent broadcasters opposed to its creating a viable fourth television network.

Future foreign investment in U.S. telecommunications thus implies a potent form of new competition that will benefit American consumers. It is therefore important that any future revision of the foreign ownership restrictions following the 1997 WTO agreement not impose a regime that sacrifices the fruits of greater domestic competition in the name of opening markets overseas. Indeed, economic analysis does not give us confidence that any sacrifice of the former would produce any increment of the latter. Such a sacrifice is the greatest harm that American trade negotiators could produce for U.S. consumers in the aftermath of the WTO accord on telecommunications services. If pursued in the post-1997 environment, the reciprocity proposals advanced in the Senate and embraced at the FCC in 1995 would be unlikely to achieve their market-opening objectives but *would be* likely to shield incumbent U.S. interexchange carriers from competition in the domestic and U.S. outbound markets. Economic analysis provides strong reasons, particularly after the successful completion of the WTO agreement, for not erecting a U.S. policy for foreign direct investment that is premised on bilateral reciprocity.

Whether Congress uses section 310(b) to promote national security or to secure market access overseas, it must recognize that the statute, and the interpretation given it by the FCC, restrict the freedom of electronic speech. The Supreme Court may soon find in the First Amendment the musculature necessary to protect speech that increasingly is conveyed by electronic means rather than by the printed word. That jurisprudential breakthrough seems inevitable and imminent. Congress and the FCC should therefore approach their revision of the foreign ownership restrictions following the 1997 WTO accord with the foresight that the First Amendment will eventually demand, if it does not already, that the purposes of those restrictions be clear, compelling, and narrowly tailored to accomplishing their goals.

In sum, the foreign ownership restrictions in the Communications Act epitomize the worst of American regulation. They are arbitrary and irrational. They rest on faulty historical analysis, faulty economic analysis, and faulty legal analysis. The agency charged with enforcing those restrictions has read them to aggrandize its own power over international transactions and to claim the authority to dabble in trade policy where no such authority has been delegated by Congress. The FCC has regulated foreign investment in a manner that has been indifferent to the harm that such regulation has caused the American consumer, to the costs that it has imposed on investors in telecommunication ventures, and to the degree to which it has infringed the rights to freedom of speech protected by the Constitution. There is nothing worth salvaging here. The foreign ownership restrictions should be scrapped. The 1997 WTO agreement on telecommunications services has supplied Congress with the opportunity and obligation to act accordingly.

Appendix

Selected WTO Documents Concerning Trade in Telecommunications Services

Reproduced below are selected official documents of the World Trade Organization concerning the negotiations leading to the agreement on telecommunications services concluded in Geneva on February 15, 1997. These and other documents can be found on the WTO's web site on the Internet, www.wto.org.

* * *

DECISION ON NEGOTIATIONS
ON BASIC TELECOMMUNICATIONS

Adopted 15 April 1994 at Marrakesh

Ministers decide as follows:

1.   Negotiations shall be entered into on a voluntary basis with a view to the progressive liberalization of trade in telecommunications transport networks and services (hereinafter referred to as "basic telecommunications") within the framework of the General Agreement on Trade in Services.

2.   Without prejudice to their outcome, the negotiations shall be comprehensive in scope, with no basic telecommunications excluded *a priori*.

3.   A Negotiating Group on Basic Telecommunications (hereinafter referred to as the "NGBT") is established to carry out this mandate. The NGBT shall report periodically on the progress of these negotiations.

4.   The negotiations in the NGBT shall be open to all governments and the European Communities which announce their intention to participate. To date, the following have announced their intention to take part in the negotiations:

> Australia, Austria, Canada, Chile, Cyprus, European Communities and their member States, Finland, Hong Kong, Hungary, Japan, Korea, Mexico, New Zealand, Norway, Slovak Republic, Sweden, Switzerland, Turkey, United States.

Further notifications of intention to participate shall be addressed to the depositary of the Agreement Establishing the World Trade Organization.

5.   The NGBT shall hold its first negotiating session no later than 16 May 1994. It shall conclude these negotiations and make a final report no later than 30 April 1996. The final report of the NGBT shall include a date for the implementation of results of these negotiations.

6.   Any commitments resulting from the negotiations, including the date of their entry into force, shall be inscribed in the Schedules annexed to the General Agreement on Trade in Services and shall be subject to all the provisions of the Agreement.

7.   Commencing immediately and continuing until the implementation date to be determined under paragraph 5, it is understood that no participant shall apply any measure affecting trade in basic telecommunications in such a manner as would improve its negotiating position and leverage. It is understood that this provision shall not prevent the pursuit of commercial and governmental arrangements regarding the provision of basic telecommunications services.

8.   The implementation of paragraph 7 shall be subject to surveillance in the NGBT. Any participant may bring to the attention of the NGBT any action or omission which it believes to be relevant to the fulfillment of paragraph 7. Such notifications shall be deemed to have been submitted to the NGBT upon their receipt by the Secretariat.

*   *   *

Negotiating Group on Basic Telecommunications

24 April 1996

REFERENCE PAPER

Scope

The following are definitions and principles on the regulatory framework for the basic telecommunications services.

Definitions

Users mean service consumers and service suppliers.

Essential facilities mean facilities of a public telecommunications transport network or service that

(a) are exclusively or predominantly provided by a single or limited number of suppliers; and

(b) cannot feasibly be economically or technically substituted in order to provide a service.

A major supplier is a supplier which has the ability to materially affect the terms of participation (having regard to price and supply) in the relevant market for basic telecommunications services as a result of:

(a) control over essential facilities; or

(b) use of its position in the market.

1. Competitive safeguards

1.1 Prevention of anti-competitive practices in telecommunications

Appropriate measures shall be maintained for the purpose of preventing suppliers who, alone or together, are a major supplier from engaging in or continuing anti-competitive practices.

1.2 Safeguards

The anti-competitive practices referred to above shall include in particular:

(a) engaging in anti-competitive cross-subsidization;

(b) using information obtained from competitors with anti-competitive results; and

(c) not making available to other services suppliers on a timely basis technical information about essential facilities and commercially relevant information which are necessary for them to provide services.

2.   Interconnection

2.1   This section applies to linking with suppliers providing public telecommunications transport networks or services in order to allow the users of one supplier to communicate with users of another supplier and to access services provided by another supplier, where specific commitments are undertaken.

2.2   Interconnection to be ensured

Interconnection with a major supplier will be ensured at any technically feasible point in the network. Such interconnection is provided

(a) under non-discriminatory terms, conditions (including technical standards and specifications) and rates and of a quality no less favourable than that provided for its own like services or for like services of non-affiliated service suppliers or for its subsidiaries or other affiliates;

(b) in a timely fashion, on terms, conditions (including technical standards and specifications) and cost-oriented rates that are transparent, reasonable, having regard to economic feasibility, and sufficiently unbundled so that the supplier need not pay for network components or facilities that it does not require for the service to be provided; and

(c) upon request, at points in addition to the network termination points offered to the majority of users, subject to charges that reflect the cost of construction of necessary additional facilities.

2.3   Public availability of the procedures for interconnection negotiations

The procedures applicable for interconnection to a major supplier will be made publicly available.

2.4   Transparency of interconnection arrangements

It is ensured that a major supplier will make publicly available either its interconnection agreements or a reference interconnection offer.

2.5   Interconnection: dispute settlement

A service supplier requesting interconnection with a major supplier will have recourse, either:

(a) at any time or

(b) after a reasonable period of time which has been made publicly known

to an independent domestic body, which may be a regulatory body as referred to in paragraph 5 below, to resolve disputes regarding appropriate terms, conditions and rates for interconnection within a reasonable period of time, to the extent that these have not been established previously.

3.   Universal service

Any Member has the right to define the kind of universal service obligation it wishes to maintain. Such obligations will not be regarded as anti-competitive *per se*, provided they are administered in a transparent, non-discriminatory and competitively neutral manner and are not more burdensome than necessary for the kind of universal service defined by the Member.

4.   Public availability of licensing criteria

Where a licence is required, the following will be made publicly available:

(a) all the licensing criteria and the period of time normally required to reach a decision concerning an application for a licence and

(b) the terms and conditions of individual licences.

The reasons for the denial of a licence will be made known to the applicant upon request.

5.   Independent regulators

The regulatory body is separate from, and not accountable to, any supplier of basic telecommunications services. The decisions of and the procedures used by regulators shall be impartial with respect to all market participants.

6.   Allocation and use of scarce resources

Any procedures for the allocation and use of scarce resources, including frequencies, numbers and rights of way, will be carried out in an objective, timely, transparent and non-discriminatory manner. The current state of allocated frequency bands will be made publicly available, but detailed identification of frequencies allocated for specific government uses is not required.

*   *   *

30 April 1996

Trade in Services

FOURTH PROTOCOL TO THE GENERAL
AGREEMENT ON TRADE IN SERVICES

Members of the World Trade Organization (hereinafter referred to as the "WTO") whose Schedules of Specific Commitments and Lists of Exemptions from Article II of the General Agreement on Trade in Services concerning basic telecommunications are annexed to this Protocol (hereinafter referred to as "Members concerned"),

Having carried out negotiations under the terms of the Ministerial Decision on Negotiations on Basic Telecommunications adopted at Marrakesh on 15 April 1994,

Having regard to the Annex on Negotiations on Basic Telecommunications,

Agree as follows:

Upon the entry into force of this Protocol, a Schedule of Specific Commitments and a List of Exemptions from Article II concerning basic telecommunications annexed to this Protocol relating to a Member shall, in accordance with the terms specified therein, supplement or modify the Schedule of Specific Commitments and the List of Article II Exemptions of that Member.

This Protocol shall be open for acceptance, by signature or otherwise, by the Members concerned until 30 November 1997.

The Protocol shall enter into force on 1 January 1998 provided it has been accepted by all Members concerned. If by 1 December 1997 the Protocol has not been accepted by all Members concerned, those Members which have accepted it by that date may decide, prior to 1 January 1998, on its entry into force.

This Protocol shall be deposited with the Director-General of the WTO. The Director-General of the WTO shall promptly furnish to each Member of the WTO a certified copy of this Protocol and notifications of acceptances thereof.

This Protocol shall be registered in accordance with the provisions of Article 102 of the Charter of the United Nations.

Done at Geneva this [— day of month] one thousand nine hundred and ninety-seven, in a single copy in the English, French and Spanish languages, each text being authentic, except as otherwise provided for in respect of the Schedules annexed hereto.

*   *   *

30 April 1996

Trade in Services

DECISION ON COMMITMENTS
IN BASIC TELECOMMUNICATIONS

Adopted by the Council for Trade in Services on 30 April 1996

The Council for Trade in Services,

Having regard to the Annex on Negotiations on Basic Telecommunications,

Having regard to the final Report of the Negotiating Group on Basic Telecommunications on the negotiations conducted under the terms of the Decision on Negotiations on Basic Telecommunications adopted at Marrakesh on 15 April 1994,

Decides as follows:

To adopt the text of the "Fourth Protocol to the General Agreement on Trade in Services" (hereinafter referred to as the Protocol) and to take note of the Schedules of Commitments and Lists of Exemptions from Article II listed in the Attachment to the final Report of the Negotiating Group on Basic Telecommunications.

Commencing immediately and continuing until the date of entry into force of the Protocol Members concerned shall, to the fullest extent consistent with their existing legislation and regulations, not take measures which would be inconsistent with their undertakings resulting from these negotiations.

During the period from 15 January to 15 February 1997, a Member which has a Schedule of Commitments annexed to the Protocol, may supplement or modify such Schedule or its List of Article II Exemptions. Any such Member which has not annexed to the Protocol a List of Article II Exemptions may submit such a list during the same period.

A Group on basic telecommunications reporting to the Council for Trade in Services shall conduct consultations on the implementation of paragraph 3 above commencing its work no later than 90 days from the adoption of the Decision.

The Council for Trade in Services shall monitor the acceptance of the Protocol by Members concerned and shall, at the request of a Member, examine any concerns raised regarding the application of paragraph 2 above.

Members of the World Trade Organization which have not annexed to the Protocol Schedules of Commitments or Lists of Exemptions from Article II may submit, for approval by the Council, Schedules of Commitments and Lists of Exemptions from Article II relating to basic telecommunications prior to 1 January 1998.

*   *   *

15 February 1997

Group on Basic Telecommunications

REPORT OF THE GROUP
ON BASIC TELECOMMUNICATIONS

[1]   This report is made in accordance with paragraph 4 of the Decision on Commitments in Basic Telecommunications, adopted by the Council for Trade in Services on 30 April 1996 (S/L/19). In paragraph 1 of this Decision, the Council also adopted the text of the Fourth Protocol to the General Agreement on Trade in Services and took note of the Schedules of Commitments and Lists of Exemptions from Article II listed in the Attachment to the final Report of the Negotiating Group on Basic Telecommunications (S/NGBT/18).

[2]   The Decision on Commitments on Basic Telecommunications established the Group on Basic Telecommunications to "conduct consultations on the implementation of paragraph 3 of the Decision." Paragraph 3 states that "during the period from 15 January to 15 February 1997, a Member which has a Schedule of Commitments annexed to the Protocol, may supplement or modify such Schedule or its List of Article II Exemptions"; and that "any such Member which has not annexed to the Protocol a List of Article II Exemptions may submit such a list during the same period."

[3]   At the Group's first meeting in July 1996, participants suggested that the principal issues before the GBT included the desirability of improving the quantity and quality of Schedules offered, and the need to address certain issues which had been left unresolved in April. Subsequently, the Group sponsored frequent rounds of bilateral negotiations on offers and regularly included discussion of outstanding issues in its meetings. In November participants began submitting revised draft offers of commitments on basic telecommunications for consideration. The Group's Report to the Council on Trade in Services (S/GBT/2), which formed part of the Report to the Singapore Ministerial Conference, recommended that Ministers "stress their commitment to bring the negotiations on basic telecommunications to a successful conclusion by 15 February 1997, urge all WTO Members to strive for significant, balanced and non-discriminatory liberalization commitments on basic telecommunications by that date and recognize the importance of resolving the principal issues

before the GBT." The Declaration adopted by Ministers in Singapore (WT/MIN(96)/DEC) contained a commitment to "achieve a successful conclusion to the negotiations on basic telecommunications in February 1997." Ministers also stated "We are determined to obtain a progressively higher level of liberalization in services on a mutually advantageous basis with appropriate flexibility for individual developing country members, as envisaged in the agreement, in the continuing negotiations and those scheduled to begin no later than 1 January 2000. In this context, we look forward to full MFN agreements based on improved market access commitments and national treatment."

[4]   In its discussions on outstanding issues, the Group considered the following matters: ways to ensure accurate scheduling of commitments—particularly with respect to supply of services over satellites and to the management of radio spectrum; potential anti-competitive distortion of trade in international services; the status of intergovernmental satellite organizations in relation to GATS provisions; and the extent to which basic telecommunications commitments include transport of video and/or broadcast signals within their scope.

[5]   The Chairman issued notes reflecting his understanding of the position reached in discussion of the scheduling of commitments and management of radio spectrum. The first such Note set out a number of assumptions applicable to the scheduling of commitments and was intended to assist in ensuring the transparency of commitments (S/GBT/W/2/Rev.1 of 16 January 1997). The second addressed the allocation of radio spectrum, suggesting that the inclusion of references to the availability of spectrum in schedules was unnecessary and that such references should be deleted (S/GBT/W/3 of 3 February 1997). These Notes are attached to this Report.

[6]   By 15 February 1997 the total number of schedules submitted had reached 55 (counting as one the offer of the European Communities and their Member States). Nine governments had submitted lists of Article II Exemptions.

[7]   The Group noted that five countries had taken Article II exemptions in respect of the application of differential accounting rates to services and service suppliers of other Members. In the light of the fact that the accounting rate system established under the International Telecommunications Regulations is the usual method of terminating international traffic and by its nature involves differential rates, and in order to avoid the submission of further such exemptions, it is the understanding of the Group that:

— the application of such accounting rates would not give rise to action by Members under dispute settlement under the WTO; and

— that this understanding will be reviewed not later than the commencement of the further Round of negotiations on Services Commitments due to begin not later than 1 January 2000.

[8]   The Group also recalled paragraph 6 of the Decision of 30 April, which stated

that Members of the World Trade Organization which have not annexed to the Protocol Schedules of Commitments or Lists of Exemptions from Article II may submit, for approval by the Council, Schedules of Commitments and Lists of Exemptions from Article II relating to basic telecommunications prior to 1 January 1998.

[9]   At its meeting of 15 February 1997, the Group adopted this report and the attached list of the Schedules of Commitments and Lists of Article II Exemptions, which, in accordance with paragraph 3 of the Decision on Commitments in Basic Telecommunications, will be attached to the Fourth Protocol to the General Agreement on Trade in Services in replacement of those attached on 30 April 1996.

* * *

17 February 1997

MARKET ACCESS COMMITMENTS
ON BASIC TELECOMMUNICATIONS

List of governments which submitted offers as of 15 February 1997

[1] ARGENTINA*
[2] **ANTIGUA & BARBUDA+**
[3] AUSTRALIA*
[4] **THE BAHAMAS+***1***
[5] **BANGLADESH+**
[6] **BELIZE+**
[7] **BOLIVIA+**
[8] **BRAZIL***
[9] **BRUNEI DARUSSALAM+**
[10] **BULGARIA+**
[11] CANADA*
[12] CHILE*
[13] COLOMBIA*
[14] CÔTE D'IVOIRE
[15] CZECH REP.*
[16] **DOMINICA+**
[17] DOMINICAN REP.*
[18] ECUADOR
[19] **EL SALVADOR+**
[20] EUROPEAN COMMUNITIES*
    AUSTRIA
    BELGIUM
    DENMARK
    FINLAND
    FRANCE

GERMANY
GREECE
IRELAND
ITALY
LUXEMBOURG
NETHERLANDS
PORTUGAL
SPAIN
SWEDEN
UNITED KINGDOM
[21] **GHANA+**
[22] **GRENADA+**
[23] **GUATEMALA+**
[24] HONG KONG*
[25] HUNGARY*
[26] ICELAND*
[27] INDIA*
[28] **INDONESIA+**
[29] ISRAEL*
[30] **JAMAICA+**
[31] JAPAN*
[32] KOREA*
[33] **MALAYSIA+**
[34] MAURITIUS*
[35] MEXICO*

[36] MOROCCO*
[37] NEW ZEALAND*
[38] NORWAY*
[39] PAKISTAN*
[40] **PAPUA NEW GUINEA**+
[41] PERU*
[42] PHILIPPINES*
[43] POLAND*
[44] **ROMANIA**+
[45] **ST. VINCENT & THE GREN-ADINES**+*1*
[46] **SENEGAL**+

[47] SINGAPORE*
[48] SLOVAK REP.*
[49] **SOUTH AFRICA**+
[50] **SRI LANKA**+
[51] SWITZERLAND*
[52] THAILAND*
[53] **TRINIDAD & TOBAGO**+
[54] **TUNISIA**+
[55] TURKEY
[56] UNITED STATES*
[57] VENEZUELA

[*] = Revisions to April 30 1996 offers submitted: 32 (46 governments)
[+] and bolded = New submissions after 30 April 1996: 23 (21 of which to be annexed to the Protocol)

Total schedules to be annexed to the GATS Protocol: 55 (69 governments).
1 To be finalized after 15 February, so will not be annexed to the Protocol.

Bibliography

Airtouch Communications, Inc., *1993 SEC Form 10-K* (1994).

Airtouch Communications, Inc., *1994 Annual Report* (1995).

Alchian, Armen A., "Uncertainty, Evolution, and Economic Theory," 58 *Journal of Political Economy* 211 (1950).

Alchian, Armen A., and Harold Demsetz, "Production, Information Costs, and Economic Organization," 62 *American Economic Review* 777 (1972).

Alchian, Armen A., and Susan E. Woodward, "The Firm Is Dead; Long Live the Firm," 26 *Journal of Economic Literature* 65 (1988).

Aleinikoff, T. Alexander, "Federal Regulation of Aliens and the Constitution," 38 *American Journal of International Law* 862 (1989).

American Bar Association, Section of Public Utility, Communications and Transportation Law, *1995 Annual Report: Infrastructure in Transition* (American Bar Association 1995).

Ameritech Corp., *1994 Form 10-K* (1995).

Archer, Gleason T., *History of Radio to 1926* (American Historical Society 1938; reprinted Arno Press 1971).

Arrow, Kenneth J., "Insurance, Risk and Resource Allocation," in *Essays in the Theory of Risk-Bearing* (North-Holland Publishing Co. 1970).

Baldwin, Richard, and Paul R. Krugman, "Industrial Policy and International Competition in Wide-Bodied Jet Aircraft," in *Trade Policy Issues and Empirical Analysis* (Robert E. Baldwin ed., University of Chicago Press 1988).

Baldwin, Robert E., and Richard K. Green, "The Effects of Protection on Domestic Output," in *Trade Policy Issues and Empirical Analysis* (Robert E. Baldwin ed., University of Chicago Press 1988).

Baldwin, Robert E., and Paul R. Krugman, "Market Access and International Competition: A Simulation Study of 16K Random Access Memories," in *Empirical Methods for International Trade* (Robert C. Feenstra ed., MIT Press 1987).

Barnouw, Erik, *A Tower in Babel: A History of Broadcasting in the United States to 1933* (Oxford University Press 1966).

Baumol, William J., and Ralph E. Gomory, "On Efficiency and Comparative Advantage in Trade Equilibria under Scale Economies," C. V. Starr Center for Applied Economics, New York University, Working Paper RR#94-13, Apr. 1994.

Bayard, Thomas O., and Kimberly Ann Elliott, *Reciprocity and Retaliation in U.S. Trade Policy* (Institute for International Economics 1994).

BCE Inc., *1993 SEC Form 20-F* (1993).

BCE Inc., *1994 Annual Report* (1995).

BCE Inc., *1994 SEC Form 20-F* (1994).

BCE Mobile Inc., *1994 Annual Report* (1995).

Bell Atlantic Corp., *1994 Annual Report* (1995).

Bell Atlantic Corp., *1994 SEC Form 10-K* (1995).

BellSouth Corp., *1994 SEC Form 10-K* (1995).

Bergsten, C. Fred, and Marcus Noland, *Reconcilable Differences? United States–Japan Economic Conflict* (Institute for International Economics 1993).

Bernard, Keith, "Global Telecommunications: Policy Implications in the U.S.A.," 16 *Telecommunications Policy* 371 (1992).

Berniker, Mark, "Sprint, Cable Partners Plan Phone Service," *Broadcasting & Cable*, Apr. 3, 1995, at 39.

Berry, Tyler, *Communications by Wire and Radio* (Callaghan & Co. 1937).

Bhagwati, Jagdish, "Aggressive Unilateralism: An Overview," in *Aggressive Unilateralism: America's 301 Trade Policy and the World Trading System* (Jagdish Bhagwati & Hugh Patrick eds., University of Michigan Press 1990).

Bhagwati, Jagdish, "Free Trade: Old and New Challenges," 104 *Economic Journal* 231 (1994).

Bhagwati, Jagdish, "Is Free Trade Passé after All?" in *Political Economy and International Economics* 26 (Douglas A. Irwin ed., MIT Press 1991).

Bhagwati, Jagdish, "U.S. Trade Policy at the Crossroads," in *Political Economy and International Economics* (Douglas A. Irwin ed., MIT Press 1991).

Biographical Directory of the American Congress, 1774–1989 (Government Printing Office 1989).

Blasi, Vincent, "The Checking Value in First Amendment Theory," 1977 *American Bar Foundation Research Journal* 521 (1977).

Blomström, Magnus, "Host Country Benefits of Foreign Investment," National Bureau of Economic Research Working Paper No. 3615, 1991.

Blomström, Magnus, Robert E. Lipsey, and Ksenia Kulchycky, "U.S. and Swedish Direct Investment and Exports," in *Trade Policy Issues and Empirical Analysis* (Robert E. Baldwin ed., University of Chicago Press 1988).

Bork, Robert H., *The Antitrust Paradox: A Policy at War with Itself* (Free Press 1978; rev. ed. 1993).

Brander, James A., and Barbara J. Spencer, "Export Subsidies and International Market Share Rivalry," 16 *Journal of International Economics* 83 (1985).

Branson, William H., and Alvin K. Klevorick, "Strategic Behavior and Trade Policy," in *Strategic Trade Policy* 241 (Paul R. Krugman ed., MIT Press 1992).

Brealey, Richard A., and Stewart C. Myers, *Principles of Corporate Finance* (McGraw Hill, 4th ed. 1991).

Brenner, Daniel L., *Law and Regulation of Common Carriers in the Communications Industry* (Westview Press 1992).

Brenner, Daniel L., Monroe E. Price, and Michael Meyerson, *Cable Television and Other Nonbroadcast Video: Law and Policy* (Clark Boardman Callaghan, rev. ed. 1996).

British Telecommunications plc, *1994 SEC Form 20-F* (1994).

Broadcasting Yearbook (Broadcasting Publications 1991).

Broadman, Harry, and Carol Balassa, "Liberalizing International Trade in Telecommunications Services," 28 *Columbia Journal of World Business* 30 (1993).

Brock, Gerald W., *The Telecommunications Industry: The Dynamics of Market Structure* (Harvard University Press 1981).

Brotman, Stuart N., "Communications Policy Making at the FCC: Past Practices, Future Direction," 7 *Cardozo Arts & Entertainment Law Journal* 55 (1985).

Bruncor, Inc., *1994 Annual Report* (1995).

BT-MCI Fact Sheet, Concert Web Site, <http://www.concert.com/deal/merger. htm.>.

Bungert, Hartwin, "Equal Protection for Foreign and Alien Corporations: Towards Intermediate Scrutiny for a Quasi-Suspect Classification," 59 *Missouri Law Review* 569 (1994).

Butler, Henry N., and Larry E. Ribstein, *The Corporation and the Constitution* (AEI Press 1995).

Cable & Wireless plc, *1994 SEC Form 20-F* (1994).

Calhoun, George C., *Digital Cellular Radio* (Artech House 1988).

Canadian Pacific Limited, *1994 Annual Report* (1995).

Carlton, Dennis W., and Jeffrey M. Perloff, *Modern Industrial Organization* (HarperCollins, 2d ed. 1994).

Case, Josephine Young, and Everett Needham Case, *Owen D. Young and American Enterprise* (David R. Godine 1982).

Cate, Fred H., "The First Amendment and the International 'Free Flow' of Information," 30 *Virginia Journal of International Law* 371 (1990).

Clark, Robert C., *Corporate Law* (Little, Brown & Co. 1986).

Cline, William R., *"Reciprocity": A New Approach to World Trade Policy?* (Institute for International Economics 1982).

Coase, Ronald H., "The Federal Communications Commission," 2 *Journal of Law and Economics* 1 (1959).

Coase, Ronald H., "The Market for Goods and the Market for Ideas," 64 *American Economic Review Papers and Proceedings* 384 (1974).

Coase, Ronald H., "The Nature of the Firm," 4 *Economica* (n.s.) (1937).

Coates, Vary T., Todd La Porte, and Mark G. Young, "Global Telecommunications and the Export of Services: The Promise and the Risk," 36 *Business Horizons* 23 (1993).

"Colossus at Bay," *Economist*, Dec. 10, 1994, at 63.

Commissioner of Navigation, *1924 Annual Report to the Secretary of Commerce* (1924).

Comcast Corp., *1994 SEC Form 10-K* (1995).

"Comcast Joins Bid for 100 Franchises," *New Media Markets,* Feb. 10, 1994.

Commissioner of Navigation, *1924 Annual Report to the Secretary of Commerce* (1924).

Cooter, Robert, and Thomas Ulen, *Law and Economics* (Scott, Foresman & Co. 1988).

Coughlin, Cletus, "Foreign-Owned Companies in the United States: Malign or Benign?" 74 *Federal Reserve Bank of St. Louis Bulletin* 17 (1992).

Crandall, Robert W., and J. Gregory Sidak, "Competition and Regulatory Policies for Interactive Broadband Networks," 68 *Southern California Law Review* 1203 (1995).

Crandall, Robert W., and Leonard Waverman, *Talk Is Cheap: The Promise of Regulatory Reform in North American Telecommunications* (Brookings Institution 1996).

Cranston, Richard, *Liberalising Telecommunications in Western Europe* (Financial Times Business Information 1995).

Culbertson, William Smith, *Commercial Policy in War Time and After* (D. Appleton & Co. 1923).

Damrosch, Lori Fisler, "Foreign States and the Constitution," 73 *Virginia Law Review* 483 (1987).

Davis, Stephen Brooks, *The Law of Radio Communication* (McGraw-Hill 1927).

Demsetz, Harold, and Kenneth M. Lehn, "The Structure of Corporate Ownership: Causes and Consequences," 93 *Journal of Political Economy* 1155 (1985).

De Soto, Clinton B., *Two Hundred Meters and Down: The Story of Amateur Radio* (American Radio Relay League 1936).

De Vany, Arthur S., Ross D. Eckert, Charles J. Meyers, Donald J. O'Hara, and Richard C. Scott., "A Property System for Market Allocation of the Electromagnetic Spectrum: A Legal-Economic-Engineering Study," 21 *Stanford Law Review* 1499 (1969).

Devins, Neal, "*Metro Broadcasting, Inc.* v. *FCC*: Requiem for a Heavyweight," 69 *Texas Law Review* 125 (1990).

Director, Aaron, "The Parity of the Economic Market Place," 7 *Journal of Law and Economics* 1 (1964).

Dixit, Avinash K., "Optimal Trade and Industrial Policies for the U.S. Automobile Industry," in *Empirical Methods for International Trade* (Robert C. Feenstra ed., MIT Press 1987).

Dixit, Avinash K., and Gene M. Grossman, "Targeted Export Promotion with Several Oligopolistic Industries," 21 *Journal of International Economics* 233 (1986).

Documents in Telecommunications Policy (John M. Kittross ed., Arno Press 1977).

Douglas, Susan J., *Inventing American Broadcasting* (Johns Hopkins University Press 1987).

Dupuy, R. Ernest, and Trevor N. Dupuy, *The Harper Encyclopedia of Military History: From 3500 B.C. to the Present* (HarperCollins, 4th ed. 1993).

Easterbrook, Frank H., and Daniel R. Fischel, *The Economic Structure of Corporate Law* (Harvard University Press 1991).

Eaton, Jonathan, and Gene M. Grossman, "Optimal Trade and Industrial Policy under Oligopoly," 101 *Quarterly Journal of Economics* 383 (1986).

Edmondson, Gail, "A Feeding Frenzy in European Telecom," *Business Week*, Nov. 21, 1994, at 119.

Edmondson, Gail, and Julia Flynn, "Missing the Wake-up Call," *Business Week*, July 10, 1995, at 18.

Emery, Henry C., "The Problem of Anti-Dumping Legislation," in *Official Report of the Third National Foreign Trade Convention* (1916).

Emord, Jonathan W., "The First Amendment Invalidity of FCC Ownership Regulations," 38 *Catholic University Law Review* 401 (1989).

Emord, Jonathan W., *Freedom, Technology, and the First Amendment* (Pacific Research Institute for Public Policy 1991).

Ennis, James G., and David N. Roberts, "Foreign Ownership in U.S. Communications Industry: The Impact of Section 310," 19 *International Business Law* 243 (1991).

Fama, Eugene F., "Agency Problems and the Theory of the Firm," 88 *Journal of Political Economy* 288 (1980).

Fama, Eugene F., and Michael C. Jensen, "Agency Problems and Residual Claims," 26 *Journal of Law and Economics* 327 (1983).

Fama, Eugene F., and Michael C. Jensen, "Separation of Ownership and Control," 26 *Journal of Law and Economics* 301 (1983).

Federal Communications Bar Association, International Practice Committee, *1993 International Communications Practice Handbook* (Paul J. Berman and Ellen K. Snyder eds., Federal Communications Bar Association 1993).

Federal Communications Commission, *Sixth Annual Report* (1940).

Federal Communications Commission, International Bureau, Global Communications Alliances, *Forms and Characteristics of Emerging Organizations* (Feb. 1996).

Federal Trade Commission, *Report on the Radio Industry* (Government Printing Office 1924).

The Federalist No. 66 (Alexander Hamilton) (Jacob E. Cooke ed., Wesleyan University Press 1961).

The Foreign Investment Debate: Opening Markets Abroad or Closing Markets at Home? (Cynthia A. Beltz ed., AEI Press 1995).

Friedlaender, Ann F., Ernst R. Berndt, and Gerard McCullough, "Governance Structure, Managerial Characteristics, and Firm Performance in the Deregulated Rail Industry," 1992 *Brookings Papers on Economic Activity: Microeconomics* 95.

Friendly, Henry J., *The Federal Administrative Agencies: The Need for Better Definition of Standards* 63 (Harvard University Press 1962).

Funabashi, Yoichi, *Managing the Dollar: From the Plaza to the Louvre* (Institute for International Economics, 2d ed. 1989).

Gavillet, Ronald W., Jill M. Foehrkolb, and Simone Wu, "Structuring Foreign Investments in FCC Licensees under Section 310(b) of the Communications Act," 27 *California Western Law Review* 7 (1990).

Gehrig, Anette, and Klaus F. Zimmermann, "Recent Developments in Strategic Trade Policy and Empirical Evidence," in *Export Activity and Strategic Trade Policy* (Horst Kräger & Klaus Zimmermann eds., Springer-Verlag-Heidelberg 1992).

General Accounting Office, *Airline Competition: Impact of Changing Foreign Investment and Control Limits on U.S. Airlines* (Government Printing Office 1993).

General Accounting Office, *Foreign Investment Implementation of Exon-Florio and Related Amendments* (GAO/NSIAD-96-12-1995).

Geotek, Inc., *1993 SEC Form 10-K* (1994).

Ginn, Sam, "Restructuring the Wireless Industry and the Information Skyway," 4 *Journal of Economics and Management Strategy* 139 (1995).

Ginsburg, Douglas H., *Regulation of Broadcasting* (West Publishing Co. 1979).

Giunta, Tara Kalagher, "Foreign Participation in Telecommunications Projects," in Federal Communications Bar Association, International Practice Committee, *1993 International Communications Practice Handbook* 43 (Paul J. Berman and Ellen K. Snyder eds., Federal Communications Bar Association 1993).

Glickman, Norman, and Dennis Woodward, "Industry Location and Public Policy," in *Regional and Local Determinants of Foreign Firm Location in the United States* (Henry W. Herzog and Alan M. Schlottman eds., University of Tennessee Press 1991).

Globerman, Steven, "Foreign Ownership in Telecommunications: A Policy Perspective," 19 *Telecommunications Policy* 21 (1995).

Globerman, Steven, "Trade Liberalization and Competitive Behavior: A Note Assessing Evidence and the Public Policy Implications," 9 *Journal of Policy Analysis and Management* 80 (1990).

Goldberg, Lee, and Charles Kolstad, "Foreign Direct Investment, Exchange Rate Variability and Demand Uncertainty in the United States," National Bureau of Economic Research Working Paper No. 4815, 1994.

Gomory, Ralph E., and William J. Baumol, "Share of World Output, Economies of Scale, and Regions Filled with Equilibria," C. V. Starr Center for Applied Economics, New York University, Working Paper RR#94-29, Oct. 1994.

Graham, Edward M. "Strategic Management and Transnational Firm Behavior: A Formal Approach," in *The Nature of the Transnational Firm* (C. N. Pitelis and R. Suqden eds., Routledge 1991).

Graham, Edward M., and Paul R. Krugman, *Foreign Direct Investment in the United States* (Institute for International Economics, 2d ed. 1991).

Graham, Edward M., and Paul R. Krugman, *Foreign Direct Investment in the United States* (Institute for International Economics, 3d ed. 1995).

Grover, Ronald, "TCI's Endless Morning After," *Business Week*, Apr. 10, 1995, at 60.

Haberler, Gottfried, *Theory of International Trade* (William Hodge & Co. 1933).

Hayek, Friedrich A., *The Fatal Conceit: The Errors of Socialism* (W. W. Bartley III ed., Routledge 1988).

Hayek, Friedrich A., "The Use of Knowledge in Society," 35 *American Economic Review* 519 (1945).

Hazlett, Thomas W., "The Rationality of U.S. Regulation of the Broadcast Spectrum," 33 *Journal of Law and Economics* 133 (1990).

Hazlett, Thomas W., and Matthew L. Spitzer, "Public Policy toward Cable Television: Regulation and the First Amendment," AEI Working Paper in Telecommunications Deregulation, 1996.

Helpman, Elhanan, and Paul R. Krugman, *Trade Policy and Market Structure* (MIT Press 1989).

Henkin, Louis, *Foreign Affairs and the Constitution* (Columbia University Press 1972).

Herzel, Leo, "'Public Interest' and the Market in Color Television Regulation," 18 *University of Chicago Law Review* 802 (1951).

Holmer, Alan F., Judith H. Bello, and Jeremy O. Preiss, "The Final Exon-Florio Regulations on Foreign Direct Investment: The Final Word or Prelude to Tighter Controls?" 23 *Law and Policy of International Business* 593 (1992).

Horstmann, Ignatius J., and James Markusen, "Up Your Average Cost Curves: Inefficient Entry and New Protectionism," 20 *Journal of International Economics* 225 (1986).

Howenstice, Ned, and William J. Zeile, "Characteristics of U.S. Manufacturing Establishments," 74 *Survey of Current Business*, no. 8, at 34 (Aug. 1994).

Howeth, L. S., *History of Communications—Electronics in the United States Navy XIV* (Bureau of Ships and Office of Naval History 1963).

Hymer, Steven H., *The International Operations of National Firms: A Study of Direct Foreign Investment* (MIT Press 1976).

Inglis, Andrew F., *Behind the Tube: A History of Broadcasting Technology and Business* (Focal Press/Butterworth Publishers 1990).

International Telecommunications Union, *World Telecommunication Development Report* (1994).

International Telecommunications Union, "Cross-Ownership in the Telecommunication Service Sector," Mar. 23, 1995.

Irwin, Douglas A, *Against the Tide: An Intellectual History of Free Trade* (Princeton University Press 1996).

Irwin, Douglas A., *Managed Trade: The Case against Import Targets* (AEI Press 1994).

Jaffe, Louis, "The Editorial Responsibility of the Broadcaster: Reflections on Fairness and Access," 85 *Harvard Law Review* 768 (1972).

Jensen, Michael C., "Takeovers: Their Causes and Consequences," 2 *Journal of Economic Perspectives* 21 (1988).

Jensen, Michael C., and William H. Meckling, "Theory of the Firm: Managerial Behavior, Agency Costs and Ownership Structure," 3 *Journal of Financial Economics* 305 (1976).

Jensen, Michael C., and Kevin M. Murphy, "Performance Pay and Top-Management Incentives," 98 *Journal of Political Economy* 225 (1990).

Johnson, Leland L., *Toward Competition in Cable Television* (MIT Press and AEI Press 1994).

Jolly, W. P., *Marconi* (Stein & Day 1972).

Jones Intercable, Inc., *1994 Annual Report* (1995).

Kahn, David, *The Codebreakers* (Weidenfeld & Nicolson 1974).

Kellogg, Michael K., John Thorne, and Peter W. Huber, *Federal Telecommunications Law* (Little, Brown & Co. 1992).

Klein, Benjamin, Robert. G. Crawford, and Armen A. Alchian, "Vertical Integration, Appropriable Rents, and the Competitive Contracting Process," 21 *Journal of Law and Economics* 297 (1978).

Koh, Harold Hongju, "The Coase Theorem and the War Power: A Response," 41 *Duke Law Journal* 122 (1991).

Krattenmaker, Thomas G., and Lucas A. Powe, Jr., "Converging First Amendment Principles for Converging Communications Media," 104 *Yale Law Journal* 1719 (1995).

Krattenmaker, Thomas G., and Lucas A. Powe, Jr., *Regulating Broadcast Programming* (MIT Press and AEI Press 1994).

Krenzler, Horst G., Director-General I, European Commission, letter to Mr. Oliver, EC Commission of the American Chamber of Commerce in Belgium (Jan. 18, 1995).

Kreps, David R., *A Course in Microeconomic Theory* (Princeton University Press 1990).

Krueger, Anne O., *American Trade Policy: A Tragedy in the Making* (AEI Press 1995).

Krueger, Anne O., "Free Trade Is the Best Policy," in *An American Trade Strategy: Options for the 1990s* (Robert Z. Lawrence and Charles L. Schultze eds., Brookings Institution 1990).

Krugman, Paul R., "Introduction," in *Empirical Studies of Strategic Trade Policy* (Paul R. Krugman ed., National Bureau of Economic Research 1994).

Krugman, Paul R., "Introduction: New Thinking about Trade Policy," in *Strategic Trade Policy and the New International Economics* (Paul R. Krugman ed., MIT Press 1986).

Krugman, Paul R., *Peddling Prosperity* (MIT Press 1994).

Krugman, Paul R., and Maurice Obstfeld, *International Economics: Theory and Practice* (HarperCollins, 3d ed. 1994).

Lackey, Raymond J., and Donald W. Upmal, "Speakeasy: The Military Software Radio," *IEEE Communications Magazine*, May 1995, at 56.

Landes, William E., and Richard A. Posner, "Market Power in Antitrust Cases," 94 *Harvard Law Review* 937 (1981).

Laster, Robert, and Martin McCauley, "Making Sense of the Profits of Foreign Firms in the U.S.," 19 *Federal Reserve Bank of New York Quarterly Review* 44 (1994).

Lehman Brothers, *The UK Cable Market: Breaking New Ground* (Feb. 17, 1994).

Levin, Harvey J., *The Invisible Resource: Use and Regulation of the Radio Spectrum* (Johns Hopkins University Press 1971).

Lewis, Tom, *Empire of the Air: The Men Who Made Radio* (Edward Burlingame Books 1991).

Lipsey, Robert, "Outbound Direct Investment and the U.S. Economy," National Bureau of Economic Research Working Paper No. 4691, 1994.

Lipsky, Abbott B., Jr., "Reconciling *Red Lion* and *Tornillo*: A Consistent Theory of Media Regulation," 28 *Stanford Law Review* 563 (1976).

Lucky, Robert W., Affidavit, Motion of Bell Atlantic Corporation, BellSouth Corporation, NYNEX Corporation, and Southwestern Bell Corporation to Vacate the Decree, *United States* v. *Western Electric Co.,* No. 82-0192 (filed D.D.C., July 6, 1994).

MacAvoy, Paul W., *The Failure of Antitrust and Regulation to Establish Competition in Long-Distance Telephone Services* (MIT Press and AEI Press 1996).

MacAvoy, Paul W., "Tacit Collusion under Regulation in the Pricing of Interstate Long-Distance Telephone Services," 4 *Journal of Economics and Management Strategy* 147 (1995).

Manne, Henry G., "Mergers and the Market for Corporate Control," 73 *Journal of Political Economy* 110 (1965).

Markey, Edward J., "Telecommunications and Financial Services Trade Hangs on NAFTA Thread," 1 *San Diego Justice Journal* 281 (1993).

Markusen, James R., "The Boundaries of Multinational Enterprises and the Theory of International Trade," 9 *Journal of Economic Perspectives* 169 (1995).

"A Marriage of Convenience," *Economist*, Nov. 9, 1996, at 72.

Martin, Donald L., *Economic Benefits to Puerto Rico from Vigorous Telecommunications Competition* (Glassman-Oliver Economic Consultants, Inc. June 27, 1994).

Mayton, William T., "The Illegitimacy of the Public Interest Standard at the FCC," 38 *Emory Law Journal* 715 (1989).

McClellan, Steve, "MCI/British Telecom Reducing ASkyB Stake; TCI Said to Be Talking about DBS Venture," *Broadcasting and Cable*, Nov. 11, 1996, at 54.

McCulloch, Rachel, "Foreign Investment in the U.S.," 30 *Finance and Development* 13 (1993).

MCI Communications Corp., *1993 Annual Report* (1994).

MCI Communications Corp., *1995 SEC Form 10-K* (1996).

"The Message in the Medium: The First Amendment on the Information Superhighway," 107 *Harvard Law Review* 1062 (1994).

Milgrom, Paul R., and John Roberts, *Economics, Organization and Management* (Prentice-Hall 1992).

Minasian, Jora R., "Property Rights in Radiation: An Alternative Approach to Radio Frequency Allocation," 18 *Journal of Law and Economics* 221 (1975).

Moore, Thomas G., "An Economic Analysis of the Concept of Freedom," 77 *Journal of Political Economy* 532 (1969).

Morgan, Kenneth O., "The Twentieth Century," in *The Oxford History of Britain* 582 (Kenneth O. Morgan ed., Oxford University Press 1988).

NYNEX Corp., *1994 Annual Report* (1995).

Olbeter, Erik, and Lawrence Chimerine, *Crossed Wires: How Foreign Policies and U.S. Regulators Are Holding Back the U.S. Telecommunications Services Industry* (Economic Strategy Institute 1994).

Ondrich, Jan, and Michael Wasylenko, *Direct Foreign Investment in the United States* (Upjohn Institute 1993).

Oster, Sharon M., *Modern Competitive Analysis* (Free Press 1990).

Owen, Bruce M., *Economics and Freedom of Expression: Media Structure and the First Amendment* (1975).

Paglin, Max, *A Legislative History of the Communications Act of 1934* (Oxford University Press 1989).

Petrazinni, Ben, *Global Telecom Talks: A Trillion Dollar Deal* (Institute for International Economics 1996).

Polsby, Daniel D., "Candidate Access to the Air: The Uncertain Future of Broadcaster Discretion," 1981 *Supreme Court Review* 223.

Pool, Ithiel de Sola, *Technologies of Freedom* (Belknap Press and Harvard University Press 1983).

Porter, Michael E., *The Competitive Advantage of Nations* (Oxford University Press, 2d ed. 1993).

Posner, Richard A., *Economic Analysis of Law* (Little, Brown & Co., 4th ed. 1992).

Powe, Lucas A., Jr., *American Broadcasting and the First Amendment* (University of California Press 1987).

Prestowitz, Clyde V., Jr., *The Future of the Airline Industry* (Economic Strategy Institute 1993).

Public Papers and Addresses of Franklin D. Roosevelt, 1934 (Samuel Rosenman ed., 1950).

Reuters Holdings plc, *1991 Annual Report* (1992).

Ricardo, David, *On the Principles of Political Economy and Taxation* (Guernsey Press Co. 1992) (1817).

Richardson, Elliot, "United States Policy toward Foreign Investment: We Can't Have It Both Ways," 4 *American University Journal of Law and Public Policy* 281 (1989).

Robinson, John O., *Spectrum Management Policy in the United States: An Historical Account* (Federal Communications Commission, Office of Plans and Policy, 1985).

Romano, Roberta, *The Genius of American Corporate Law* (AEI Press 1993).

Romano, Roberta, "A Guide to Takeovers: Theory, Evidence and Regulation," 9 *Yale Journal on Regulation* 119 (1992).

Rose, Ian M., "Barring Foreigners from Our Airwaves: An Anachronistic Pothole on the Global Information Highway," 95 *Columbia Law Review* 1188 (1995).

Sampson, Anthony, *The Sovereign State of ITT* (Stein & Day 1973).

SBC Communications, Inc., *1994 SEC Form 10-K* (1995).

Schelling, Thomas C., *The Strategy of Conflict* (Oxford University Press 1960).

Shuck, Peter, "The Transformation of Immigration Law," 84 *Columbia Law Review* 1 (1984).

Sidak, J. Gregory, "Antitrust Preliminary Injunctions in Hostile Tender Offers," 30 *Kansas Law Review* 491 (1982).

Sidak, J. Gregory, "A Framework for Administering the 1916 Antidumping Act: Lessons from Antitrust Economics," 18 *Stanford Journal of International Law* 377 (1982).

Sidak, J. Gregory, "The Recommendation Clause," 77 *Georgetown Law Journal* 2079 (1989).

Sidak, J. Gregory, "Telecommunications in Jericho," 82 *California Law Review* 1209 (1993).

Sidak, J. Gregory, "To Declare War," 41 *Duke Law Journal* 27 (1991).

Sidak, J. Gregory, "War, Liberty, and Enemy Aliens," 67 *New York University Law Review* 1402 (1992).

"Silencing the Speech of Strangers: Constitutional Values and the First Amendment Rights of Aliens," 81 *Georgetown Law Journal* 2073 (1993).

Smith, Adam, *An Inquiry into the Nature and Causes of the Wealth of Nations* (Oxford University Press 1976) (1776).

Smolla, Rodney A., and Stephen A. Smith, "Propaganda, Xenophobia, and the First Amendment," 67 *Oregon Law Review* 253 (1988).

Socolow, A. Walter, *The Law of Radio Broadcasting* (Baker, Voorhis & Co. 1939).

Southwestern Bell Corporation, *1993 SEC Form 10-K* (1994).

Spak, Michael I., "America for Sale: When Well-Connected Former Federal Officials Peddle Their Influence to the Highest Foreign Bidder—A Statutory Analysis and Proposals for Reform of the Foreign Agents Registration Act and the Ethics in Government Act," 78 *Kentucky Law Journal* 237 (1990).

Spitzer, Matthew L., "The Constitutionality of Licensing Broadcasters," 64 *New York University Law Review* 990 (1989).

Spitzer, Matthew L., "Controlling the Content of Print and Broadcast," 58 *Southern California Law Review* 1349 (1985).

Spitzer, Matthew L., "Justifying Minority Preferences in Broadcasting," 64 *Southern California Law Review* 293 (1991).

Spitzer, Matthew L., *Seven Dirty Words and Six Other Stories* (Yale University Press 1986).

Sprint Corp. *1994 SEC Form 10-K* (1995).

Spulber, Daniel F., *Regulation and Markets* (MIT Press 1989).

Stone, Geoffrey, "Imagining a Free Press," 90 *Michigan Law Review* 1246 (1992).

Taylor, Lester D., *Telecommunications Demand in Theory and Practice* (Kluwer Academic Publishers, rev. ed. 1994).

"Telecoms Liberalisation Reaches Final Stages in EU," *Business Europe*, Nov. 28, 1994.

Telefónica de España, S.A. *1993 SEC Form 20-F* (1993).

Temin, Peter, *The Fall of the Bell System: A Study in Prices and Politics* (Cambridge University Press 1987).

Terraine, John, *The U-Boat Wars, 1916–1945* (G. P. Putnam's Sons 1989).

Tirole, Jean, *The Theory of Industrial Organization* (MIT Press 1988).

Torres, Amelia, "EU Has Hard Time Policing the Telecoms Market," *Reuter European Community Report,* July 11, 1995.

Tuchman, Barbara W., *The Zimmermann Telegram* (Macmillan 1966).

U.S. Congress, Office of Technology Assessment, *U.S. Telecommunications Services in European Markets* (Government Printing Office 1993).

U.S. Congress, Office of Technology Assessment, *U.S. Telecommunications Services in European Markets* (Government Printing Office 1994).

U.S. Department of Commerce, *Foreign Direct Investment in the United States: An Update* (Government Printing Office 1993).

U.S. Department of Commerce, Bureau of Economic Analysis, 74 *Survey of Current Business*, no. 8 (Aug. 1994).

U S West Inc., *1994 Annual Report* (1995).

Varian, Hal R., *Microeconomic Analysis* (W. W. Norton & Co., 3d ed. 1992).

Vickers, John, and George Yarrow, *Privatization: An Economic Analysis* (MIT Press 1988).

Viner, Jacob, *Dumping: A Problem in International Trade* (1923).

Wallin, Homer N., *Pearl Harbor: Why, How, Fleet Salvage and Final Appraisal* (Government Printing Office 1968).

Waterman, David, "Vertical Integration and Program Access in the Cable Television Industry," 47 *Federal Communications Law Journal* 511 (1995),

Waterman, David, and Andrew Weiss, *Vertical Integration in Cable Television* (MIT Press and AEI Press 1997).

Watkins, John J., "Alien Ownership and the Communications Act," 33 *Federal Communications Law Journal* 1 (1980).

White, Llewellyn, *The American Radio* (University of Chicago Press 1971).

Williamson, Oliver E., *The Economic Institutions of Capitalism: Firms, Markets, Relational Contracting* (Free Press 1985).

Williamson, Oliver E., "Franchise Bidding for National Monopolies—In General and with Respect to CATV," 7 *Bell Journal of Economics* 73 (1976).

Williamson, Oliver E., *The Mechanisms of Governance* (Oxford University Press 1996).

World Trade Organization, Negotiating Group on Basic Telecommunications, "Reference Paper," Apr. 24, 1996.

Zeile, William J., "Foreign Direct Investment in the United States: 1992 Benchmark Survey Results," 74 *Survey of Current Business*, no. 7, at 154 (July 1994).

Case and Regulatory Proceeding Index

United States *v.* Sprint Corp. and
Joint Venture Co., 60 Fed. Reg.
44,049 (1995), 2
United States *v.* Western Elec.
Co., No. 82-0192 (D.D.C.,
filed July 6, 1994), 170
United States *v.* Zenith Radio
Corp., 12 F.2d 614 (N.D. Ill.
1926), 59
United States Postal Service *v.*
Council of Greenburgh Civic
Ass'ns, 453 U.S. 114 (1981),
308
United States Tel. Ass'n *v.* United
States, No. 1:94-CV-0196-1
(D.D.C. Jan. 27, 1995), 329
Univision Holdings, Inc., 7 F.C.C.
Rcd. 6672 (1992), *recons.
denied*, 8 F.C.C. Rcd. 3931
(1993), 153
Upsouth Corp., 9 F.C.C. Rcd.
2130 (1994), 142, 357
U S WEST, Inc. *v.* United States,
48 F.3d 1092 (9th Cir. 1994),
329

Vermont Tel. Co., 10 F.C.C. Rcd.
9337 (1995), 132

Viereck *v.* United States, 318 U.S.
236 (1943), 112

Western Union Corp., 4 F.C.C.
Rcd. 2219 (1989), 105
WHDH, Inc., 17 F.C.C.2d 856
(1969), 152
White *v.* Johnson, 282 U.S. 367
(1931), 309
Whitney *v.* California, 274 U.S.
357 (1927), 289
Wickard *v.* Filburn, 317 U.S. 111
(1942), 307
Widmar *v.* Vincent, 454 U.S. 263
(1981), 294
William S. Paley, 61 Rad. Reg. 2d
(P & F) 413 (1986), *recons.
denied*, 62 Rad. Reg. 2d (P &
F) 852 (1987), *aff'd sub. nom.*,
Fairness in Media, 851 F.2d
1500 (D.C. Cir. 1988), 146
WWIZ, Inc. 2 Rad. Reg. 2d (P&F)
169 (1964), *aff'd sub nom.*
Lorain Journal Co. *v.* FCC, 351
F.2d 824 (D.C. Cir. 1964),
cert. denied, 383 U.S. 967
(1966), 152
WWOR-TV, Inc., 6 F.C.C. Rcd.
193 (1990), 153

Subject Index